ELEMENTS OF SET THEORY

ELEMENTS OF SET THEORY

Herbert B. Enderton

DEPARTMENT OF MATHEMATICS
UNIVERSITY OF CALIFORNIA, LOS ANGELES
LOS ANGELES, CALIFORNIA

Academic Press
An Imprint of Elsevier
San Diego New York Boston
London Sydney Tokyo Toronto

Academic Press
An Imprint of Elsevier
525 B Street, Suite 1900, San Diego, California 92101-4495, USA
http://www.academicpress.com

Academic Press
84 Theobalds Road, London WC1X 8RR, UK
http://www.academicpress.com

Library of Congress Cataloging-in-Publication Data

Enderton, Herbert, B
Elements of Set Theory.

Bibliography: p.
Includes index.

1. Set theory I. Title.

QA248.E5 511'.3 76-27438

ISBN-13: 978-0-12-238440-0

AMS (MOS) 1970 Subject Classifications: 04-01, 04A05, 04A10

Transferred to Digital Printing 2009

To my teachers with gratitude
To my students with hope

CONTENTS

PREFACE

This is an introductory undergraduate textbook in set theory. In mathematics these days, essentially everything is a set. Some knowledge of set theory is a necessary part of the background everyone needs for further study of mathematics. It is also possible to study set theory for its own interest—it is a subject with intriguing results about simple objects. This book starts with material that nobody can do without. There is no end to what can be learned of set theory, but here is a beginning.

The author of a book always has a preferred manner for using the book: A reader should simply study it from beginning to end. But in practice, the users of a book have their own goals. I have tried to build into the present book enough flexibility to accommodate a variety of goals.

The axiomatic material in the text is marked by a stripe in the margin. The purpose of the stripe is to allow a user to deemphasize the axiomatic material, or even to omit it entirely.

A course in axiomatic set theory might reasonably cover the first six or seven chapters, omitting Chapter 5. This is the amount of set theory that everyone with an interest in matters mathematical should know. Those with a special interest in set theory itself are encouraged to continue to the end of the book (and beyond). A very different sort of course might emphasize the set-theoretic construction of the number systems. This course might cover the first five chapters, devoting only as much attention to the axiomatic material as desired. The book presupposes no specific background. It does give real proofs. The first difficult proof is not met until part way through Chapter 4.

The hierarchical view of sets, constructed by transfinite iteration of the power set operation, is adopted from the start. The axiom of regularity is not added until it can be proved to be equivalent to the assertion that every set has a rank.

The exercises are placed at the end of each (or nearly each) section. In addition, Chapters 2, 3, and 4 have "Review Exercises" at the ends of the chapters. These are comparatively straightforward exercises for the reader wishing additional review of the material. There are, in all, close to 300 exercises.

There is a brief appendix dealing with some topics from logic, such as truth tables and quantifiers. This appendix also contains an example of how one might discover a proof.

At the end of this text there is an annotated list of books recommended for further study. In fact it includes diverse books for several further studies in a variety of directions. Those wishing to track down the source of particular results or historical points are referred to the books on the list that provide specific citations.

There are two stylistic matters that require mention. The end of a proof is marked by a reversed turnstile (⊣). This device is due to C. C. Chang and H. J. Keisler. In definitions, I generally pass up the traditionally correct "if" in favor of the logically correct "iff" (meaning "if and only if").

Two preliminary editions of the text have been used in my courses at UCLA. I would be pleased to receive comments and corrections from further users of the book.

LIST OF SYMBOLS

The number indicates the page on which the symbol first occurs in the text or the page on which it is defined.

CHAPTER 1

INTRODUCTION

BABY SET THEORY

We shall begin with an informal discussion of some basic concepts of set theory. In these days of the "new math," much of this material will be already familiar to you. Indeed, the practice of beginning each mathematics course with a discussion of set theory has become widespread, extending even to the elementary schools. But we want to review here elementary-school set theory (and do it in our notation). Along the way we shall be able to point out some matters that will become important later. We shall not, in these early sections, be particularly concerned with rigor. The more serious work will start in Chapter 2.

A *set* is a collection of things (called its *members* or *elements*), the collection being regarded as a single object. We write "$t \in A$" to say that t is a member of A, and we write "$t \notin A$" to say that t is not a member of A.

For example, there is the set whose members are exactly the prime numbers less than 10. This set has four elements, the numbers 2, 3, 5, and 7. We can name the set conveniently by listing the members within braces (curly brackets):

$$\{2, 3, 5, 7\}.$$

Call this set A. And let B be the set of all solutions to the polynomial equation

$$x^4 - 17x^3 + 101x^2 - 247x + 210 = 0.$$

Now it turns out (as the industrious reader can verify) that the set B has exactly the same four members, 2, 3, 5, and 7. For this reason A and B are the same set, i.e., $A = B$. It matters not that A and B were defined in different ways. Because they have exactly the same elements, they are equal; that is, they are one and the same set. We can formulate the general principle:

Principle of Extensionality If two sets have exactly the same members, then they are equal.

Here and elsewhere, we can state things more concisely and less ambiguously by utilizing a modest amount of symbolic notation. Also we abbreviate the phrase "if and only if" as "iff." Thus we have the restatement:

Principle of Extensionality If A and B are sets such that for every object t,

$$t \in A \quad \text{iff} \quad t \in B,$$

then $A = B$.

For example, the set of primes less than 10 is the same as the set of solutions to the equation $x^4 - 17x^3 + 101x^2 - 247x + 210 = 0$. And the set $\{2\}$ whose only member is the number 2 is the same as the set of even primes.

Incidentally, we write "$A = B$" to mean that A and B are the same object. That is, the expression "A" on the left of the equality symbol names the same object as does the expression "B" on the right. If $A = B$, then automatically (i.e., by logic) anything that is true of the object A is also true of the object B (it being the same object). For example, if $A = B$, then it is automatically true that for any object t, $t \in A$ iff $t \in B$. (This is the converse to the principle of extensionality.) As usual, we write "$A \neq B$" to mean that it is not true that $A = B$.

A small set would be a set $\{0\}$ having only one member, the number 0. An even smaller set is the empty set $\varnothing$. The set $\varnothing$ has no members at all. Furthermore it is the only set with no members, since extensionality tells us that any two such sets must coincide. It might be thought at first that the empty set would be a rather useless or even frivolous set to mention, but, in fact, from the empty set by various set-theoretic operations a surprising array of sets will be constructed.

For any objects x and y, we can form the pair set $\{x, y\}$ having just the members x and y. Observe that $\{x, y\} = \{y, x\}$, as both sets have exactly the same members. As a special case we have (when $x = y$) the set $\{x, x\} = \{x\}$.

For example, we can form the set $\{\varnothing\}$ whose only member is $\varnothing$. Note that $\{\varnothing\} \neq \varnothing$, because $\varnothing \in \{\varnothing\}$ but $\varnothing \notin \varnothing$. The fact that $\{\varnothing\} \neq \varnothing$ is reflected in the fact that a man with an empty container is better off than a man with nothing—at least he has the container. Also we can form $\{\{\varnothing\}\}$, $\{\{\{\varnothing\}\}\}$, and so forth, all of which are distinct (Exercise 2).

Similarly for any objects x, y, and z we can form the set $\{x, y, z\}$. More generally, we have the set $\{x_1, \ldots, x_n\}$ whose members are exactly the objects $x_1, \ldots, x_n$. For example,

$$\{\varnothing, \{\varnothing\}, \{\{\varnothing\}\}\}$$

is a three-element set.

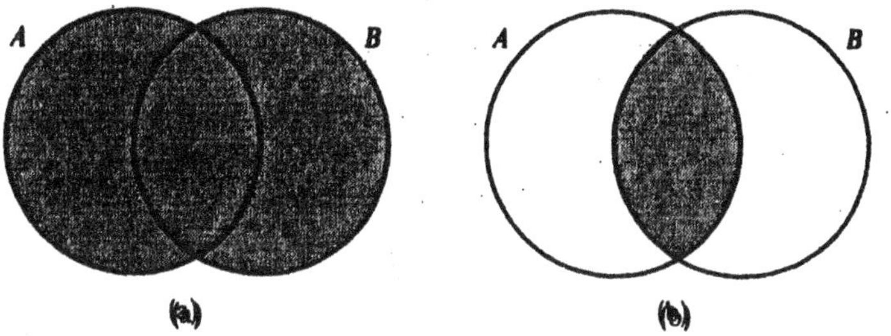

Fig. 1. The shaded areas represent (a) $A \cup B$ and (b) $A \cap B$.

Two other familiar operations on sets are union and intersection. The *union* of sets A and B is the set $A \cup B$ of all things that are members of A or B (or both). Similarly the *intersection* of A and B is the set $A \cap B$ of all things that are members of both A and B. For example,

$$\{x, y\} \cup \{z\} = \{x, y, z\}$$

and

$$\{2, 3, 5, 7\} \cap \{1, 2, 3, 4\} = \{2, 3\}.$$

Figure 1 gives the usual pictures illustrating these operations. Sets A and B are said to be *disjoint* when they have no common members, i.e., when $A \cap B = \varnothing$.

A set A is said to be a *subset* of a set B (written $A \subseteq B$) iff all the members of A are also members of B. Note that any set is a subset of itself. At the other extreme, $\varnothing$ is a subset of every set. This fact (that $\varnothing \subseteq A$ for any A) is "vacuously true," since the task of verifying, for every member of $\varnothing$, that it also belongs to A, requires doing nothing at all.

If $A \subseteq B$, then we also say that A is *included* in B or that B *includes* A. The inclusion relation ($\subseteq$) is not to be confused with the membership relation ($\in$). If we want to know whether $A \in B$, we look at the set A as a single object, and we check to see if this single object is among the members of B. By contrast, if we want to know whether $A \subseteq B$, then we must open up the set A, examine its various members, and check whether its various members can be found among the members of B.

Examples 1. $\varnothing \subseteq \varnothing$, but $\varnothing \notin \varnothing$.

2. $\{\varnothing\} \in \{\{\varnothing\}\}$ but $\{\varnothing\} \nsubseteq \{\{\varnothing\}\}$. $\{\varnothing\}$ is not a subset of $\{\{\varnothing\}\}$ because there is a member of $\{\varnothing\}$, namely $\varnothing$, that is not a member of $\{\{\varnothing\}\}$.

3. Let Us be the set of all people in the United States, and let Un be the set of all countries belonging to the United Nations. Then

$$\text{John Jones} \in Us \in Un.$$

But John Jones $\notin Un$ (since he is not even a country), and hence $Us \nsubseteq Un$.

Any set A will have one or more subsets. (In fact, if A has n elements, then A has 2^n subsets. But this is a matter we will take up much later.) We can gather all of the subsets of A into one collection. We then have the set of all subsets of A, called the *power*[1] *set* $\mathscr{P}A$ of A. For example,

$$\mathscr{P}\varnothing = \{\varnothing\},$$
$$\mathscr{P}\{\varnothing\} = \{\varnothing, \{\varnothing\}\},$$
$$\mathscr{P}\{0, 1\} = \{\varnothing, \{0\}, \{1\}, \{0, 1\}\}.$$

A very flexible way of naming a set is the method of *abstraction*. In this method we specify a set by giving the condition—the entrance requirement—that an object must satisfy in order to belong to the set. In this way we obtain the set of all objects x such that x meets the entrance requirement. The notation used for the set of all objects x such that the condition $\_\_ x \_\_$ holds is

$$\{x \mid \_\_ x \_\_\}.$$

For example:

1. $\mathscr{P}A$ is the set of all objects x such that x is a subset of A. Here "x is a subset of A" is the entrance requirement that x must satisfy in order to belong to $\mathscr{P}A$. We can write

$$\begin{aligned}\mathscr{P}A &= \{x \mid x \text{ is a subset of } A\} \\ &= \{x \mid x \subseteq A\}.\end{aligned}$$

[1] The reasons for using the word "power" in this context are not very convincing, but the usage is now well established.

2. $A \cap B$ is the set of all objects y such that $y \in A$ and $y \in B$. We can write

$$A \cap B = \{y \mid y \in A \text{ and } y \in B\}.$$

It is unimportant whether we use "x" or "y" or another letter as the symbol (which is used as a pronoun) here.

3. The set $\{z \mid z \neq z\}$ equals $\varnothing$, because the entrance requirement "$z \neq z$" is not satisfied by any object z.

4. The set $\{n \mid n \text{ is an even prime number}\}$ is the same as the set $\{2\}$.

There are, however, some dangers inherent in the abstraction method. For certain bizarre choices of the entrance requirement, it may happen that there is no set containing exactly those objects meeting the entrance requirement. There are two ways in which disaster can strike.

One of the potential disasters is illustrated by

$$\{x \mid x \text{ is a positive integer definable in one line of type}\}.$$

The tricky word here is "definable." Some numbers are easy to define in one line. For example, the following lines each serve to define a positive integer:

12,317,
the millionth prime number,
the least number of the form $2^{2^n} + 1$ that is not prime,
the 23rd perfect[2] number.

Observe that there are only finitely many possible lines of type (because there are only finitely many symbols available to the printer, and there is a limit to how many symbols will fit on a line). Consequently

$$\{x \mid x \text{ is a positive integer definable in one line of type}\}$$

is only a finite set of integers. Consider the least positive integer not in this set; that is, consider

the least positive integer not definable in one line of type.

The preceding line defines a positive integer in one line, but that number is, by its construction, not definable in one line! So we are in trouble, and the trouble can be blamed on the entrance requirement of the set, i.e., on the phrase "is a positive integer definable in one line of type." While it may have

[2] A positive integer is *perfect* if it equals the sum of its smaller divisors, e.g., $6 = 1 + 2 + 3$. It is *deficient* (or *abundant*) if the sum of its smaller divisors is less than (or greater than, respectively) the number itself. This terminology is a vestigial trace of numerology, the study of the mystical significance of numbers. The first four perfect numbers are 6, 28, 496, and 8128.

appeared originally to be a meaningful entrance requirement, it now appears to be gravely defective. (This example was given by G. G. Berry in 1906. A related example was published in 1905 by Jules Richard.)

There is a second disaster that can result from an overly free-swinging use of the abstraction method. It is exemplified by

$$\{x \mid x \notin x\},$$

this is, by the set of all objects that are not members of themselves. Call this set A, and ask "is A a member of itself?" If $A \notin A$, then A meets the entrance requirement for A, whereupon $A \in A$. But on the other hand, if $A \in A$, then A fails to meet the entrance requirement and so $A \notin A$. Thus both "$A \in A$" and "$A \notin A$" are untenable. Again, we are in trouble. The phrase "is not a member of itself" appears to be an illegal entrance requirement for the abstraction method. (This example is known as Russell's paradox. It was communicated by Bertrand Russell in 1902 to Gottlob Frege, and was published in 1903. The example was independently discovered by Ernst Zermelo.)

These two sorts of disaster will be blocked in precise ways in our axiomatic treatment, and less formally in our nonaxiomatic treatment. The first sort of disaster (the Berry example) will be avoided by adherence to entrance requirements that can be stated in a totally unambiguous form, to be specified in the next chapter. The second sort of disaster will be avoided by the distinction between *sets* and *classes*. Any collection of sets will be a *class*. Some collections of sets (such as the collections $\varnothing$ and $\{\varnothing\}$) will be sets. But some collections of sets (such as the collection of all sets not members of themselves) will be too large to allow as sets. These oversize collections will be called *proper classes*. The distinction will be discussed further presently.

In practice, avoidance of disaster will not really be an oppressive or onerous task. We will merely avoid ambiguity and avoid sweepingly vast sets. A prudent person would not want to do otherwise.

Exercises

1. Which of the following become true when "$\in$" is inserted in place of the blank? Which become true when "$\subseteq$" is inserted?
   (a) $\{\varnothing\}$ __ $\{\varnothing, \{\varnothing\}\}$.
   (b) $\{\varnothing\}$ __ $\{\varnothing, \{\{\varnothing\}\}\}$.
   (c) $\{\{\varnothing\}\}$ __ $\{\varnothing, \{\varnothing\}\}$.
   (d) $\{\{\varnothing\}\}$ __ $\{\varnothing, \{\{\varnothing\}\}\}$.
   (e) $\{\{\varnothing\}\}$ __ $\{\varnothing, \{\varnothing, \{\varnothing\}\}\}$.

2. Show that no two of the three sets $\varnothing$, $\{\varnothing\}$, and $\{\{\varnothing\}\}$ are equal to each other.

3. Show that if $B \subseteq C$, then $\mathscr{P}B \subseteq \mathscr{P}C$.

4. Assume that x and y are members of a set B. Show that $\{\{x\}, \{x, y\}\} \in \mathscr{P}\mathscr{P}B$.

SETS—AN INFORMAL VIEW

We are about to present a somewhat vague description of how sets are obtained. (The description will be repeated much later in precise form.) None of our later work will actually depend on this informal description, but we hope it will illuminate the motivation behind some of the things we will do.

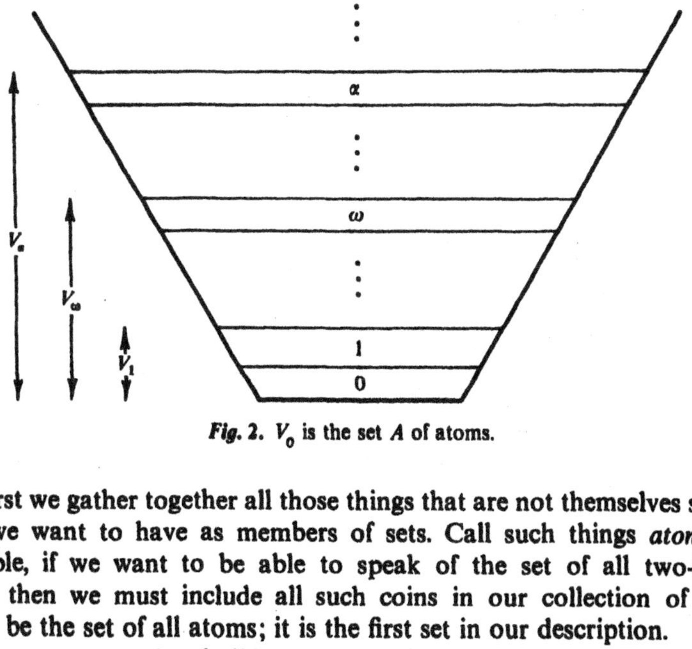

Fig. 2. V_0 is the set A of atoms.

First we gather together all those things that are not themselves sets but that we want to have as members of sets. Call such things *atoms*. For example, if we want to be able to speak of the set of all two-headed coins, then we must include all such coins in our collection of atoms. Let A be the set of all atoms; it is the first set in our description.

We now proceed to build up a hierarchy

$$V_0 \subseteq V_1 \subseteq V_2 \subseteq \cdots$$

of sets. At the bottom level (in a vertical arrangement as in Fig. 2) we take $V_0 = A$, the set of atoms. The next level will also contain all sets of atoms:

$$V_1 = V_0 \cup \mathscr{P}V_0 = A \cup \mathscr{P}A.$$

The third level contains everything that is in a lower level, plus all sets of things from lower levels:

$$V_2 = V_1 \cup \mathscr{P}V_1.$$

And in general

$$V_{n+1} = V_n \cup \mathscr{P}V_n.$$

Thus we obtain successively V_0, V_1, V_2, ... But even this infinite hierarchy does not include enough sets. For example, $\varnothing \in V_1$, $\{\varnothing\} \in V_2$, $\{\{\varnothing\}\} \in V_3$, etc., but we do not yet have the infinite set

$$\{\varnothing, \{\varnothing\}, \{\{\varnothing\}\}, \ldots\}.$$

To remedy this lack, we take the infinite union

$$V_\omega = V_0 \cup V_1 \cup \cdots,$$

and then let $V_{\omega+1} = V_\omega \cup \mathscr{P}V_\omega$, and we continue. In general for any α,

$$V_{\alpha+1} = V_\alpha \cup \mathscr{P}V_\alpha,$$

and this goes on "forever." Whenever you might think that the construction is finished, you instead take the union of all the levels obtained so far, take the power set of that union, and continue.

A better explanation of the "forever" idea must be delayed until we discuss (in Chapter 7) the "numbers" being used as subscripts in the preceding paragraphs. These are the so-called "ordinal numbers." The ordinal numbers begin with 0, 1, 2, ...; then there is the infinite number ω, then $\omega + 1$, $\omega + 2$, ...; and this goes on "forever."

A fundamental principle is the following: Every set appears somewhere in this hierarchy. That is, for every set a there is some α with $a \in V_{\alpha+1}$. That is what the sets are; they are the members of the levels of our hierarchy.

Examples Suppose that a and b are sets. Say that $a \in V_{\alpha+1}$ and $b \in V_{\beta+1}$ and suppose that $V_{\beta+1}$ is "higher" in the hierarchy than $V_{\alpha+1}$. Then both a and b are in $V_{\beta+1}$, since each level includes all lower levels. Consequently in $V_{\beta+2}$ we have the pair set $\{a, b\}$. On the other hand at no point do we obtain a set of all sets, i.e., a set having all sets as members. There simply is no such set.

There is one way in which we can simplify our picture. We were very indefinite about just what was in the set A of atoms. The fact of the matter is that the atoms serve no mathematically necessary purpose, so we banish them; we take $A = \varnothing$. In so doing, we lose the ability to form sets of flowers or sets of people. But this is no cause for concern; we do not need set theory to talk about people and we do not need people in our set theory. But we definitely do want to have sets of numbers, e.g., $\{2, 3 + i\pi\}$. Numbers do not appear at first glance to be sets. But as we shall discover (in Chapters 4 and 5), we can find sets that serve perfectly well as numbers.

Our theory then will ignore all objects that are not sets (as interesting and real as such objects may be). Instead we will concentrate just on "pure" sets that can be constructed without the use of such external objects. In

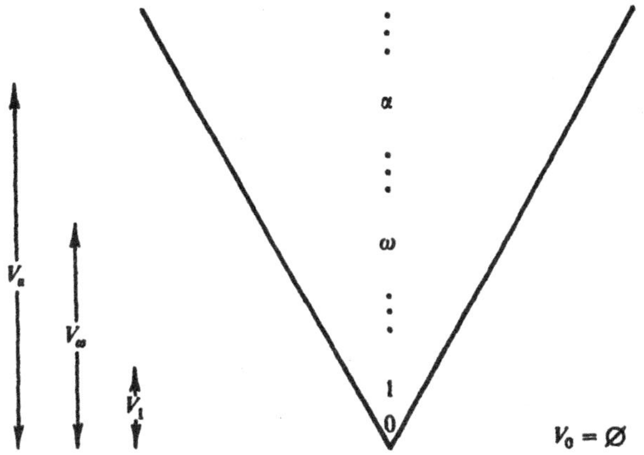

Fig. 3. The ordinals are the backbone of the universe.

particular, any member of one of our sets will itself be a set, and each of *its* members, if any, will be a set, and so forth. (This does not produce an infinite regress, because we stop when we reach $\varnothing$.)

Now that we have banished atoms, the picture becomes narrower (Fig. 3). The construction is also simplified. We have defined $V_{\alpha+1}$ to be $V_\alpha \cup \mathscr{P}V_\alpha$. Now it turns out that this is the same as $A \cup \mathscr{P}V_\alpha$ (see Exercise 6). With $A = \varnothing$, we have simply $V_{\alpha+1} = \mathscr{P}V_\alpha$.

Exercises

5. Define the rank of a set c to be the least α such that $c \subseteq V_\alpha$. Compute the rank of $\{\{\varnothing\}\}$. Compute the rank of $\{\varnothing, \{\varnothing\}, \{\varnothing, \{\varnothing\}\}\}$.

6. We have stated that $V_{\alpha+1} = A \cup \mathscr{P}V_\alpha$. Prove this at least for $\alpha < 3$.

7. List all the members of V_3. List all the members of V_4. (It is to be assumed here that there are no atoms.)

CLASSES

There is no "set of all sets," i.e., there is no set having all sets as members. This is in accordance with our informal image of the hierarchical way sets are constructed. Later, the nonexistence of a set of all sets will become a theorem (Theorem 2A), provable from the axioms.

Nonetheless, there is some mild inconvenience that results if we forbid ourselves even to speak of the collection of all sets. The collection cannot be a set, but what status can we give it? Basically there are two alternatives:

The Zermelo-Fraenkel alternative The collection of all sets need have no ontological status at all, and we need never speak of it. When tempted to speak of it, we can seek a rephrasing that avoids it.

The von Neumann–Bernays alternative The collection of all sets can be called a *class*. Similarly any other collection of sets can be called a class. In particular, any set is a class, but some classes are too large to be sets. Informally, a class A is a set if it is included in some level V_α of our hierarchy (and then is a member of $V_{\alpha+1}$). Otherwise it is not a set, and can never be a member of a set.

For advanced works in set theory, the Zermelo–Fraenkel alternative seems to be the better of the two. It profits from the simplicity of having to deal with only one sort of object (sets) instead of two (classes and sets). And the circumlocutions it requires (in order to avoid reference to classes that are not sets) are things to which set-theorists learn early to adapt.

For introductory works in set theory (such as this book), the choice between the two alternatives is less clear. The prohibition against mentioning any class that fails to be a set seems unnatural and possibly unfair. Desiring to have our cake and eat it too, we will proceed as follows. We officially adopt the Zermelo–Fraenkel alternative. Consequently the axioms and theorems shall make no mention of any class that is not a set. But in the expository comments, we will not hesitate to mention, say, the class of all sets if it appears helpful to do so. To avoid confusion, we will reserve upper-case sans serif letters (A, B, ...) for classes that are not guaranteed to be sets.

AXIOMATIC METHOD

In this book we are going to state the axioms of set theory, and we are going to show that our theorems are consequences of those axioms. The great advantage of the axiomatic method is that it makes totally explicit just what our initial assumptions are.

It is sometimes said that "mathematics can be embedded in set theory." This means that mathematical objects (such as numbers and differentiable functions) can be defined to be certain sets. And the theorems of mathematics (such as the fundamental theorem of calculus) then can be viewed as statements about sets. Furthermore, these theorems will be provable from our axioms. Hence our axioms provide a sufficient collection of assumptions for the development of the whole of mathematics—a remarkable fact. (In Chapter 5 we will consider further the procedure for embedding mathematics in set theory.)

The axiomatic method has been useful in other subjects as well as in set theory. Consider plane geometry, for example. It is quite possible to talk about lines and triangles without using axioms. But the advantages of axiomatizing geometry were seen very early in the history of the subject.

The nonaxiomatic approach to set theory is often referred to as "naive set theory," a terminology that does not hide its bias. Historically, set theory originated in nonaxiomatic form. But the paradoxes of naive set theory (the most famous being Russell's paradox) forced the development of axiomatic set theory, by showing that certain assumptions, apparently plausible, were inconsistent and hence totally untenable. It then became mandatory to give explicit assumptions that could be examined by skeptics for possible inconsistency. Even without the foundational crises posed by the paradoxes of naive set theory, the axiomatic approach would have been developed to cope with later controversy over the truth or falsity of certain principles, such as the axiom of choice (Chapter 6). Of course our selection of axioms will be guided by the desire to reflect as accurately as possible our informal (preaxiomatic) ideas regarding sets and classes.

It is nonetheless quite possible to study set theory from the nonaxiomatic viewpoint. We have therefore arranged the material in the book in a "two-tier" fashion. The passages dealing with axioms will henceforth be marked by a stripe in the left margin. The reader who omits such passages and reads only the unstriped material will thereby find a nonaxiomatic development of set theory. Perhaps he will at some later time wish to look at the axiomatic underpinnings. On the other hand, the reader who omits nothing will find an axiomatic development. Notice that most of the striped passages appear early in the book, primarily in the first three chapters. Later in the book the nonaxiomatic and the axiomatic approaches largely converge.

Our axiom system begins with two *primitive notions*, the concepts of "set" and "member." In terms of these concepts we will define others, but the primitive notions remain undefined. Instead we adopt a list of *axioms* concerning the primitive notions. (The axioms can be thought of as divulging partial information regarding the meaning of the primitive notions.)

Having adopted a list of axioms, we will then proceed to derive sentences that are *logical consequences* (or *theorems*) of the axioms. Here a sentence σ is said to be a logical consequence of the axioms if any assignment of meaning to the undefined notions of set and member making the axioms true also makes σ true.

We have earlier sketched, in an informal and sometimes vague way, what "set" and "member" are *intended* to mean. But for a sentence to be a logical consequence of the axioms, it must be true *whatever* "set" and "member" mean, provided only that the axioms are true. The sentences that appear, on the basis of our informal viewpoint, as if they ought to be true, must still be shown to be logical consequences of the axioms before being accepted as theorems. In return for adopting this restriction, we escape any drawbacks of the informality and vagueness of the nonaxiomatic viewpoint.

(There is an interesting point here concerning the foundations of mathematics. If σ is a logical consequence of a list of axioms, is there then a finitely long *proof* of σ from the axioms? The answer is affirmative, under a very reasonable definition of "proof." This is an important result in mathematical logic. The topic is treated, among other places, in our book *A Mathematical Introduction to Logic*, Academic Press, 1972.)

For example, the first of our axioms is the axiom of extensionality, which is almost as follows: Whenever A and B are sets such that exactly the same things are members of one as are members of the other, then $A = B$. Imagine for the moment that this were our *only* axiom. We can then consider some of the logical consequences of this one axiom.

For a start, take the sentence: "There cannot be two different sets, each of which has no members." This sentence *is* a logical consequence of extensionality, for we claim that any assignment of meaning to "set" and "member" making extensionality true also makes the above sentence true. To prove this, we argue as follows. Let A and B be any sets, each of which has no members. Then exactly the same things belong to A as to B, since none belong to either. Hence by extensionality, $A = B$. (The validity of this argument, while independent of the meaning of "set" or "member," does depend on the meaning of the logical terms such as "each," "no," "equal," etc.)

On the other hand, consider a simple sentence such as σ: "There are two sets, one of which is a member of the other." This sentence is *not* a logical consequence of extensionality. To see this, let the word "set" mean "a number equal to 2," and let "member of" mean "unequal to." Under this interpretation, extensionality is true but σ is false. Of course we will soon add other axioms of which σ will be a logical consequence.

NOTATION

To denote sets we will use a variety of letters, both lowercase (a, b, ...), uppercase (A, B, ...), and even script letters and Greek letters. Where feasible, we will attempt to have the larger and fancier letters denote sets higher in our hierarchy of levels than those denoted by smaller and plainer letters. In addition, letters can be embellished with subscripts, primes, and the like. This assures us of an inexhaustible supply of symbols for sets. In other words, when it comes to naming sets, anything goes.

It will often be advantageous to exploit the symbolic notation of mathematical logic. This symbolic language, when used in judicious amounts to replace the English language, has the advantages of both conciseness (so that expressions are shorter) and preciseness (so that expressions are less ambiguous).

The following symbolic expressions will be used to abbreviate the corresponding English expressions:

$\forall x$	for every set x
$\exists x$	there exists a set x such that
$\neg$	not
&	and
or	or (in the sense "one or the other *or both*")
$\Rightarrow$	implies ("__ $\Rightarrow$ __" abbreviates "if __, then __")
$\Leftrightarrow$	if and only if, also abbreviated "iff"

You have probably seen these abbreviations before; they are discussed in more detail in the Appendix.

We also have available the symbols $\in$ and $=$ (and $\notin$ and $\neq$, although we could economize by eliminating, e.g., "$a \notin B$" in favor of "$\neg a \in B$"). With all these symbols, variables (a, b, ...), and parentheses we *could* avoid the English language altogether in the statement of axioms and theorems. We will not actually do so, or at least not all at once. But the fact that we have this splendid formal language available is more than a theoretical curiosity. When we come to stating certain axioms, the formal language will be a necessity.

Notice that we read $\forall x$ as "for all *sets* x," rather than "for all *things* x." This is due to our decision to eliminate atoms from our theory. A result of this elimination is that everything we consider is a set; e.g., every member of a set will itself be a set.

Example The principle of extensionality can be written as

$$\forall A\ \forall B[(A \text{ and } B \text{ have exactly the same members}) \Rightarrow A = B].$$

Then "A and B have exactly the same members" can be written as

$$\forall x(x \in A \iff x \in B)$$

so that extensionality becomes

$$\forall A\, \forall B[\forall x(x \in A \iff x \in B) \Rightarrow A = B].$$

Example "There is a set to which nothing belongs" can be written

$$\exists B\, \forall x\, x \notin B.$$

These two examples constitute our first two axioms. It is not really necessary for us to state the axioms in symbolic form. But we will seize the opportunity to show how it *can* be done, on the grounds that the more such examples you see, the more natural the notation will become to you.

HISTORICAL NOTES

The concept of a set is very basic and natural, and has been used in mathematical writings since ancient times. But the theory of abstract sets, as objects to be studied for their own interest, was originated largely by Georg Cantor (1845–1918). Cantor was a German mathematician, and his papers on set theory appeared primarily during the period from 1874 to 1897.

Cantor was led to the study of set theory in a very indirect way. He was studying trigonometric series (Fourier series), e.g., series of the form

$$a_1 \sin x + a_2 \sin 2x + a_3 \sin 3x + \cdots.$$

Such series had been studied throughout the nineteenth century as solutions to differential equations representing physical problems. Cantor's work on Fourier series led him to consider more and more general sets of real numbers. In 1871 he realized that a certain operation on sets of real numbers (the operation of forming the set of limit points) could be iterated more than a finite number of times; starting with a set P_0 one could form $P_0, P_1, P_2, \ldots, P_\omega, P_{\omega+1}, \ldots, P_{\omega+\omega}, \ldots$. In December of 1873 Cantor proved that the set of all real numbers could not be put into one-to-one correspondence with the integers (Theorem 6B); this result was published in 1874. In 1879 and subsequent years, he published a series of papers setting forth the general concepts of abstract sets and "transfinite numbers."

Cantor's work was well received by some of the prominent mathematicians of his day, such as Richard Dedekind. But his willingness to regard infinite sets as objects to be treated in much the same way as finite sets was bitterly attacked by others, particularly Kronecker. There was no objection to a "potential infinity" in the form of an unending process, but an "actual infinity" in the form of a completed infinite set was harder to accept.

About the turn of the century, attempts were made to present the principles of set theory as being principles of logic—as self-evident truths of deductive thought. The foremost work in this direction was done by Gottlob Frege. Frege was a German mathematician by training, who contributed to both mathematics and philosophy. In 1893 and 1903 he published a two-volume work in which he indicated how mathematics could be developed from principles that he regarded as being principles of logic. But just as the second volume was about to appear, Bertrand Russell informed Frege of a contradiction derivable from the principles (Russell's paradox).

Russell's paradox had a tremendous impact on the ideas of that time regarding the foundations of mathematics. It was not, to be sure, the first paradox to be noted in set theory. Cantor himself had observed that some collections, such as the collection of all sets, had to be regarded as "inconsistent totalities," in contrast to the more tractable "consistent totalities," such as the set of numbers. In this he foreshadowed the distinction between proper classes and sets, which was introduced by John von Neumann in 1925. Also in 1897, Cesare Burali-Forti had observed a paradoxical predicament in Cantor's theory of transfinite ordinal numbers. But the simplicity and the directness of Russell's paradox seemed to destroy utterly the attempt to base mathematics on the sort of set theory that Frege had proposed.

The first axiomatization of set theory was published by Ernst Zermelo in 1908. His axioms were essentially the ones we state in Chapters 2 and 4. (His *Aussonderung* axioms were later made more precise by Thoralf Skolem and others.) It was observed by several people that for a satisfactory theory of ordinal numbers, Zermelo's axioms required strengthening. The axiom of replacement (Chapter 7) was proposed by Abraham Fraenkel (in 1922) and others, giving rise to the list of axioms now known as the "Zermelo–Fraenkel" (ZF) axioms. The axiom of regularity or foundation (Chapter 7) was at least implicit in a 1917 paper by Dmitry Mirimanoff and was explicitly included by von Neumann in 1925.

An axiomatization admitting proper classes as legitimate objects was formulated by von Neumann in the 1925 paper just mentioned. Some of his ideas were utilized by Paul Bernays in the development of a more satisfactory axiomatization, which appeared in a series of papers published in 1937 and later years. A modification of Bernays's axioms was used by Kurt Gödel in a 1940 monograph. This approach is now known as "von Neumann–Bernays" (VNB) or "Gödel–Bernays" (GB) set theory.

The use of the symbol ∈ (a stylized form of the Greek epsilon) to denote membership was initiated by the Italian mathematician Giuseppe Peano in 1889. It abbreviates the Greek word ἐστί, which means "is." The underlying rationale is illustrated by the fact that if B is the set of all blue objects, then we write "$x \in B$" in order to assert that x *is* blue.

Present-day research in set theory falls for the most part into two branches. One branch involves investigating the consequences of new and stronger axioms. The other branch involves the "metamathematics" of set theory, which is the study not of sets but of the workings of set *theory* itself: its proofs, its theorems, and its nontheorems.

CHAPTER 2

AXIOMS AND OPERATIONS

In this chapter we begin by introducing the first six of our ten axioms. Initially the axiomatization might appear to be like cumbersome machinery to accomplish simple tasks. But we trust that it will eventually prove itself to be powerful machinery for difficult tasks. Of course the axioms are not chosen at random, but must ultimately reflect our informal ideas about what sets are.

In addition to the introduction of basic concepts, this chapter provides practice in using the symbolic notation introduced in Chapter 1. Finally the chapter turns to the standard results on the algebra of sets.

AXIOMS

The first of our axioms is the principle of extensionality.

Extensionality Axiom If two sets have exactly the same members, then they are equal:

$$\forall A\, \forall B[\forall x(x \in A \iff x \in B) \implies A = B].$$

The above symbolic rendering of extensionality is the one we developed previously. We will supply similar symbolizations for the other axioms. The more practice the better!

Next we need some axioms assuring the existence of some basic sets that were encountered informally in the preceding chapter.

Empty Set Axiom There is a set having no members:

$$\exists B\, \forall x\; x \notin B.$$

Pairing Axiom For any sets u and v, there is a set having as members just u and v:

$$\forall u\, \forall v\, \exists B\, \forall x(x \in B \iff x = u \text{ or } x = v).$$

Union Axiom, Preliminary Form For any sets a and b, there is a set whose members are those sets belonging either to a or to b (or both):

$$\forall a\, \forall b\, \exists B\, \forall x(x \in B \iff x \in a \text{ or } x \in b).$$

Power Set Axiom For any set a, there is a set whose members are exactly the subsets of a:

$$\forall a\, \exists B\, \forall x(x \in B \iff x \subseteq a).$$

Here we can, if we wish, rewrite "$x \subseteq a$" in terms of the definition of $\subseteq$:

$$\forall t(t \in x \implies t \in a).$$

Later we will expand this list to include

subset axioms, replacement axioms,
infinity axiom, regularity axiom,
choice axiom.

Also the union axiom will be restated in a stronger form. (Not all of these axioms are really necessary; some will be found to be redundant.)

The set existence axioms can now be used to justify the definition of symbols used informally in Chapter 1. First of all, we want to define the symbol "$\varnothing$."

Definition $\varnothing$ is the set having no members.

This definition bestows the name "$\varnothing$" on a certain set. But when we write down such a definition there are two things of which we must be sure: We must know that there exists a set having no members, and we must know that there cannot be more than one set having no members. The empty set axiom provides the first fact, and extensionality provides the second. Without both facts the symbol "$\varnothing$" would not be well defined.

Severe logical difficulties arise from introducing symbols when either there is no object for the symbol to name, or (even worse) the symbol names ambiguously more than one object.

The other set existence axioms justify the definition of the following symbols.

Definition. (i) For any sets u and v, the *pair set* $\{u, v\}$ is the set whose only members are u and v.

(ii) For any sets a and b, the *union* $a \cup b$ is the set whose members are those sets belonging either to a or to b (or both).

(iii) For any set a, the *power set* $\mathscr{P}a$ is the set whose members are exactly the subsets of a.

As with the empty set, our set existence axioms assure us that the sets being named exist, and extensionality assures us that the sets being named are unique.

We can use pairing and union together to form other finite sets. First of all, given any x we have the *singleton* $\{x\}$, which is defined to be $\{x, x\}$. And given any x_1, x_2, and x_3 we can define

$$\{x_1, x_2, x_3\} = \{x_1, x_2\} \cup \{x_3\}.$$

Similarly we can define $\{x_1, x_2, x_3, x_4\}$, and so forth.

Having defined the union operation, we should accompany it by the intersection operation. But to justify the definition of intersection we need new axioms, to which we now turn. In the next few paragraphs we shall use our informal view of sets to motivate the formulation of these axioms.

Observe that our set existence axioms contain expressions like "there is a set B whose members are those sets x satisfying the condition __," where the blank is filled by some condition specifying which sets we want. In symbols this becomes

$$\exists B\, \forall x(x \in B \quad \Leftrightarrow \quad \_\_).$$

If the axiom mentions some other sets $t_1, \ldots, t_k$, then the full version becomes

$$\forall t_1 \cdots \forall t_k\, \exists B\, \forall x(x \in B \quad \Leftrightarrow \quad \_\_)$$

with the blank filled by some expression involving $t_1, \ldots, t_k$, and x. The empty set axiom is not quite in this form, but it can be rewritten as

$$\exists B\, \forall x(x \in B \quad \Leftrightarrow \quad x \neq x),$$

which is in the above form (with $k = 0$). The set B whose existence is asserted by such an axiom is (by extensionality) uniquely determined by $t_1, \ldots, t_k$, so we can give it a name (in which the symbols $t_1, \ldots, t_k$ appear). This is just what we have done.

Now let us try to be more general and consider any sentence σ of the form

$$\forall t_1 \cdots \forall t_k \exists B \forall x(x \in B \iff \_\_),$$

where the blank is filled by some expression involving at most $t_1, \ldots, t_k$, and x. If this sentence is true, then the set B could be named by use of the abstraction notation of Chapter 1:

$$B = \{x \mid \_\_\}.$$

The sets recently defined can be named by use of the abstraction notation:

$$\varnothing = \{x \mid x \neq x\},$$
$$\{u, v\} = \{x \mid x = u \text{ or } x = v\},$$
$$a \cup b = \{x \mid x \in a \text{ or } x \in b\},$$
$$\mathscr{P}a = \{x \mid x \subseteq a\}.$$

One might be tempted to think that *any* sentence σ of the form

$$\forall t_1 \cdots \forall t_k \exists B \forall x(x \in B \iff \_\_)$$

should be adopted as true. But this is wrong; some sentences of this form are false in our informal view of sets (Chapter 1). For example,

$$\exists B \forall x(x \in B \iff x = x)$$

is false, since it asserts the existence of a set B to which every set belongs. The most that can be said is that there is a *class* A (but not necessarily a set) whose members are those *sets* x such that __:

$$\mathsf{A} = \{x \mid \_\_\}.$$

In order for the class A to be a set it must be included in some level V_α of the hierarchy. In fact it is enough for A to be included in any set c, for then

$$\mathsf{A} \subseteq c \subseteq V_\alpha$$

for some α, and from this it follows that $\mathsf{A} \in V_{\alpha+1}$.

All this is to motivate the adoption of the subset axioms. These axioms say, very roughly, that any class A included in some *set* c must in fact be a set. But the axioms can refer (in the Zermelo–Fraenkel alternative) only to sets. So instead of direct reference to the class A, we refer instead to the expression __ that defined A.

Subset Axioms For each formula __ not containing B, the following is an axiom:

$$\forall t_1 \cdots \forall t_k \forall c\, \exists B\, \forall x(x \in B \iff x \in c \;\&\; \_\_).$$

In English, the axiom asserts (for any $t_1, \ldots, t_k$ and c) the existence of a set B whose members are exactly those sets x in c such that __. It then follows automatically that B is a subset of c (whence the name "subset axiom"). The set B is uniquely determined (by $t_1, \ldots, t_k$ and c), and can be named by use of a variation on the abstraction notation:

$$B = \{x \in c \mid \_\_\}.$$

Example One of the subset axioms is:

$$\forall a\, \forall c\, \exists B\, \forall x(x \in B \iff x \in c \;\&\; x \in a).$$

This axiom asserts the existence of the set we define to be the ***intersection*** $c \cap a$ of c and a.

We are not tied to one particular choice of letters. For example, we will also allow as a subset axiom:

$$\forall A\, \forall B\, \exists S\, \forall t[t \in S \iff t \in A \;\&\; t \notin B].$$

This set S is the ***relative complement*** of B in A, denoted $A - B$.

Note on Terminology The subset axioms are often known by the name Zermelo gave them, ***Aussonderung axioms***. The word *Aussonderung* is German, and is formed from *sonderen* (to separate) and *aus* (out).

Example In Chapter 4 we shall construct the set ω of natural numbers:

$$\omega = \{0, 1, 2, \ldots\}.$$

We will then be able to use the subset axioms to form the set of even numbers and the set of primes:

$$\{x \in \omega \mid x \text{ is even}\} \qquad \text{and} \qquad \{y \in \omega \mid y \text{ is prime}\}.$$

(But to do this we must be able to express "x is even" by means of a legal formula; we will return to this point shortly.)

Example Let s be some set. Then there is a set Q whose members are the one-element subsets of s:

$$Q = \{a \in \mathscr{P}s \mid a \text{ is a one-element subset of } s\}.$$

We can now use the argument of Russell's paradox to show that the *class* V of all sets is not itself a set.

Theorem 2A There is no set to which every set belongs.

Proof Let A be a set; we will construct a set not belonging to A. Let

$$B = \{x \in A \mid x \notin x\}.$$

We claim that $B \notin A$. We have, by the construction of B,

$$B \in B \iff B \in A \;\&\; B \notin B.$$

If $B \in A$, then this reduces to

$$B \in B \iff B \notin B,$$

which is impossible, since one side must be true and the other false. Hence $B \notin A$. ⊣

One might ask whether a set can ever be a member of itself. We will argue much later (in Chapter 7) that it cannot. And consequently in the preceding proof, the set B is actually the same as the set A.

At this point we need to say just what a formula is. After all, it would be most unfortunate to have as one of the subset axioms

$$\exists B\, \forall x(x \in B \iff x \in \omega \;\&\; x \text{ is an integer definable in one line of type}).$$

We are saved from this disaster by our logical symbols. By insisting that the formula be expressible in the formal language these symbols give us, we can eliminate "x is an integer definable in one line of type" from the list of possible formulas. (Moral: Those symbols are your friends!)

The simplest formulas are expressions such as

$$a \in B \qquad \text{and} \qquad a = b$$

(and similarly with other letters). More complicated formulas can then be built up from these by use of the expressions

$$\forall x, \quad \exists x, \quad \neg, \quad \&, \quad \text{or}, \quad \Rightarrow, \quad \iff,$$

together with enough parentheses to avoid ambiguity. That is, from formulas φ and ψ we can construct longer formulas $\forall x\ \varphi$, $\exists x\ \varphi$ (and similarly $\forall y\ \varphi$, etc.), $(\neg\varphi)$, $(\varphi \;\&\; \psi)$, $(\varphi \text{ or } \psi)$, $(\varphi \Rightarrow \psi)$, and $(\varphi \iff \psi)$. We define a *formula* to be a string of symbols constructed from the simplest formulas by use of the above-listed methods. For example,

$$\exists x(x \in A \;\&\; \forall t(t \in x \Rightarrow (\neg t \in A)))$$

is a formula. In practice, however, we are likely to abbreviate it by something a little more readable, such as

$$(\exists x \in A)(\forall t \in x)\, t \notin A.$$

An ungrammatical string of symbols such as $)) \Rightarrow A$ is not a formula, nor is

$$x \text{ is an integer definable in one line of type.}$$

Example Let s be some set. In a previous example we formed the set of one-element subsets of s:

$$\{a \in \mathscr{P}s \mid a \text{ is a one-element subset of } s\}.$$

Now "a is a one-element subset of s" is not itself a formula, but it can be rewritten as a formula. As a first step, it can be expressed as

$$a \subseteq s \;\&\; a \neq \varnothing \;\&\; \text{any two members of } a \text{ coincide.}$$

This in turn becomes the formula

$$((\forall x(x \in a \quad \Rightarrow \quad x \in s) \;\&\; \exists y\, y \in a) \;\&\; \forall u\, \forall v((u \in a \;\&\; v \in a) \quad \Rightarrow \quad u = v)).$$

In applications of subset axioms we generally will not write out the formula itself. And this example shows why; "a is a one-element subset of s" is much easier to read than the legal formula. But in every case it will be possible (for a person with unbounded patience) to eliminate the English words and the defined symbols (such as $\varnothing$, $\cup$, and so forth) in order to arrive at a legal formula. The procedure for eliminating defined symbols is discussed further in the Appendix.

ARBITRARY UNIONS AND INTERSECTIONS

The union operation previously described allows us to form the union $a \cup b$ of two sets. By repeating the operation, we can form the union of three sets or the union of forty sets. But suppose we want the union of infinitely many sets; suppose we have an infinite collection of sets

$$A = \{b_0, b_1, b_2, \ldots\}$$

and we want to take the union of all the b_i's. For this we need a more general union operation:

$$\begin{aligned} \bigcup A &= \bigcup_i b_i \\ &= \{x \mid x \text{ belongs to some member } b_i \text{ of } A\}. \end{aligned}$$

This leads us to make the following definition. For any set A, the *union* $\bigcup A$ of A is the set defined by

$$\begin{aligned} \bigcup A &= \{x \mid x \text{ belongs to some member of } A\} \\ &= \{x \mid (\exists b \in A)\, x \in b\}. \end{aligned}$$

Thus $\bigcup A$ is a melting pot into which all members of A are dumped. For example, suppose Un is (as on p. 4) the set of countries belonging to the United Nations. Then $\bigcup Un$ is the set of all people that are citizens of some country belonging to the United Nations. A smaller example (and one that avoids sets of people) is

$$\bigcup\{\{2, 4, 6\}, \{6, 16, 26\}, \{0\}\}.$$

You should evaluate this expression to make certain that you understand the union operation. If you do it correctly, you will end up with a set of six numbers.

We need an improved version of the union axiom in order to know that a set exists containing the members of the members of A.

Union Axiom For any set A, there exists a set B whose elements are exactly the members of the members of A:

$$\forall x[x \in B \iff (\exists b \in A)\, x \in b].$$

We can state the definition of $\bigcup A$ in the following form:

$$x \in \bigcup A \iff (\exists b \in A)\, x \in b.$$

For example,

$$\begin{aligned}\bigcup\{a, b\} &= \{x \mid x \text{ belongs to some member of } \{a, b\}\}\\ &= \{x \mid x \text{ belongs to } a \text{ or to } b\}\\ &= a \cup b.\end{aligned}$$

This example shows that our preliminary form of the union axiom can be discarded in favor of the new form. That is, the set $a \cup b$ produced by the preliminary form can also be obtained from pairing and the revised form of the union axiom.

Similarly we have

$$\bigcup\{a, b, c, d\} = a \cup b \cup c \cup d \qquad \text{and} \qquad \bigcup\{a\} = a.$$

An extreme case is $\bigcup\varnothing = \varnothing$.

We also want a corresponding generalization of the intersection operation. Suppose we want to take the intersection of infinitely many sets $b_0, b_1, \ldots$. Then where

$$A = \{b_0, b_1, \ldots\}$$

the desired intersection can be informally characterized as

$$\begin{aligned}\bigcap A &= \bigcap_i b_i\\ &= \{x \mid x \text{ belongs to every } b_i \text{ in } A\}.\end{aligned}$$

In general, we define for every nonempty set A, the *intersection* $\bigcap A$ of A by the condition

$$x \in \bigcap A \quad \Leftrightarrow \quad x \text{ belongs to every member of } A.$$

In contrast to the union operation, no special axiom is needed to justify the intersection operation. Instead we have the following theorem.

Theorem 2B For any nonempty set A, there exists a unique set B such that for any x,

$$x \in B \quad \Leftrightarrow \quad x \text{ belongs to every member of } A.$$

This theorem permits defining $\bigcap A$ to be that unique set B.

Proof We are given that A is nonempty; let c be some fixed member of A. Then by a subset axiom there is a set B such that for any x,

$$\begin{aligned} x \in B \quad &\Leftrightarrow \quad x \in c \mathbin{\&} x \text{ belongs to every other member of } A \\ &\Leftrightarrow \quad x \text{ belongs to every member of } A. \end{aligned}$$

Uniqueness, as always, follows from extensionality. ⊣

Examples

$$\bigcap\{\{1, 2, 8\}, \{2, 8\}, \{4, 8\}\} = \{8\},$$
$$\bigcup\{\{1, 2, 8\}, \{2, 8\}, \{4, 8\}\} = \{1, 2, 4, 8\}.$$

Examples

$$\bigcap\{a\} = a,$$
$$\bigcap\{a, b\} = a \cap b,$$
$$\bigcap\{a, b, c\} = a \cap b \cap c.$$

In these last examples, as A becomes larger, $\bigcap A$ gets smaller. More precisely: Whenever $A \subseteq B$, then $\bigcap B \subseteq \bigcap A$. There is one troublesome extreme case. What happens if $A = \varnothing$? For any x at all, it is vacuously true that x belongs to every member of $\varnothing$. (There can be no member of $\varnothing$ to which x fails to belong.) Thus it looks as if $\bigcap\varnothing$ should be the class V of all sets. By Theorem 2A, there is no set C such that for all x,

$$x \in C \quad \Leftrightarrow \quad x \text{ belongs to every member of } \varnothing$$

since the right side is true of every x. This presents a mild notational problem: How do we define $\bigcap\varnothing$? The situation is analogous to division by zero in arithmetic. How does one define $a \div 0$? One option is to leave $\bigcap\varnothing$ undefined, since there is no very satisfactory way of defining it. This option works perfectly well, but some logicians dislike it. It leaves $\bigcap\varnothing$ as an untidy loose end, which they may later trip over. The other option is to select some arbitrary scapegoat (the set $\varnothing$ is always used for this) and define $\bigcap\varnothing$ to equal that object. Either way, whenever one forms $\bigcap A$ one must beware the possibility that perhaps $A = \varnothing$. Since it makes no difference which of the two options one follows, we will not bother to make a choice between them at all.

Example If $b \in A$, then $b \subseteq \bigcup A$.

Example If $\{\{x\}, \{x, y\}\} \in A$, then $\{x, y\} \in \bigcup A$, $x \in \bigcup\bigcup A$, and $y \in \bigcup\bigcup A$.

Example $\bigcap\{\{a\}, \{a, b\}\} = \{a\} \cap \{a, b\} = \{a\}$. Hence

$$\bigcup\bigcap\{\{a\}, \{a, b\}\} = \bigcup\{a\} = a.$$

On the other hand,

$$\bigcap\bigcup\{\{a\}, \{a, b\}\} = \bigcap\{a, b\} = a \cap b.$$

Exercises

See also the Review Exercises at the end of this chapter.

1. Assume that A is the set of integers divisible by 4. Similarly assume that B and C are the sets of integers divisible by 9 and 10, respectively. What is in $A \cap B \cap C$?

2. Give an example of sets A and B for which $\bigcup A = \bigcup B$ but $A \neq B$.

3. Show that every member of a set A is a subset of $\bigcup A$. (This was stated as an example in this section.)

4. Show that if $A \subseteq B$, then $\bigcup A \subseteq \bigcup B$.

5. Assume that every member of $\mathscr{A}$ is a subset of B. Show that $\bigcup\mathscr{A} \subseteq B$.

6. (a) Show that for any set A, $\bigcup\mathscr{P}A = A$.
   (b) Show that $A \subseteq \mathscr{P}\bigcup A$. Under what conditions does equality hold?

7. (a) Show that for any sets A and B,

$$\mathscr{P}A \cap \mathscr{P}B = \mathscr{P}(A \cap B).$$

   (b) Show that $\mathscr{P}A \cup \mathscr{P}B \subseteq \mathscr{P}(A \cup B)$. Under what conditions does equality hold?

8. Show that there is no set to which every singleton (that is, every set of the form $\{x\}$) belongs. [*Suggestion*: Show that from such a set, we could construct a set to which every set belonged.]

9. Give an example of sets a and B for which $a \in B$ but $\mathscr{P}a \notin \mathscr{P}B$.

10. Show that if $a \in B$, then $\mathscr{P}a \in \mathscr{P}\mathscr{P}\bigcup B$. [*Suggestion*: If you need help, look in the Appendix.]

ALGEBRA OF SETS

Two basic operations on sets are the operations of union and intersection:

$$A \cup B = \{x \mid x \in A \text{ or } x \in B\},$$
$$A \cap B = \{x \mid x \in A \;\&\; x \in B\}.$$

Also we have for any sets A and B the *relative complement* $A - B$ *of* B *in* A:

$$A - B = \{x \in A \mid x \notin B\}.$$

The usual diagram for $A - B$ is shown in Fig. 4. In some books the minus sign is needed for other uses, and the relative complement is then denoted $A \setminus B$.

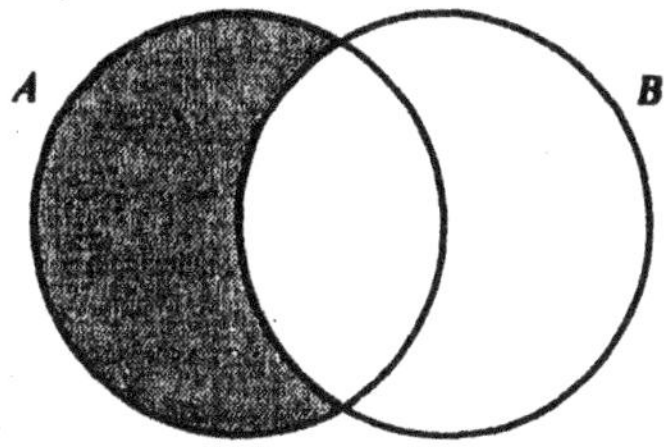

Fig. 4. The shaded area represents $A - B$.

The union axiom was used to give us $A \cup B$. But $A \cap B$ and $A - B$ were both obtained from subset axioms.

We cannot form (as a set) the "absolute complement" of B, i.e., $\{x \mid x \notin B\}$. This class fails to be a set, for its union with B would be the class of all sets. In any event, the absolute complement is unlikely to be an interesting object of study.

For example, suppose one is studying sets of real numbers. Let $\mathbb{R}$ be the set of all real numbers, and suppose that $B \subseteq \mathbb{R}$. Then the relative complement $\mathbb{R} - B$ consists of these real numbers not in B. On the other hand, the absolute complement of B would be a huge class containing all manner of irrelevant things; it would contain *any* set that was not a real number.

Example Let A be the set of all left-handed people, let B be the set of all blond people, and let C be the set of all male people. (We choose to suppress in this example, as in others, the fact that we officially banned people from our sets.) Then $A \cup (B - C)$ is the set of all people who either are left-handed or are blond nonmales (or both). On the other hand $(A \cup B) - C$ is the set of all nonmales who are either left-handed or blond. These two sets are different; Joe (who is a left-handed male) belongs to the first set but not the second. The set

$$(A - C) \cup (B - C)$$

is the same as one of the two sets mentioned above. Which one?

The study of the operations of union ($\cup$), intersection ($\cap$), and relative complementation ($-$), together with the inclusion relation ($\subseteq$), goes by the name of the *algebra of sets*. In some ways, the algebra of sets obeys laws reminiscent of the algebra of real numbers (with $+$, $\cdot$, $-$, and $\leq$), but there are significant differences.

The following identities, which hold for any sets, are some of the elementary facts of the algebra of sets.

Commutative laws

$$A \cup B = B \cup A \qquad \text{and} \qquad A \cap B = B \cap A.$$

Associative laws

$$A \cup (B \cup C) = (A \cup B) \cup C,$$
$$A \cap (B \cap C) = (A \cap B) \cap C.$$

Distributive laws

$$A \cap (B \cup C) = (A \cap B) \cup (A \cap C),$$
$$A \cup (B \cap C) = (A \cup B) \cap (A \cup C).$$

De Morgan's laws

$$C - (A \cup B) = (C - A) \cap (C - B),$$
$$C - (A \cap B) = (C - A) \cup (C - B).$$

Identities involving $\varnothing$

$$A \cup \varnothing = A \qquad \text{and} \qquad A \cap \varnothing = \varnothing,$$
$$A \cap (C - A) = \varnothing.$$

Often one considers sets, all of which are subsets of some large set or "space" S. A common example is the study of subsets of the space $\mathbb{R}$ of real numbers. Assume then that A and B are subsets of S. Then we can abbreviate $S - A$ as simply $-A$, the set S being understood as fixed. In this abbreviation, De Morgan's laws become

$$-(A \cup B) = -A \cap -B,$$
$$-(A \cap B) = -A \cup -B.$$

Further, we have (still under the assumption that $A \subseteq S$)

$$A \cup S = S \quad \text{and} \quad A \cap S = A,$$
$$A \cup -A = S \quad \text{and} \quad A \cap -A = \varnothing.$$

Now we should say something about how one proves all these facts. Let us take as a sample the distributive law:

$$A \cap (B \cup C) = (A \cap B) \cup (A \cap C).$$

One way to check this is to draw the picture (Fig. 5). After shading

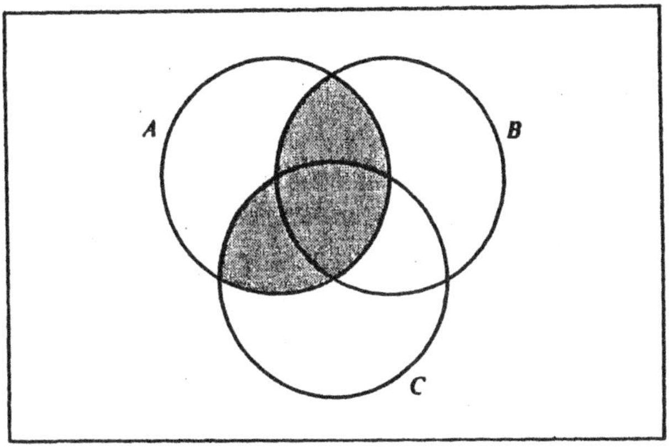

Fig. 5. Diagram for three sets.

the region representing $A \cap (B \cup C)$ and the region representing $(A \cap B) \cup (A \cap C)$, one discovers that these regions are the same.

Is the foregoing proof, which relies on a picture, really trustworthy? Let us run through it again without the picture. To prove the desired equation, it suffices (by extensionality) to consider an arbitrary x, and to show that x belongs to $A \cap (B \cup C)$ iff it belongs to $(A \cap B) \cup (A \cap C)$. So consider this arbitrary x. We do not know for sure whether $x \in A$ or not, whether $x \in B$ or not, etc., but we can list all eight possibilities:

$x \in A$	$x \in B$	$x \in C$
$x \in A$	$x \in B$	$x \notin C$
$x \in A$	$x \notin B$	$x \in C$
$x \in A$	$x \notin B$	$x \notin C$
$x \notin A$	$x \in B$	$x \in C$
$x \notin A$	$x \in B$	$x \notin C$
$x \notin A$	$x \notin B$	$x \in C$
$x \notin A$	$x \notin B$	$x \notin C$.

(These cases correspond to the eight regions of Fig. 5.) We can then verify that in each of the eight cases,

$$x \in A \cap (B \cup C) \quad \text{iff} \quad x \in (A \cap B) \cup (A \cap C).$$

For example, in the fifth case we find that

$$x \notin A \cap (B \cup C) \qquad \text{and} \qquad x \notin (A \cap B) \cup (A \cap C).$$

(What region of Fig. 5 does this case represent?) When verification for the seven other cases has been made, the proof of the equation is complete.

This method of proof is applicable to all the equations listed thus far; in fact it works for any equation or inclusion of this sort. If the equation involves n letters, then there will be 2^n cases. In the distributive law we had three letters and eight cases. But less mechanical methods of proof will be needed for some of the facts listed below.

For the inclusion relation, we have the following monotonicity properties:

$$\begin{aligned} A \subseteq B &\Rightarrow A \cup C \subseteq B \cup C, \\ A \subseteq B &\Rightarrow A \cap C \subseteq B \cap C, \\ A \subseteq B &\Rightarrow \bigcup A \subseteq \bigcup B, \end{aligned}$$

and the "antimonotone" results:

$$\begin{aligned} A \subseteq B &\Rightarrow C - B \subseteq C - A, \\ \varnothing \neq A \subseteq B &\Rightarrow \bigcap B \subseteq \bigcap A. \end{aligned}$$

In each case, the proof is straightforward. For example, in the last case, we assume that every member of A is also a member of B. Hence if $x \in \bigcap B$, i.e., if x belongs to every member of B, then *a fortiori* x belongs to every member of the smaller collection A. And consequently $x \in \bigcap A$.

Next we list some more identities involving arbitrary unions and intersections.

Distributive laws

$$\begin{aligned} A \cup \bigcap \mathscr{B} &= \bigcap \{A \cup X \mid X \in \mathscr{B}\} \qquad \text{for} \quad \mathscr{B} \neq \varnothing, \\ A \cap \bigcup \mathscr{B} &= \bigcup \{A \cap X \mid X \in \mathscr{B}\}. \end{aligned}$$

The notation used on the right side is an extension of the abstraction notation. The set $\{A \cup X \mid X \in \mathscr{B}\}$ (read "the set of all $A \cup X$ such that $X \in \mathscr{B}$") is the unique set $\mathscr{D}$ whose members are exactly the sets of the form $A \cup X$ for some X in $\mathscr{B}$; i.e.,

$$t \in \mathscr{D} \iff t = A \cup X \quad \text{for some } X \text{ in } \mathscr{B}.$$

The existence of such a set $\mathscr{D}$ can be proved by observing that $A \cup X \subseteq A \cup \bigcup \mathscr{B}$. Hence the set $\mathscr{D}$ we seek is a subset of $\mathscr{P}(A \cup \bigcup \mathscr{B})$. A subset axiom produces

$$\{t \in \mathscr{P}(A \cup \bigcup \mathscr{B}) \mid t = A \cup X \text{ for some } X \text{ in } \mathscr{B}\},$$

and this is exactly the set $\mathscr{D}$.

For another example of this notation, suppose that sets $\mathscr{A}$ and C are under consideration. Then

$$\{C - X \mid X \in \mathscr{A}\}$$

is the set of relative complements of members of $\mathscr{A}$, i.e., for any t,

$$t \in \{C - X \mid X \in \mathscr{A}\} \quad \Leftrightarrow \quad t = C - X \quad \text{for some } X \text{ in } \mathscr{A}.$$

Similarly, $\{\mathscr{P}X \mid X \in \mathscr{A}\}$ is the set for which

$$t \in \{\mathscr{P}X \mid X \in \mathscr{A}\} \quad \Leftrightarrow \quad t = \mathscr{P}X \quad \text{for some } X \text{ in } \mathscr{A}.$$

It is not entirely obvious that any such set can be proved to exist, but see Exercise 10.

De Morgan's laws (for $\mathscr{A} \neq \varnothing$)

$$C - \bigcup \mathscr{A} = \bigcap \{C - X \mid X \in \mathscr{A}\},$$
$$C - \bigcap \mathscr{A} = \bigcup \{C - X \mid X \in \mathscr{A}\}.$$

If $\bigcup \mathscr{A} \subseteq S$, then these laws can be written as

$$-\bigcup \mathscr{A} = \bigcap \{-X \mid X \in \mathscr{A}\},$$
$$-\bigcap \mathscr{A} = \bigcup \{-X \mid X \in \mathscr{A}\},$$

where it is understood that $-X$ is $S - X$.

To prove, for example, that for nonempty $\mathscr{A}$ the equation

$$C - \bigcup \mathscr{A} = \bigcap \{C - X \mid X \in \mathscr{A}\}$$

holds, we can argue as follows:

$$\begin{aligned} t \in C - \bigcup \mathscr{A} \quad &\Rightarrow \quad t \in C \quad \text{but } t \text{ belongs to no member of } \mathscr{A} \\ &\Rightarrow \quad t \in C - X \quad \text{for every } X \text{ in } \mathscr{A} \\ &\Rightarrow \quad t \in \bigcap \{C - X \mid X \in \mathscr{A}\}. \end{aligned}$$

Furthermore every step reverses, so that "$\Rightarrow$" can become "$\Leftrightarrow$." (A question for the alert reader: Where do we use the fact that $\mathscr{A} \neq \varnothing$?)

A Final Remark on Notation There is another style of writing some of the unions and intersections with which we have been working. We can write, for example,

$$\bigcap_{X \in \mathscr{B}} (A \cup X) \qquad \text{for} \qquad \bigcap\{A \cup X \mid X \in \mathscr{B}\}$$

and

$$\bigcup_{X \in \mathscr{A}} (C - X) \qquad \text{for} \qquad \bigcup\{C - X \mid X \in \mathscr{A}\}.$$

But for the most part we will stick to our original notation.

Exercises

11. Show that for any sets A and B,

$$A = (A \cap B) \cup (A - B) \qquad \text{and} \qquad A \cup (B - A) = A \cup B.$$

12. Verify the following identity (one of De Morgan's laws):

$$C - (A \cap B) = (C - A) \cup (C - B).$$

13. Show that if $A \subseteq B$, then $C - B \subseteq C - A$.

14. Show by example that for some sets A, B, and C, the set $A - (B - C)$ is different from $(A - B) - C$.

15. Define the symmetric difference $A + B$ of sets A and B to be the set $(A - B) \cup (B - A)$.

(a) Show that $A \cap (B + C) = (A \cap B) + (A \cap C)$.

(b) Show that $A + (B + C) = (A + B) + C$.

16. Simplify:

$$[(A \cup B \cup C) \cap (A \cup B)] - [(A \cup (B - C)) \cap A].$$

17. Show that the following four conditions are equivalent.

(a) $A \subseteq B$, (b) $A - B = \varnothing$,

(c) $A \cup B = B$, (d) $A \cap B = A$.

18. Assume that A and B are subsets of S. List all of the different sets that can be made from these three by use of the binary operations $\cup$, $\cap$, and $-$.

19. Is $\mathscr{P}(A - B)$ always equal to $\mathscr{P}A - \mathscr{P}B$? Is it ever equal to $\mathscr{P}A - \mathscr{P}B$?

20. Let A, B, and C be sets such that $A \cup B = A \cup C$ and $A \cap B = A \cap C$. Show that $B = C$.

21. Show that $\bigcup(A \cup B) = \bigcup A \cup \bigcup B$.

22. Show that if A and B are nonempty sets, then $\bigcap(A \cup B) = \bigcap A \cap \bigcap B$.

23. Show that if $\mathscr{B}$ is nonempty, then $A \cup \bigcap\mathscr{B} = \bigcap\{A \cup X \mid X \in \mathscr{B}\}$.

24. (a) Show that if $\mathscr{A}$ is nonempty, then $\mathscr{P}\bigcap\mathscr{A} = \bigcap\{\mathscr{P}X \mid X \in \mathscr{A}\}$.
(b) Show that

$$\bigcup\{\mathscr{P}X \mid X \in \mathscr{A}\} \subseteq \mathscr{P}\bigcup\mathscr{A}.$$

Under what conditions does equality hold?

25. Is $A \cup \bigcup\mathscr{B}$ always the same as $\bigcup\{A \cup X \mid X \in \mathscr{B}\}$? If not, then under what conditions does equality hold?

EPILOGUE

This chapter was entitled "Axioms and Operations." Some of the operations featured in the chapter are union, intersection, the power set axiom, and relative complementation. You should certainly know what all of these operations are and be able to work with them. You should also be able to prove various properties of the operations.

Also in this chapter we introduced six of our ten axioms. Or more accurately, we introduced five axioms and one axiom schema. (An *axiom schema* is an infinite bundle of axioms, such as the subset axioms.) From these axioms we can justify the definitions of the operations mentioned above. And the proofs of properties of the operations are, ultimately, proofs from our list of axioms.

Review Exercises

26. Consider the following sets: $A = \{3, 4\}$, $B = \{4, 3\} \cup \varnothing$, $C = \{4, 3\} \cup \{\varnothing\}$, $D = \{x \mid x^2 - 7x + 12 = 0\}$, $E = \{\varnothing, 3, 4\}$, $F = \{4, 4, 3\}$, $G = \{4, \varnothing, \varnothing, 3\}$. For each pair of sets (e.g., A and E) specify whether or not the sets are equal.

27. Give an example of sets A and B for which $A \cap B$ is nonempty and

$$\bigcap A \cap \bigcap B \neq \bigcap(A \cap B).$$

28. Simplify:

$$\bigcup\{\{3, 4\}, \{\{3\}, \{4\}\}, \{3, \{4\}\}, \{\{3\}, 4\}\}.$$

29. Simplify:
(a) $\bigcap\{\mathscr{P}\mathscr{P}\mathscr{P}\varnothing, \mathscr{P}\mathscr{P}\varnothing, \mathscr{P}\varnothing, \varnothing\}$.
(b) $\bigcap\{\mathscr{P}\mathscr{P}\mathscr{P}\{\varnothing\}, \mathscr{P}\mathscr{P}\{\varnothing\}, \mathscr{P}\{\varnothing\}\}$.

30. Let A be the set $\{\{\varnothing\}, \{\{\varnothing\}\}\}$. Evaluate the following:
(a) $\mathscr{P}A$, (b) $\bigcup A$, (c) $\mathscr{P}\bigcup A$, (d) $\bigcup\mathscr{P}A$.

31. Let B be the set $\{\{1, 2\}, \{2, 3\}, \{1, 3\}, \{\varnothing\}\}$. Evaluate the following sets.
(a) $\bigcup B$, (b) $\bigcap B$, (c) $\bigcap\bigcup B$, (d) $\bigcup\bigcap B$.

32. Let S be the set $\{\{a\}, \{a, b\}\}$. Evaluate and simplify:
(a) $\bigcup\bigcup S$, (b) $\bigcap\bigcap S$, (c) $\bigcap\bigcup S \cup (\bigcup\bigcup S - \bigcup\bigcap S)$.

33. With S as in the preceding exercise, evaluate $\bigcup(\bigcup S - \bigcap S)$ when $a \neq b$ and when $a = b$.

34. Show that $\{\varnothing, \{\varnothing\}\} \in \mathscr{P}\mathscr{P}\mathscr{P}S$ for every set S.

35. Assume that $\mathscr{P}A = \mathscr{P}B$. Prove that $A = B$.

36. Verify that for all sets the following are correct.
(a) $A - (A \cap B) = A - B$.
(b) $A - (A - B) = A \cap B$.

37. Show that for all sets the following equations hold.
(a) $(A \cup B) - C = (A - C) \cup (B - C)$.
(b) $A - (B - C) = (A - B) \cup (A \cap C)$.
(c) $(A - B) - C = A - (B \cup C)$.

38. Prove that for all sets the following are valid.
(a) $A \subseteq C \;\&\; B \subseteq C \iff A \cup B \subseteq C$.
(b) $C \subseteq A \;\&\; C \subseteq B \iff C \subseteq A \cap B$.

CHAPTER 3

RELATIONS AND FUNCTIONS

In this chapter we introduce some concepts that are important throughout mathematics. The correct formulation (and understanding) of the definitions will be a major goal. The theorems initially will be those needed to justify the definitions, and those verifying some properties of the defined objects.

ORDERED PAIRS

The pair set $\{1, 2\}$ can be thought of as an unordered pair, since $\{1, 2\} = \{2, 1\}$. We will need another object $\langle 1, 2\rangle$ that will encode more information: that 1 is the first component and 2 is the second. In particular, we will demand that $\langle 1, 2\rangle \neq \langle 2, 1\rangle$.

More generally, we want to define a set $\langle x, y\rangle$ that uniquely encodes both what x and y are, and also what order they are in. In other words, if an ordered pair can be represented in two ways

$$\langle x, y\rangle = \langle u, v\rangle,$$

then the representations are identical in the sense that $x = u$ and $y = v$. And in fact any way of defining $\langle x, y\rangle$ that satisfies this property of unique decomposition will suffice. It will be instructive to consider first some examples of definitions *lacking* this property.

Example 1 If we define $\langle x, y\rangle_1 = \{x, y\}$, then (as noted above), $\langle 1, 2\rangle_1 = \langle 2, 1\rangle_1$.

Example 2 Let $\langle x, y\rangle_2 = \{x, \{y\}\}$. Again the desired property fails, since $\langle\{\varnothing\}, \{\varnothing\}\rangle_2 = \langle\{\{\varnothing\}\}, \varnothing\rangle_2$, both sides being equal to $\{\{\varnothing\}, \{\{\varnothing\}\}\}$.

The first successful definition was given by Norbert Wiener in 1914, who proposed to let

$$\langle x, y\rangle_3 = \{\{\{x\}, \varnothing\}, \{\{y\}\}\}.$$

A simpler definition was given by Kazimierz Kuratowski in 1921, and is the definition in general use today:

Definition $\langle x, y\rangle$ is defined to be $\{\{x\}, \{x, y\}\}$.

We must prove that this definition succeeds in capturing the desired property: The ordered pair $\langle x, y\rangle$ uniquely determines both what x and y are, and the order upon them.

Theorem 3A $\langle u, v\rangle = \langle x, y\rangle$ iff $u = x$ and $v = y$.

Proof One direction is trivial; if $u = x$ and $v = y$, then $\langle u, v\rangle$ is the same thing as $\langle x, y\rangle$.

To prove the interesting direction, assume that $\langle u, v\rangle = \langle x, y\rangle$, i.e.,

$$\{\{u\}, \{u, v\}\} = \{\{x\}, \{x, y\}\}.$$

Then we have

$$\{u\} \in \{\{x\}, \{x, y\}\} \qquad \text{and} \qquad \{u, v\} \in \{\{x\}, \{x, y\}\}.$$

From the first of these, we know that either

$$\text{(a)}\quad \{u\} = \{x\} \qquad \text{or} \qquad \text{(b)}\quad \{u\} = \{x, y\},$$

and from the second we know that either

$$\text{(c)}\quad \{u, v\} = \{x\} \qquad \text{or} \qquad \text{(d)}\quad \{u, v\} = \{x, y\}.$$

First suppose (b) holds; then $u = x = y$. Then (c) and (d) are equivalent, and tell us that $u = v = x = y$. In this case the conclusion of the theorem holds. Similarly if (c) holds, we have the same situation.

There remains the case in which (a) and (d) hold. From (a) we have $u = x$. From (d) we get either $u = y$ or $v = y$. In the first case (b) holds; that case has already been considered. In the second case we have $v = y$, as desired. ⊣

The preceding theorem lets us unambiguously define the *first coordinate* of $\langle x, y\rangle$ to be x, and the *second coordinate* to be y.

Example Let $\mathbb{R}$ be the set of all real numbers. The pair $\langle x, y\rangle$ can be visualized as a point in the plane (Fig. 6), where coordinate axes have been established. This representation of points in the plane is attributed to Descartes.

Now suppose that we have two sets A and B, and we form ordered pairs $\langle x, y\rangle$ with $x \in A$ and $y \in B$. The collection of *all* such pairs is called the *Cartesian product* $A \times B$ of A and B:

$$A \times B = \{\langle x, y\rangle \mid x \in A \ \&\ y \in B\}.$$

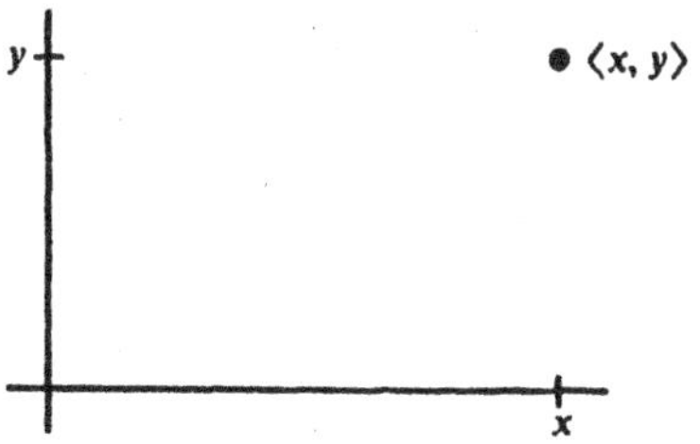

Fig. 6. The pair $\langle x, y\rangle$ as a point in the plane.

We must verify that this collection is actually a set before the definition is legal. When we use the abstraction notation $\{t \mid \_\ t\ \_\}$ for a set, we must verify that there does indeed exist a set D such that

$$t \in D \quad \Leftrightarrow \quad \_\ t\ \_$$

for every t. For example, just as strings of symbols, the expressions

$$\{x \mid x = x\} \qquad \text{and} \qquad \{x \mid x \neq x\}$$

look similar. But by Theorem 2A the first does not name a set, whereas (by the empty set axiom) the second does.

The strategy we follow in order to show that $A \times B$ is a set (and not a proper class) runs as follows. If we can find a large set that already contains all of the pairs $\langle x, y\rangle$ we want, then we can use a subset axiom to cut things down to $A \times B$. A suitable large set to start with is provided by the next lemma.

Lemma 3B If $x \in C$ and $y \in C$, then $\langle x, y\rangle \in \mathscr{P}\mathscr{P}C$.

Proof As the following calculation demonstrates, the fact that the braces in "$\{\{x\}, \{x, y\}\}$" are nested to a depth of 2 is responsible for the two applications of the power set operation:

$$\begin{array}{ccc} x \in C & \text{and} & y \in C, \\ \{x\} \subseteq C & \text{and} & \{x, y\} \subseteq C, \\ \{x\} \in \mathscr{P}C & \text{and} & \{x, y\} \in \mathscr{P}C, \end{array}$$

$$\{\{x\}, \{x, y\}\} \subseteq \mathscr{P}C,$$

$$\{\{x\}, \{x, y\}\} \in \mathscr{P}\mathscr{P}C.$$

Corollary 3C For any sets A and B, there is a set whose members are exactly the pairs $\langle x, y\rangle$ with $x \in A$ and $y \in B$.

Proof From a subset axiom we can construct

$$\{w \in \mathscr{P}\mathscr{P}(A \cup B) \mid w = \langle x, y\rangle \text{ for some } x \text{ in } A \text{ and some } y \text{ in } B\}.$$

Clearly this set contains only pairs of the desired sort; by the preceding lemma it contains them all. ⊣

This corollary justifies our earlier definition of the Cartesian product $A \times B$.

As you have probably observed, our decision to use the Kuratowski definition

$$\langle x, y\rangle = \{\{x\}, \{x, y\}\}$$

is somewhat arbitrary. There are other definitions that would serve as well. The essential fact is that satisfactory ways exist of defining ordered pairs in terms of other concepts of set theory.

Exercises

See also the Review Exercises at the end of this chapter.

1. Suppose that we attempted to generalize the Kuratowski definitions of ordered pairs to ordered triples by defining

$$\langle x, y, z\rangle^* = \{\{x\}, \{x, y\}, \{x, y, z\}\}.$$

Show that this definition is unsuccessful by giving examples of objects u, v, w, x, y, z with $\langle x, y, z\rangle^* = \langle u, v, w\rangle^*$ but with either $y \neq v$ or $z \neq w$ (or both).

2. (a) Show that $A \times (B \cup C) = (A \times B) \cup (A \times C)$.
   (b) Show that if $A \times B = A \times C$ and $A \neq \varnothing$, then $B = C$.
3. Show that $A \times \bigcup\mathscr{B} = \bigcup\{A \times X \mid X \in \mathscr{B}\}$.
4. Show that there is no set to which every ordered pair belongs.

5. (a) Assume that A and B are given sets, and show that there exists a set C such that for any y,

$$y \in C \iff y = \{x\} \times B \quad \text{for some } x \text{ in } A.$$

In other words, show that $\{\{x\} \times B \mid x \in A\}$ is a set.

(b) With A, B, and C as above, show that $A \times B = \bigcup C$.

RELATIONS

Before attempting to say what, in general, a relation is, it would be prudent to contemplate a few examples.

The ordering relation $<$ on the set $\{2, 3, 5\}$, is one example. We might

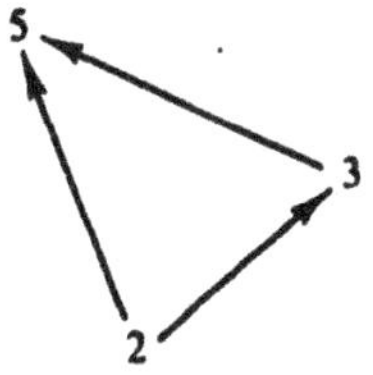

Fig. 7. The ordering relation $<$ on $\{2, 3, 5\}$.

say that $<$ relates each number to each of the larger numbers. Thus $3 < 5$, so $<$ relates 3 to 5. Pictorially we can represent this by drawing an arrow from 3 to 5. Altogether we get three arrows in this way (Fig. 7).

What *set* adequately encodes this ordering relation? In place of the arrows, we take the ordered pairs $\langle 2, 3\rangle$, $\langle 2, 5\rangle$, and $\langle 3, 5\rangle$. The set of these pairs

$$R = \{\langle 2, 3\rangle, \langle 2, 5\rangle, \langle 3, 5\rangle\}$$

completely captures the information in Fig. 7. At one time it was fashionable to refer to the set R as the *graph* of the relation, a terminology that seems particularly appropriate if we think of R as a subset of the coordinate plane. But nowadays an even simpler viewpoint has become dominant: R *is* the ordering relation on $\{2, 3, 5\}$. It consists of the pairs tying each number to the larger numbers; a relation *is* this collection of "ties."

A homier example might be the relation of marriage. (We ignore for the moment the fact that we banished people from our set theory.) This relation is the aggregate total of individual ties between each married person and his or her spouse. Or to say it more mathematically, the relation is

$$\{\langle x, y\rangle \mid x \text{ is married to } y\}.$$

You have probably guessed that for us a relation will be a set of ordered pairs. And there will be no further restrictions; *any* set of ordered pairs is some relation, even if a peculiar one.

Definition A *relation* is a set of ordered pairs.

For a relation R, we sometimes write xRy in place of $\langle x, y\rangle \in R$. For example, in the case of the ordering relation $<$ on the set $\mathbb{R}$ of real numbers

$$< \,=\{\langle x, y\rangle \in \mathbb{R} \times \mathbb{R} \mid x \text{ is less than } y\},$$

the notation "$x < y$" is preferred to "$\langle x, y\rangle \in <$."

Examples Let ω be the set $\{0, 1, 2, \ldots\}$, which is introduced more formally in the next chapter. Then the divisibility relation is

$$\{\langle m, n\rangle \in \omega \times \omega \mid (\exists p \in \omega) m \cdot p = n\}.$$

The identity relation on ω is

$$I_\omega = \{\langle n, n\rangle \mid n \in \omega\}.$$

And *any* subset of $\omega \times \omega$ (of which there are a great many) is some sort of relation.

Of course some relations are much more interesting than others. In the coming pages we shall look at functions, equivalence relations, and ordering relations. At this point we make some very general definitions.

Definition We define the *domain* of R (dom R), the *range* of R (ran R), and the *field* of R (fld R) by

$$\begin{aligned} x \in \operatorname{dom} R &\iff \exists y\, \langle x, y\rangle \in R, \\ x \in \operatorname{ran} R &\iff \exists t\, \langle t, x\rangle \in R, \\ \operatorname{fld} R &= \operatorname{dom} R \cup \operatorname{ran} R. \end{aligned}$$

For example, let $\mathbb{R}$ be the set of all real numbers (a set we construct officially in Chapter 5) and suppose that $R \subseteq \mathbb{R} \times \mathbb{R}$. Then R is a subset of the coordinate plane (Fig. 8). The projection of R onto the horizontal axis is dom R, and the projection onto the vertical axis is ran R.

To justify the foregoing definition, we must be sure that for any given R, there exists a set containing all first coordinates and second coordinates of pairs of R. The problem here is analogous to the recent problem of justifying the definition of $A \times B$, which was accomplished by Corollary 3C. The crucial fact needed now is that there exists a large set already containing all of the elements we seek. This fact is provided by the following lemma, which is related to Lemma 3B. (Lemma 3D was stated as an example in the preceding chapter.)

Lemma 3D If $\langle x, y\rangle \in A$, then x and y belong to $\bigcup\bigcup A$.

Proof We assume that $\{\{x\}, \{x, y\}\} \in A$. Consequently, $\{x, y\} \in \bigcup A$ since it belongs to a member of A. And from this we conclude that $x \in \bigcup\bigcup A$ and $y \in \bigcup\bigcup A$. ⊣

This lemma indicates how we can use subset axioms to construct the domain and range of R:

$$\operatorname{dom} R = \{x \in \bigcup\bigcup R \mid \exists y\, \langle x, y\rangle \in R\},$$

$$\operatorname{ran} R = \{x \in \bigcup\bigcup R \mid \exists t\, \langle t, x\rangle \in R\}.$$

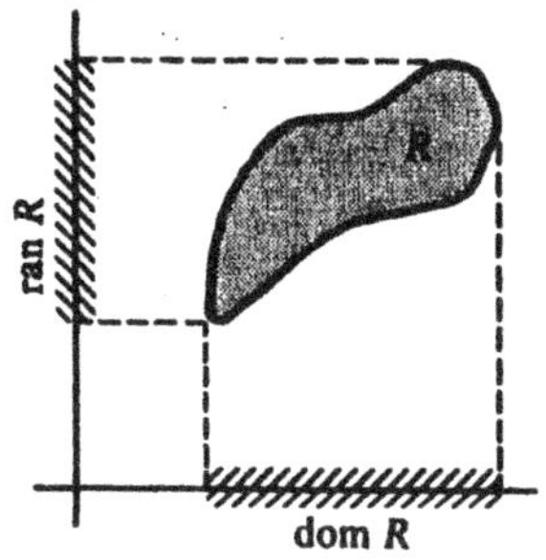

Fig. 8. A relation as a subset of the plane.

Exercises

6. Show that a set A is a relation iff $A \subseteq \operatorname{dom} A \times \operatorname{ran} A$.
7. Show that if R is a relation, then $\operatorname{fld} R = \bigcup\bigcup R$.
8. Show that for any set $\mathscr{A}$:

$$\operatorname{dom} \bigcup\mathscr{A} = \bigcup\{\operatorname{dom} R \mid R \in \mathscr{A}\},$$

$$\operatorname{ran} \bigcup\mathscr{A} = \bigcup\{\operatorname{ran} R \mid R \in \mathscr{A}\}.$$

9. Discuss the result of replacing the union operation by the intersection operation in the preceding problem.

n-ARY RELATIONS

We can extend the ideas behind ordered pairs to the case of ordered triples and, more generally, to ordered n-tuples. For triples we define

$$\langle x, y, z\rangle = \langle\langle x, y\rangle, z\rangle.$$

Similarly we can form ordered quadruples:

$$\begin{aligned}\langle x_1, x_2, x_3, x_4\rangle &= \langle\langle x_1, x_2, x_3\rangle, x_4\rangle \\ &= \langle\langle\langle x_1, x_2\rangle, x_3\rangle, x_4\rangle.\end{aligned}$$

Clearly we could continue in this way to define ordered quintuples or ordered n-tuples for any particular n. It is convenient for reasons of uniformity to define also the 1-tuple $\langle x\rangle = x$.

We define an *n-ary relation on A* to be a set of ordered n-tuples with all components in A. Thus a binary (2-ary) relation on A is just a subset of $A \times A$. And a ternary (3-ary) relation on A is a subset of $(A \times A) \times A$. There is, however, a terminological quirk here. If $n > 1$, then any n-ary relation on A is actually a relation. But a unary (1-ary) relation on A is just a subset of A; thus it may not be a relation at all.

Exercise

10. Show that an ordered 4-tuple is also an ordered m-tuple for every positive integer m less than 4.

FUNCTIONS

Calculus books often describe a function as a rule that assigns to each object in a certain set (its domain) a unique object in a possibly different set (its range). A typical example is the squaring function, which assigns to each real number x its square x^2. The action of this function on a particular number can be described by writing

$$3 \mapsto 9, \quad -2 \mapsto 4, \quad 1 \mapsto 1, \quad \tfrac{1}{2} \mapsto \tfrac{1}{4}, \quad \text{etc.}$$

Each individual action can be represented by an ordered pair:

$$\langle 3, 9\rangle, \quad \langle -2, 4\rangle, \quad \langle 1, 1\rangle, \quad \langle \tfrac{1}{2}, \tfrac{1}{4}\rangle, \quad \text{etc.}$$

The set of all these pairs (one for each real number) adequately represents the squaring function. The set of pairs has at times been called the *graph* of the function; it is a subset of the coordinate plane $\mathbb{R} \times \mathbb{R}$. But the simplest procedure is to take this set of ordered pairs to *be* the function.

Thus a function is a set of ordered pairs (i.e., a relation). But it has a special property: It is "single-valued," i.e., for each x in its domain there is a unique y such that $x \mapsto y$. We build these ideas into the following definition.

Definition A *function* is a relation F such that for each x in dom F there is only one y such that xFy.

For a function F and a point x in dom F, the unique y such that xFy is called the *value* of F *at* x and is denoted $F(x)$. Thus $\langle x, F(x)\rangle \in F$. The "$F(x)$" notation was introduced by Euler in the 1700s. We hereby resolve to use this notation *only* when F is a function and $x \in \operatorname{dom} F$. There are, however, some artificial ways of defining $F(x)$ that are meaningful for any F and x. For example, the set

$$\bigcup\{y \mid \langle x, y\rangle \in F\}$$

is equal to $F(x)$ whenever F is a function and $x \in \operatorname{dom} F$.

Functions are basic objects appearing in all parts of mathematics.[1] As a result, there is a good deal of terminology used in connection with functions. Unfortunately, no terminology has become uniformly standardized. We collect below some of this terminology.

We say that F is a function *from A into B* or that F *maps A into B* (written $F\colon A \to B$) iff F is a function, dom $F = A$, and ran $F \subseteq B$. Note the unequal treatment of A and B here; we demand only that ran $F \subseteq B$. If, in addition, ran $F = B$, then F is a function from A *onto* B. (Thus any function F maps its domain onto its range. And it maps its domain into any set B that includes ran F. The applicability of the word "onto" depends both on F and on the set B, not just on F. The word "onto" must never be used as an adjective.)

A function F is *one-to-one* iff for each $y \in \operatorname{ran} F$ there is only one x such that xFy. For example, the function defined by

$$f(x) = x^3 \qquad \text{for each real number } x$$

is one-to-one, whereas the squaring function is not, since $(-3)^2 = 3^2$. One-to-one functions are sometimes called *injections*.

It will occasionally be useful to apply the concept of "one-to-one" to relations that are not functions. Since the phrase "one-to-one" seems inappropriate in such cases, we will use the phrase "single-rooted," in analogy to "single-valued."

Definition A set R is *single-rooted* iff for each $y \in \operatorname{ran} R$ there is only one x such that xRy.

Thus for a function, it is single-rooted iff it is one-to-one.

It is entirely possible to have the domain of a function F consist of ordered pairs or n-tuples. For example, addition is a function $+\colon \mathbb{R} \times \mathbb{R} \to \mathbb{R}$. Thus the domain of addition consists of pairs of numbers, and the addition

[1] Despite this ubiquity, the general concept of "a function" emerged slowly over a period of time. There was a reluctance to separate the concept of a function itself from the idea of a written formula defining the function. There still is.

function itself consists of triples of numbers. In place of $+(\langle x, y\rangle)$ we write either $+(x, y)$ or $x + y$.

The following operations are most commonly applied to functions, sometimes are applied to relations, but can actually be defined for arbitrary sets A, F, and G.

Definition (a) The *inverse* of F is the set

$$F^{-1} = \{\langle u, v\rangle \mid vFu\}.$$

(b) The *composition* of F and G is the set

$$F \circ G = \{\langle u, v\rangle \mid \exists t(uGt \;\&\; tFv)\}.$$

(c) The *restriction* of F to A is the set

$$F \restriction A = \{\langle u, v\rangle \mid uFv \;\&\; u \in A\}.$$

(d) The *image* of A *under* F is the set

$$\begin{aligned} F[\![A]\!] &= \operatorname{ran}(F \restriction A) \\ &= \{v \mid (\exists u \in A)uFv\}. \end{aligned}$$

$F[\![A]\!]$ can be characterized more simply when F is a function and $A \subseteq \operatorname{dom} F$; in this case

$$F[\![A]\!] = \{F(u) \mid u \in A\}.$$

In each case we can easily apply a subset axiom to establish the existence of the desired set. Specifically, $F^{-1} \subseteq \operatorname{ran} F \times \operatorname{dom} F$, $F \circ G \subseteq \operatorname{dom} G \times \operatorname{ran} F$, $F \restriction A \subseteq F$, and $F[\![A]\!] \subseteq \operatorname{ran} F$. (A more detailed justification of the definition of F^{-1} would go as follows: By a subset axiom there is a set B such that for any x,

$$x \in B \iff x \in \operatorname{ran} F \times \operatorname{dom} F \;\&\; \exists u\, \exists v(x = \langle u, v\rangle \;\&\; vFu).$$

It then follows that

$$x \in B \iff \exists u\, \exists v(x = \langle u, v\rangle \;\&\; vFu).$$

This unique set B we denote as F^{-1}.)

Example Let $F\colon \mathbb{R} \to \mathbb{R}$ be defined by the equation $F(x) = x^2$. Let A be the set $\{x \in R \mid -1 \le x \le 2\}$, i.e., the closed interval $[-1, 2]$. Then $F[\![A]\!] = [0, 4]$; see Fig. 9. And $F^{-1}[\![A]\!] = [-\sqrt{2}, \sqrt{2}]$. Notice that although F here is a function, F^{-1} is *not* a function, because both $\langle 9, 3\rangle$ and $\langle 9, -3\rangle$ are in F^{-1}. The so-called "multiple-valued functions" are relations, not functions. People write "$F^{-1}(9) = \pm 3$," but it would be preferable to write: $F^{-1}[\![\{9\}]\!] = \{-3, 3\}$.

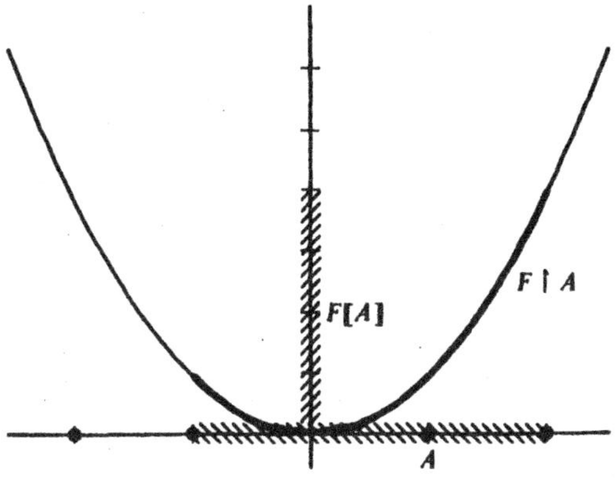

Fig. 9. $F[\![A]\!]$ is the image of the set A under F.

Example In mathematical analysis it is often necessary to consider the "inverse image" of a set A under a function F, i.e., the set $F^{-1}[\![A]\!]$. For a function F,

$$F^{-1}[\![A]\!] = \{x \in \operatorname{dom} F \mid F(x) \in A\}$$

(Exercise 24). In general, F^{-1} will not be a function.

Example Let g be the sine function of trigonometry. Then g^{-1} is not a function. (Why not?) But the restriction of g to the closed interval $[-\pi/2, \pi/2]$ is one-to-one, and its inverse

$$(g \restriction [-\pi/2, \pi/2])^{-1}$$

is the arc sine function.

Example Assume that we have the set of all people (!) in mind, and we define P to be the relation of parenthood, i.e.,

$$P = \{\langle x, y\rangle \mid x \text{ is a parent of } y\}.$$

Then

$$P^{-1} = \{\langle x, y\rangle \mid x \text{ is a son or daughter of } y\}$$

and

$$P \circ P = \{\langle x, y\rangle \mid x \text{ is a grandparent of } y\}.$$

If A is the set of people born in Poland, then

$$(P \circ P)[\![A]\!] = \{t \mid \text{a grandparent of } t \text{ was born in Poland}\}.$$

To further complicate matters, let S be the relation that holds between siblings:

$$S = \{\langle x, y\rangle \mid x \text{ is a brother or sister of } y\}$$

Then $S^{-1} = S$. To find out what $P \circ S$ is, we can calculate:

$$\begin{aligned} \langle x, y\rangle \in P \circ S \quad &\Leftrightarrow \quad xSt \;\&\; tPy \quad \text{for some } t \\ &\Leftrightarrow \quad x \text{ is a sibling of a parent of } y \\ &\Leftrightarrow \quad x \text{ is an aunt or uncle of } y. \end{aligned}$$

None of the relations in this example are functions.

Example Let

$$F = \{\langle \varnothing, a\rangle, \langle \{\varnothing\}, b\rangle\}.$$

Observe that F is a function. We have

$$F^{-1} = \{\langle a, \varnothing\rangle, \langle b, \{\varnothing\}\rangle\}.$$

F^{-1} is a function iff $a \neq b$. The restriction of F to $\varnothing$ is $\varnothing$, but

$$F \upharpoonright \{\varnothing\} = \{\langle \varnothing, a\rangle\}.$$

Consequently, $F[\![\{\varnothing\}]\!] = \{a\}$, in contrast to the fact that $F(\{\varnothing\}) = b$.

The following facts about inverses are not difficult to show; the proofs of some of them are left as exercises.

Theorem 3E For a set F, $\operatorname{dom} F^{-1} = \operatorname{ran} F$ and $\operatorname{ran} F^{-1} = \operatorname{dom} F$. For a relation F, $(F^{-1})^{-1} = F$.

Theorem 3F For a set F, F^{-1} is a function iff F is single-rooted. A relation F is a function iff F^{-1} is single-rooted.

Theorem 3G Assume that F is a one-to-one function. If $x \in \operatorname{dom} F$, then $F^{-1}(F(x)) = x$. If $y \in \operatorname{ran} F$, then $F(F^{-1}(y)) = y$.

Proof Suppose that $x \in \operatorname{dom} F$; then $\langle x, F(x)\rangle \in F$ and $\langle F(x), x\rangle \in F^{-1}$. Thus $F(x) \in \operatorname{dom} F^{-1}$. F^{-1} is a function by Theorem 3F, so $x = F^{-1}(F(x))$.

If $y \in \operatorname{ran} F$, then by applying the first part of the theorem to F^{-1} we obtain the equation $(F^{-1})^{-1}(F^{-1}(y)) = y$. But $(F^{-1})^{-1} = F$. ⊣

In place of Theorem 3G, we could have *defined* F^{-1} (for a one-to-one function F) to be that function whose value at $F(x)$ is x (and whose domain is ran F). But this would be too restrictive; F^{-1} can be a useful relation even when it is not a function. Hence we prefer a definition of F^{-1} that is applicable to any set.

Theorem 3H Assume that F and G are functions. Then $F \circ G$ is a function, its domain is

$$\{x \in \operatorname{dom} G \mid G(x) \in \operatorname{dom} F\},$$

and for x in its domain, $(F \circ G)(x) = F(G(x))$.

Proof To see that $F \circ G$ is a function, assume that $x(F \circ G)y$ and $x(F \circ G)y'$. Then for some t and t',

$$xGt \;\&\; tFy \qquad \text{and} \qquad xGt' \;\&\; t'Fy'.$$

Since G is a function, $t = t'$. Since F is a function, $y = y'$. Hence $F \circ G$ is a function.

Now suppose that $x \in \operatorname{dom} G$ and $G(x) \in \operatorname{dom} F$. We must show that $x \in \operatorname{dom}(F \circ G)$ and that $(F \circ G)(x) = F(G(x))$. We have $\langle x, G(x)\rangle \in G$ and $\langle G(x), F(G(x))\rangle \in F$. Hence $\langle x, F(G(x))\rangle \in F \circ G$, and this yields the desired facts.

Conversely, if $x \in \operatorname{dom} F \circ G$, then we know that for some y and t, xGt and tFy. Hence $x \in \operatorname{dom} G$ and $t = G(x) \in \operatorname{dom} F$. ⊣

Again we could have *defined* $F \circ G$ (for functions F and G) as the function with the properties stated in the above theorem. But we prefer to use a definition applicable to nonfunctions as well. For example, in Exercise 32, we will want $R \circ R$ for an arbitrary relation R.

Example Assume that G is some one-to-one function. Then by Theorem 3H, $G^{-1} \circ G$ is a function, its domain is

$$\{x \in \operatorname{dom} G \mid G(x) \in \operatorname{dom} G^{-1}\} = \operatorname{dom} G,$$

and for x in its domain,

$$\begin{aligned} (G^{-1} \circ G)(x) &= G^{-1}(G(x)) \\ &= x \qquad \text{by Theorem 3G.} \end{aligned}$$

Thus $G^{-1} \circ G$ is $I_{\operatorname{dom} G}$, the identity function on dom G, by Exercise 11. Similarly one can show that $G \circ G^{-1}$ is $I_{\operatorname{ran} G}$ (Exercise 25).

Theorem 3I For any sets F and G,

$$(F \circ G)^{-1} = G^{-1} \circ F^{-1}.$$

Proof Both $(F \circ G)^{-1}$ and $G^{-1} \circ F^{-1}$ are relations. We calculate:

$$\begin{aligned} \langle x, y\rangle \in (F \circ G)^{-1} &\Leftrightarrow \langle y, x\rangle \in F \circ G \\ &\Leftrightarrow yGt \;\&\; tFx \qquad \text{for some } t \\ &\Leftrightarrow xF^{-1}t \;\&\; tG^{-1}y \qquad \text{for some } t \\ &\Leftrightarrow \langle x, y\rangle \in G^{-1} \circ F^{-1}. \end{aligned}$$

⊣

In a less abstract form, Theorem 3I expresses common knowledge. In getting dressed, one first puts on socks and then shoes. But in the inverse process of getting undressed, one first removes shoes and then socks.

Theorem 3J Assume that $F: A \to B$, and that A is nonempty.

(a) There exists a function $G: B \to A$ (a "left inverse") such that $G \circ F$ is the identity function I_A on A iff F is one-to-one.
(b) There exists a function $H: B \to A$ (a "right inverse") such that $F \circ H$ is the identity function I_B on B iff F maps A *onto* B.

Proof (a) First assume that there is a function G for which $G \circ F = I_A$. If $F(x) = F(y)$, then by applying G to both sides of the equation we have

$$x = G(F(x)) = G(F(y)) = y,$$

and hence F is one-to-one.

For the converse, assume that F is one-to-one. Then F^{-1} is a function from ran F onto A (by Theorems 3E and 3F). The idea is to extend F^{-1} to a function G defined on all of B. By assumption A is nonempty, so we can fix some a in A. Then we define G so that it assigns a to every point in $B - \operatorname{ran} F$:

$$G(x) = \begin{cases} F^{-1}(x) & \text{if } x \in \operatorname{ran} F \\ a & \text{if } x \in B - \operatorname{ran} F. \end{cases}$$

In one line,

$$G = F^{-1} \cup (B - \operatorname{ran} F) \times \{a\}$$

(see Fig. 10a). This choice for G does what we want: G is a function mapping B into A, $\operatorname{dom}(G \circ F) = A$, and $G(F(x)) = F^{-1}(F(x)) = x$ for each x in A. Hence $G \circ F = I_A$.

(b) Next assume that there is a function H for which $F \circ H = I_B$. Then for any y in B we have $y = F(H(y))$, so that $y \in \operatorname{ran} F$. Thus ran F is all of B.

The converse poses a difficulty. We cannot take $H = F^{-1}$, because in general F will not be one-to-one and so F^{-1} will not be a function. Assume that F maps A onto B, so that ran $F = B$. The idea is that for each $y \in B$ we must *choose* some x for which $F(x) = y$ and then let $H(y)$ be the chosen x. Since $y \in \operatorname{ran} F$ we know that such x's exist, so there is no problem (see Fig. 10b).

Or is there? For any *one* y we know there exists an appropriate x. But that is not by itself enough to let us form a function H. We have in general no way of defining any one particular choice of x. What is needed here is the axiom of choice.

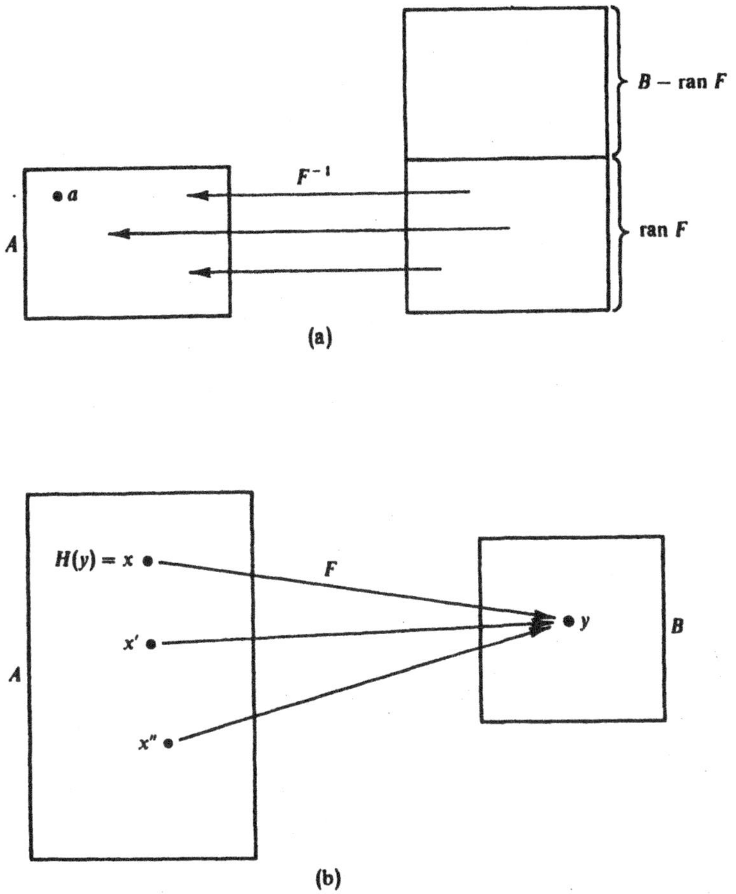

Fig. 10. The proof of Theorem 3J. In part (a), make $G(x) = a$ for $x \in B - \operatorname{ran} F$. In part (b), $H(y)$ is the chosen x for which $F(x) = y$.

Axiom of Choice (first form) For any relation R there is a function $H \subseteq R$ with dom $H =$ dom R.

With this axiom we can now proceed with the proof of Theorem 3J(b); take H to be a function with $H \subseteq F^{-1}$ and dom $H =$ dom $F^{-1} = B$. Then H does what we want: Given any y in B, we have $\langle y, H(y)\rangle \in F^{-1}$; hence $\langle H(y), y\rangle \in F$, and so $F(H(y)) = y$. ⊣

In Chapter 6 we will give a systematic discussion of the axiom of choice. It is the only axiom that we discuss without using the marginal stripe.

Theorem 3K The following hold for any sets. (F need not be a function.)

(a) The image of a union is the union of the images:

$$F[A \cup B] = F[A] \cup F[B] \quad \text{and} \quad F[\bigcup \mathscr{A}] = \bigcup \{F[A] \mid A \in \mathscr{A}\}.$$

(b) The image of an intersection is included in the intersection of the images:

$$F[A \cap B] \subseteq F[A] \cap F[B] \quad \text{and} \quad F[\bigcap \mathscr{A}] \subseteq \bigcap \{F[A] \mid A \in \mathscr{A}\}$$

for nonempty $\mathscr{A}$. Equality holds if F is single-rooted.

(c) The image of a difference includes the difference of the images:

$$F[A] - F[B] \subseteq F[A - B].$$

Equality holds if F is single-rooted.

Example Let $F: \mathbb{R} \to \mathbb{R}$ be defined by $F(x) = x^2$. Let A and B be the closed intervals $[-2, 0]$ and $[1, 2]$:

$$A = \{x \mid -2 \le x \le 0\} \quad \text{and} \quad B = \{x \mid 1 \le x \le 2\}.$$

Then $F[A] = [0, 4]$ and $F[B] = [1, 4]$. This example shows that equality does not always hold in parts (b) and (c) of Theorem 3K, for $F[A \cap B] = F[\varnothing] = \varnothing$, whereas $F[A] \cap F[B] = [1, 4]$. And $F[A] - F[B] = [0, 1)$, whereas $F[A - B] = F[A] = [0, 4]$.

Proof To prove Theorem 3K we calculate

$$\begin{aligned} y \in F[A \cup B] &\Leftrightarrow (\exists x \in A \cup B) xFy \\ &\Leftrightarrow (\exists x \in A) xFy \text{ or } (\exists x \in B) xFy \\ &\Leftrightarrow y \in F[A] \text{ or } y \in F[B]. \end{aligned}$$

This proves the first half of (a). For intersections we have the corresponding calculation, except that the middle step

$$(\exists x \in A \cap B) xFy \Rightarrow (\exists x \in A) xFy \;\&\; (\exists x \in B) xFy$$

is not always reversible. It is possible that both $x_1 \in A$ with $x_1 Fy$ and $x_2 \in B$ with $x_2 Fy$, and yet there might be no x in $A \cap B$ with xFy. But if F is single-rooted, then $x_1 = x_2$ and so it is in $A \cap B$. Thus we obtain the first half of (b).

The second halves of (a) and (b) generalize the first halves. The proofs follow the same outlines as the first halves, but we leave the details to Exercise 26.

For part (c) we also calculate:

$$\begin{aligned} y \in F[\![A]\!] - F[\![B]\!] &\Leftrightarrow (\exists x \in A)xFy \;\&\; \neg(\exists t \in B)tFy \\ &\Rightarrow (\exists x \in A - B)xFy \\ &\Leftrightarrow y \in F[\![A - B]\!]. \end{aligned}$$

Again if F is single-rooted, then there is only one x such that xFy. In this case the middle step can be reversed. ⊣

Since the inverse of a function is always single-rooted, we have as an immediate consequence of Theorem 3K that unions, intersections, and relative complements are always preserved under inverse images.

Corollary 3L For any function G and sets A, B, and $\mathscr{A}$:

$$G^{-1}[\![\bigcup\mathscr{A}]\!] = \bigcup\{G^{-1}[\![A]\!] \mid A \in \mathscr{A}\},$$

$$G^{-1}[\![\bigcap\mathscr{A}]\!] = \bigcap\{G^{-1}[\![A]\!] \mid A \in \mathscr{A}\} \qquad \text{for} \quad \mathscr{A} \neq \varnothing,$$

$$G^{-1}[\![A - B]\!] = G^{-1}[\![A]\!] - G^{-1}[\![B]\!].$$

We conclude our discussion of functions with some definitions that may be useful later. Our intent is to build a large working vocabulary of set-theoretic notations.

An infinite union is often "indexed," as when we write $\bigcup_{i \in I} A_i$. We can give a formal definition to such a union as follows. Let I be a set, called the *index* set. Let F be a function whose domain includes I. Then we define

$$\begin{aligned} \bigcup_{i \in I} F(i) &= \bigcup\{F(i) \mid i \in I\} \\ &= \{x \mid x \in F(i) \text{ for some } i \text{ in } I\}. \end{aligned}$$

For example, if $I = \{0, 1, 2, 3\}$, then

$$\begin{aligned} \bigcup_{i \in I} F(i) &= \bigcup\{F(0), F(1), F(2), F(3)\} \\ &= F(0) \cup F(1) \cup F(2) \cup F(3). \end{aligned}$$

Similar remarks apply to intersections (provided that I is nonempty):

$$\begin{aligned} \bigcap_{i \in I} F(i) &= \bigcap\{F(i) \mid i \in I\} \\ &= \{x \mid x \in F(i) \text{ for every } i \text{ in } I\}. \end{aligned}$$

If we use the alternative notation

$$F_i = F(i),$$

then we can rewrite the above equations as

$$\bigcup_{i \in I} F_i = \bigcup\{F_i \mid i \in I\}$$
$$= \{x \mid x \in F_i \text{ for some } i \text{ in } I\}$$

and

$$\bigcap_{i \in I} F_i = \bigcap\{F_i \mid i \in I\}$$
$$= \{x \mid x \in F_i \text{ for every } i \text{ in } I\}.$$

For sets A and B we can form the collection of functions F from A into B. Call the set of all such functions ${}^{A}B$:

$${}^{A}B = \{F \mid F \text{ is a function from } A \text{ into } B\}.$$

If $F: A \to B$, then $F \subseteq A \times B$, and so $F \in \mathscr{P}(A \times B)$. Consequently we can apply a subset axiom to $\mathscr{P}(A \times B)$ to construct the set of all functions from A into B.

The notation ${}^{A}B$ is read "B-pre-A." Some authors write B^A instead; this notation is derived from the fact that if A and B are finite sets and the number of elements in A and B is a and b, respectively, then ${}^{A}B$ has b^a members. (To see this, note that for each of the a elements of A, we can choose among b points in B into which it could be mapped. The number of ways of making all a such choices is $b \cdot b \cdot \dots \cdot b$, a times.) We will return to this point in Chapter 6.

Example Let $\omega = \{0, 1, 2, \dots\}$. Then ${}^{\omega}\{0, 1\}$ is the set of all possible functions $f: \omega \to \{0, 1\}$. Such an f can be thought of as an infinite sequence $f(0), f(1), f(2), \dots$ of 0's and 1's.

Example For a nonempty set A, we have ${}^{A}\varnothing = \varnothing$. This is because no function could have a nonempty domain and an empty range. On the other hand, ${}^{\varnothing}A = \{\varnothing\}$ for any set A, because $\varnothing: \varnothing \to A$, but $\varnothing$ is the only function with empty domain. As a special case, we have ${}^{\varnothing}\varnothing = \{\varnothing\}$.

Exercises

11. Prove the following version (for functions) of the extensionality principle: Assume that F and G are functions, $\operatorname{dom} F = \operatorname{dom} G$, and $F(x) = G(x)$ for all x in the common domain. Then $F = G$.

12. Assume that f and g are functions and show that

$$f \subseteq g \iff \operatorname{dom} f \subseteq \operatorname{dom} g \;\&\; (\forall x \in \operatorname{dom} f)\, f(x) = g(x).$$

13. Assume that f and g are functions with $f \subseteq g$ and dom $g \subseteq$ dom f. Show that $f = g$.

14. Assume that f and g are functions.

(a) Show that $f \cap g$ is a function.

(b) Show that $f \cup g$ is a function iff $f(x) = g(x)$ for every x in $(\operatorname{dom} f) \cap (\operatorname{dom} g)$.

15. Let $\mathscr{A}$ be a set of functions such that for any f and g in $\mathscr{A}$, either $f \subseteq g$ or $g \subseteq f$. Show that $\bigcup \mathscr{A}$ is a function.

16. Show that there is no set to which every function belongs.

17. Show that the composition of two single-rooted sets is again single-rooted. Conclude that the composition of two one-to-one functions is again one-to-one.

18. Let R be the set

$$\{\langle 0, 1\rangle, \langle 0, 2\rangle, \langle 0, 3\rangle, \langle 1, 2\rangle, \langle 1, 3\rangle, \langle 2, 3\rangle\}.$$

Evaluate the following: $R \circ R$, $R \restriction \{1\}$, $R^{-1} \restriction \{1\}$, $R[\![\{1\}]\!]$, and $R^{-1}[\![\{1\}]\!]$.

19. Let

$$A = \{\langle \varnothing, \{\varnothing, \{\varnothing\}\}\rangle, \langle \{\varnothing\}, \varnothing\rangle\}.$$

Evaluate each of the following: $A(\varnothing)$, $A[\![\varnothing]\!]$, $A[\![\{\varnothing\}]\!]$, $A[\![\{\varnothing, \{\varnothing\}\}]\!]$, A^{-1}, $A \circ A$, $A \restriction \varnothing$, $A \restriction \{\varnothing\}$, $A \restriction \{\varnothing, \{\varnothing\}\}$, $\bigcup\bigcup A$.

20. Show that $F \restriction A = F \cap (A \times \operatorname{ran} F)$.

21. Show that $(R \circ S) \circ T = R \circ (S \circ T)$ for any sets R, S, and T.

22. Show that the following are correct for any sets.

(a) $A \subseteq B \Rightarrow F[\![A]\!] \subseteq F[\![B]\!]$.

(b) $(F \circ G)[\![A]\!] = F[\![G[\![A]\!]]\!]$.

(c) $Q \restriction (A \cup B) = (Q \restriction A) \cup (Q \restriction B)$.

23. Let I_A be the identity function on the set A. Show that for any sets B and C,

$$B \circ I_A = B \restriction A \qquad \text{and} \qquad I_A[\![C]\!] = A \cap C.$$

24. Show that for a function F, $F^{-1}[\![A]\!] = \{x \in \operatorname{dom} F \mid F(x) \in A\}$.

25. (a) Assume that G is a one-to-one function. Show that $G \circ G^{-1}$ is $I_{\operatorname{ran} G}$, the identity function on ran G.

(b) Show that the result of part (a) holds for any function G, not necessarily one-to-one.

26. Prove the second halves of parts (a) and (b) of Theorem 3K.

27. Show that $\operatorname{dom}(F \circ G) = G^{-1}[\![\operatorname{dom} F]\!]$ for any sets F and G. (F and G need not be functions.)

28. Assume that f is a one-to-one function from A into B, and that G is the function with dom $G = \mathscr{P}A$ defined by the equation $G(X) = f[\![X]\!]$. Show that G maps $\mathscr{P}A$ one-to-one into $\mathscr{P}B$.

29. Assume that $f\colon A \to B$ and define a function $G\colon B \to \mathscr{P}A$ by

$$G(b) = \{x \in A \mid f(x) = b\}.$$

Show that if f maps A *onto* B, then G is one-to-one. Does the converse hold?

30. Assume that $F\colon \mathscr{P}A \to \mathscr{P}A$ and that F has the monotonicity property:

$$X \subseteq Y \subseteq A \quad\Rightarrow\quad F(X) \subseteq F(Y).$$

Define

$$B = \bigcap\{X \subseteq A \mid F(X) \subseteq X\} \qquad \text{and} \qquad C = \bigcup\{X \subseteq A \mid X \subseteq F(X)\}.$$

(a) Show that $F(B) = B$ and $F(C) = C$.
(b) Show that if $F(X) = X$, then $B \subseteq X \subseteq C$.

INFINITE CARTESIAN PRODUCTS[2]

We can form something like the Cartesian product of infinitely many sets, provided that the sets are suitably indexed. More specifically, let I be a set (which we will refer to as the *index set*) and let H be a function whose domain includes I. Then for each i in I we have the set $H(i)$; we want the product of the $H(i)$'s for all $i \in I$. We define:

$$\bigtimes_{i \in I} H(i) = \{f \mid f \text{ is a function with domain } I \text{ and } (\forall i \in I)\, f(i) \in H(i)\}.$$

Thus the members of $\bigtimes_{i \in I} H(i)$ are "I-tuples" (i.e., functions with domain I) for which the "ith coordinate" (i.e., the value at i) is in $H(i)$.

The members of $\bigtimes_{i \in I} H(i)$ are all functions from I into $\bigcup_{i \in I} H(i)$ and hence are members of ${}^{I}(\bigcup_{i \in I} H(i))$. Thus the set $\bigtimes_{i \in I} H(i)$ can be formed by applying a subset axiom to ${}^{I}(\bigcup_{i \in I} H(i))$.

Example If for every $i \in I$ we have $H(i) = A$ for some one fixed A, then $\bigtimes_{i \in I} H(i) = {}^{I}A$.

Example Assume that the index set is the set $\omega = \{0, 1, 2, \ldots\}$. Then $\bigtimes_{i \in \omega} H(i)$ consists of "ω-sequences" (i.e., functions with domain ω) that have for their ith term some member of $H(i)$. If we picture the sets $H(i)$ as shown in Fig. 11, then a typical member of $\bigtimes_{i \in \omega} H(i)$ is a "thread" that selects a point from each set.

[2] This section may be omitted if certain obvious adjustments are made in Theorem 6M.

If any one $H(i)$ is empty, then clearly the product $\times_{i \in I} H(i)$ is empty. Conversely, suppose that $H(i) \neq \varnothing$ for every i in I. Does it follow that $\times_{i \in I} H(i) \neq \varnothing$? To obtain a member f of the product, we need to select some member from each $H(i)$, and put $f(i)$ equal to that selected member. This requires the axiom of choice, and in fact this is one of the many equivalent ways of stating the axiom.

Axiom of Choice (second form) For any set I and any function H with domain I, if $H(i) \neq \varnothing$ for all i in I, then $\times_{i \in I} H(i) \neq \varnothing$.

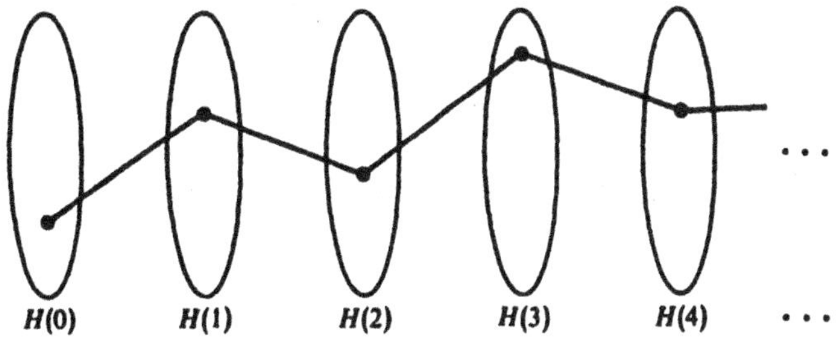

Fig. 11. The thread is a member of the Cartesian product.

Exercise

31. Show that from the first form of the axiom of choice we can prove the second form, and conversely.

EQUIVALENCE RELATIONS

Consider a set A (Fig. 12a). We might want to partition A into little boxes (Fig. 12b). For example, take $A = \omega$; we can partition ω into six parts:

$$\{0, 6, 12, \ldots\},$$
$$\{1, 7, 13, \ldots\},$$
$$\vdots$$
$$\{5, 11, 17, \ldots\}.$$

By "partition" we mean that every element of A is in exactly one little box, and that each box is a nonempty subset of A.

Now we need some mental agility. We want to think of each little box as being a single object, instead of thinking of it as a plurality of objects. (Actually we have been doing this sort of thing throughout the book, whenever we think of a set as a single object. It is really no harder than thinking of a brick house as a single object and not as a multitude of bricks.) This changes the picture (Fig. 12c); each box is now, in our mind, a single point. The set B of boxes is very different from the set A. In our example, B has only six members whereas A is infinite. (When we get around to defining "six" and "infinite" officially, we must certainly do it in a way that makes the preceding sentence true.)

The process of transforming a situation like Fig. 12a into Fig. 12c is common in abstract algebra and elsewhere in mathematics. And in Chapter 5 the process will be applied several times in the construction of the real numbers.

Suppose we now define a binary relation R on A as follows: For x and y in A,

$$xRy \Leftrightarrow x \text{ and } y \text{ are in the same little box.}$$

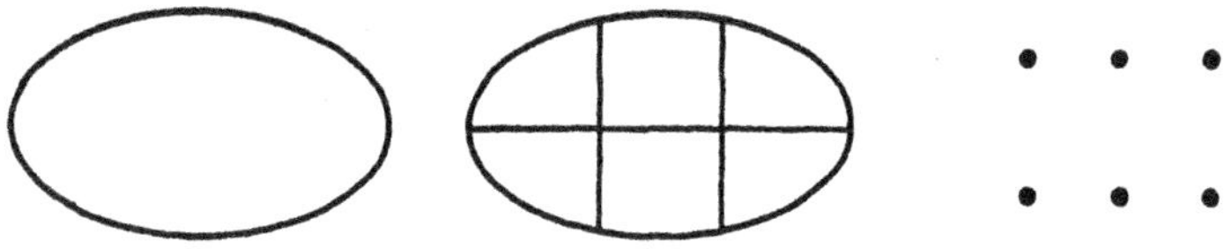

(a) (b) (c)

Fig. 12. Partitioning a set into six little boxes.

Then we can easily see that R has the following three properties.

1. R is *reflexive on A*, by which we mean that xRx for all $x \in A$.
2. R is *symmetric*, by which we mean that whenever xRy, then also yRx.
3. R is *transitive*, by which we mean that whenever xRy and yRz, then also xRz.

Definition R is an *equivalence relation on A* iff R is a binary relation on A that is reflexive on A, symmetric, and transitive.

Theorem 3M If R is a symmetric and transitive relation, then R is an equivalence relation on fld R.

Proof Any relation R is a binary relation on its field, since

$$R \subseteq \operatorname{dom} R \times \operatorname{ran} R \subseteq \operatorname{fld} R \times \operatorname{fld} R.$$

What we must show is that R is reflexive on fld R. We have

$$\begin{aligned} x \in \operatorname{dom} R \quad &\Rightarrow \quad xRy && \text{for some } y \\ &\Rightarrow \quad xRy \;\&\; yRx && \text{by symmetry} \\ &\Rightarrow \quad xRx && \text{by transitivity,} \end{aligned}$$

and a similar calculation applies to points in ran R.

This theorem deserves a precautionary note: If R is a symmetric and transitive relation on A, it does *not* follow that R is an equivalence relation on A. R is reflexive on fld R, but fld R may be a small subset of A.

We have shown how a partition of a set A induces an equivalence relation. (A more formal version of this is in Exercise 37.) Next we want to reverse the process, and show that from any equivalence relation R on A, we get a partition of A.

Definition The set $[x]_R$ is defined by

$$[x]_R = \{t \mid xRt\}.$$

If R is an equivalence relation and $x \in \text{fld } R$, then $[x]_R$ is called the *equivalence class* of x (*modulo* R). If the relation R is fixed by the context, we may write just $[x]$.

The status of $[x]_R$ as a set is guaranteed by a subset axiom, since $[x]_R \subseteq \text{ran } R$. Furthermore we can construct a set of equivalence classes such as $\{[x]_R \mid x \in A\}$, since this set is included in $\mathscr{P}(\text{ran } R)$.

Lemma 3N Assume that R is an equivalence relation on A and that x and y belong to A. Then

$$[x]_R = [y]_R \quad \text{iff} \quad xRy.$$

Proof ($\Rightarrow$) Assume that $[x]_R = [y]_R$. We know that $y \in [y]_R$ (because yRy), consequently $y \in [x]_R$ (because $[x]_R = [y]_R$). By the definition of $[x]_R$, this means that xRy.

($\Leftarrow$) Next assume that xRy. Then

$$\begin{aligned} t \in [y]_R &\Rightarrow yRt \\ &\Rightarrow xRt \qquad \text{because } xRy \text{ and } R \text{ is transitive} \\ &\Rightarrow t \in [x]_R. \end{aligned}$$

Thus $[y]_R \subseteq [x]_R$. Since R is symmetric, we also have yRx and we can reverse x and y in the above argument to obtain $[x]_R \subseteq [y]_R$. ⊣

Definition A *partition* Π of a set A is a set of nonempty subsets of A that is disjoint and exhaustive, i.e.,

(a) no two different sets in Π have any common elements, and
(b) each element of A is in some set in Π.

Theorem 3P Assume that R is an equivalence relation on A. Then the set $\{[x]_R \mid x \in A\}$ of all equivalence classes is a partition of A.

Proof Each equivalence class $[x]_R$ is nonempty (because $x \in [x]_R$) and is a subset of A (because R is a binary relation on A). The main thing that we must prove is that the collection of equivalence classes is disjoint, i.e., part (a) of the above definition is satisfied. So suppose that $[x]_R$ and $[y]_R$ have a common element t. Thus

$$xRt \quad \text{and} \quad yRt.$$

But then xRy and by Lemma 3N, $[x]_R = [y]_R$. ⊣

If R is an equivalence relation on A, then we can define the *quotient set*

$$A/R = \{[x]_R \mid x \in A\}$$

whose members are the equivalence classes. (The expression A/R is read "A modulo R.") We also have the *natural map* (or *canonical map*) $\varphi: A \to A/R$ defined by

$$\varphi(x) = [x]_R$$

for $x \in A$.

Example Let $\omega = \{0, 1, 2, \ldots\}$; define the binary relation $\sim$ on ω by

$$m \sim n \quad \Leftrightarrow \quad m - n \text{ is divisible by } 6.$$

Then $\sim$ is an equivalence relation on ω (as you should verify). The quotient set $\omega/\sim$ has six members:

$$[0], \quad [1], \quad [2], \quad [3], \quad [4], \quad [5],$$

corresponding to the six possible remainders after division by 6.

Example The relation of congruence of triangles in the plane is an equivalence relation.

Example Textbooks on linear algebra often define vectors in the plane as follows. Let A be the set of all directed line segments in the plane. Two such line segments are considered to be equivalent iff they have the same length and direction. A *vector* is then defined to be an equivalence class of directed line segments. But to avoid the necessity of dealing explicitly with equivalence relations, books use phrases like "equivalent vectors are regarded as equal even though they are located in different positions," or "we write $PQ = RS$ to say that PQ and RS have the same length and direction even though they are not identical sets of points," or simply "we identify two line segments having the same length and direction."

Example Let $F: A \to B$ and for points in A define

$$x \sim y \quad \text{iff} \quad F(x) = F(y).$$

The relation $\sim$ is an equivalence relation on A. There is a unique one-to-one function $\hat{F}: A/\sim \to B$ such that $F = \hat{F} \circ \varphi$ (where φ is the natural map as

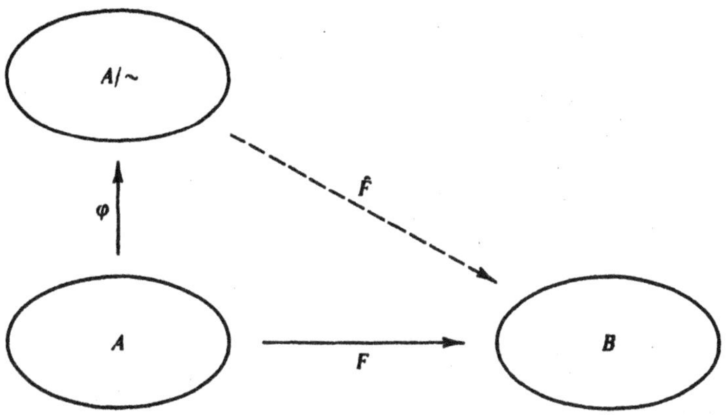

Fig. 13. F factors into the natural map followed by a one-to-one function.

shown in Fig. 13). The value of $\hat{F}$ at a particular equivalence class is the common value of F at the members of the equivalence class.

The last problem we want to examine in this section is the problem of defining functions on a quotient set. Specifically, assume that R is an equivalence relation on A and that $F\colon A \to A$. We ask whether or not there exists a corresponding function $\hat{F}\colon A/R \to A/R$ such that for all $x \in A$,

$$\hat{F}([x]_R) = [F(x)]_R$$

(see Fig. 14). Here we are attempting to define the value of $\hat{F}$ at an equivalence class by selecting a particular member x from the class and then forming $[F(x)]_R$. But suppose x_1 and x_2 are in the same equivalence class. Then $\hat{F}$ is not well defined unless $F(x_1)$ and $F(x_2)$ are in the same equivalence class.

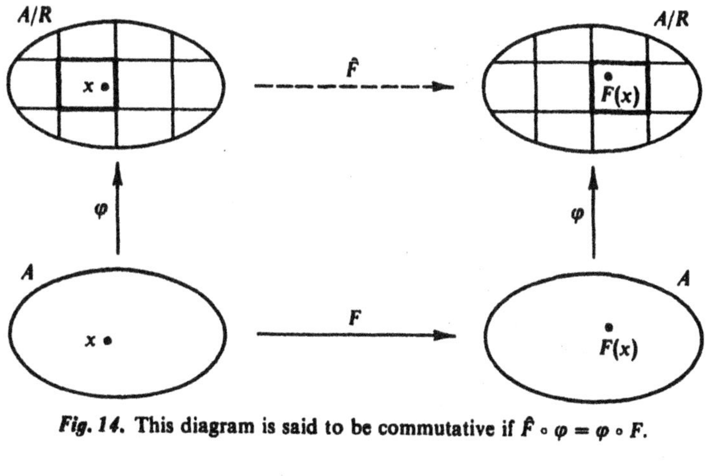

Fig. 14. This diagram is said to be commutative if $\hat{F} \circ \varphi = \varphi \circ F$.

Example Consider $\omega/\sim$ where $m \sim n$ iff $m - n$ is divisible by 6. Three functions from ω into ω are defined by

$$F_1(n) = 2n, \qquad F_2(n) = n^2, \qquad F_3(n) = 2^n.$$

In each case we can ask whether there is $\hat{F}_i: \omega/\sim \to \omega/\sim$ such that for every n in ω:

$$\hat{F}_1([n]) = [2n], \qquad \hat{F}_2([n]) = [n^2], \qquad \hat{F}_3([n]) = [2^n].$$

It is easy to see that if $m \sim n$, then $2m \sim 2n$. Because of this fact $\hat{F}_1$ is well defined; that is, there exists a function $\hat{F}_1$ satisfying the equation $\hat{F}_1([n]) = [2n]$. No matter what representative m of the equivalence class $[n]$ we look at, we always obtain the same equivalence class $[2m]$. (For further details, see the proof of Theorem 3Q below.) Similarly if $m \sim n$, then $m^2 \sim n^2$, for recall that $m^2 - n^2 = (m + n)(m - n)$. Consequently $\hat{F}_2$ is also well defined. On the other hand, $\hat{F}_3$ is *not* well defined. For example, $0 \sim 6$ but $2^0 = 1 \not\sim 64 = 2^6$. Thus although $[0] = [6]$, we have $[2^0] \neq [2^6]$. Hence there cannot possibly exist any function $\hat{F}_3$ such that the equation $\hat{F}_3([n]) = [2^n]$ holds for both $n = 0$ and $n = 6$.

In order to formulate a general theorem here, let us say that F is *compatible* with R iff for all x and y in A,

$$xRy \Rightarrow F(x)RF(y).$$

Theorem 3Q Assume that R is an equivalence relation on A and that $F: A \to A$. If F is compatible with R, then there exists a unique $\hat{F}: A/R \to A/R$ such that

$$(\text{☆}) \qquad \hat{F}([x]_R) = [F(x)]_R \qquad \text{for all } x \text{ in } A.$$

If F is not compatible with R, then no such $\hat{F}$ exists. Analogous results apply to functions from $A \times A$ into A.

Proof First assume that F is *not* compatible; we will show that there can be no $\hat{F}$ satisfying (☆) The incompatibility tells us that for certain x and y in A we have xRy (and hence $[x] = [y]$) but not $F(x)RF(y)$ (and hence $[F(x)] \neq [F(y)]$). For (☆) to hold we would need both

$$\hat{F}([x]) = [F(x)] \qquad \text{and} \qquad \hat{F}([y]) = [F(y)].$$

But this is impossible, since the left sides coincide and the right sides differ.

Now for the converse, assume that F is compatible with R. Since (☆) demands that the pair $\langle [x], [F(x)] \rangle \in \hat{F}$, we will try defining $\hat{F}$ to be the set of all such ordered pairs:

$$\hat{F} = \{\langle [x], [F(x)] \rangle \mid x \in A\}.$$

The crucial matter is whether this relation $\hat{F}$ is a function. So consider pairs $\langle [x], [F(x)] \rangle$ and $\langle [y], [F(y)] \rangle$ in $\hat{F}$. The calculation

$$\begin{array}{rlll} [x] = [y] & \Rightarrow & xRy & \text{by Lemma 3N} \\ & \Rightarrow & F(x)RF(y) & \text{by compatibility} \\ & \Rightarrow & [F(x)] = [F(y)] & \text{by Lemma 3N} \end{array}$$

shows that $\hat{F}$ is indeed a function. The remaining things to check are easier. Clearly dom $\hat{F} = A/R$ and ran $\hat{F} \subseteq A/R$, hence $\hat{F}: A/R \to A/R$. Finally (☆) holds because $\langle [x], [F(x)] \rangle \in \hat{F}$.

We leave it to you to explain why $\hat{F}$ is unique, and to formulate the "analogous results" for a binary operation (Exercise 42). ⊣

Exercises

32. (a) Show that R is symmetric iff $R^{-1} \subseteq R$.
(b) Show that R is transitive iff $R \circ R \subseteq R$.

33. Show that R is a symmetric and transitive relation iff $R = R^{-1} \circ R$.

34. Assume that $\mathscr{A}$ is a nonempty set, every member of which is a transitive relation.
(a) Is the set $\bigcap \mathscr{A}$ a transitive relation?
(b) Is $\bigcup \mathscr{A}$ a transitive relation?

35. Show that for any R and x, we have $[x]_R = R[\![\{x\}]\!]$.

36. Assume that $f: A \to B$ and that R is an equivalence relation on B. Define Q to be the set

$$\{\langle x, y \rangle \in A \times A \mid \langle f(x), f(y) \rangle \in R\}.$$

Show that Q is an equivalence relation on A.

37. Assume that Π is a partition of a set A. Define the relation R_Π as follows:

$$xR_\Pi y \quad \Leftrightarrow \quad (\exists B \in \Pi)(x \in B \;\&\; y \in B).$$

Show that R_Π is an equivalence relation on A. (This is a formalized version of the discussion at the beginning of this section.)

38. Theorem 3P shows that A/R is a partition of A whenever R is an equivalence relation on A. Show that if we start with the equivalence relation R_Π of the preceding exercise, then the partition A/R_Π is just Π.

39. Assume that we start with an equivalence relation R on A and define Π to be the partition A/R. Show that R_Π, as defined in Exercise 37, is just R.

40. Define an equivalence relation R on the set P of positive integers by

$$mRn \Leftrightarrow m \text{ and } n \text{ have the same number of prime factors.}$$

Is there a function $f: P/R \to P/R$ such that $f([n]_R) = [3n]_R$ for each n?

41. Let $\mathbb{R}$ be the set of real numbers and define the relation Q on $\mathbb{R} \times \mathbb{R}$ by $\langle u, v\rangle Q\langle x, y\rangle$ iff $u + y = x + v$.

(a) Show that Q is an equivalence relation on $\mathbb{R} \times \mathbb{R}$.

(b) Is there a function $G: (\mathbb{R} \times \mathbb{R})/Q \to (\mathbb{R} \times \mathbb{R})/Q$ satisfying the equation

$$G([\langle x, y\rangle]_Q) = [\langle x + 2y, y + 2x\rangle]_Q \,?$$

42. State precisely the "analogous results" mentioned in Theorem 3Q. (This will require extending the concept of compatibility in a suitable way.)

ORDERING RELATIONS

The first example of a relation we gave in this chapter was the ordering relation

$$\{\langle 2, 3\rangle, \langle 2, 5\rangle, \langle 3, 5\rangle\}$$

on the set $\{2, 3, 5\}$; recall Fig. 7. Now we want to consider ordering relations on other sets. In the present section we will set forth the basic concepts, which will be useful in Chapter 5. A more thorough discussion of ordering relations can be found in Chapter 7.

Our first need is for a definition. What, in general, should it mean to say that R is an ordering relation on a set A? Well, for one thing R should tell us, given any distinct x and y in A, just which one is smaller. No x should be smaller than itself. And furthermore if x is less than y and y is less than z, then x should be less than z. The following definition captures these ideas.

Definition Let A be any set. *A linear ordering* on A (also called a *total ordering* on A) is a binary relation R on A (i.e., $R \subseteq A \times A$) meeting the following two conditions:

(a) R is a transitive relation; i.e., whenever xRy and yRz, then xRz.

(b) R satisfies trichotomy on A, by which we mean that for any x and y in A exactly one of the three alternatives

$$xRy, \qquad x = y, \qquad yRx$$

holds.

To clarify the meaning of trichotomy, consider first the special case where x and y are the *same* member of A (with two names). Then trichotomy demands that exactly one of

$$xRx, \qquad x = x, \qquad xRx$$

holds. Since the middle alternative certainly holds, we can conclude that xRx *never* holds.

Next consider the case where x and y are two distinct members of A. Then the middle alternative $x = y$ fails, so trichotomy demands that either xRy or yRx (but not both). Thus we have proved the following:

Theorem 3R Let R be a linear ordering on A.

(i) There is no x for which xRx.
(ii) For distinct x and y in A, either xRy or yRx.

In fact for a transitive relation R on A, conditions (i) and (ii) are equivalent to trichotomy. A relation meeting condition (i) is called *irreflexive*; one meeting condition (ii) is said to be *connected* on A.

Note also that a linear ordering R can never lead us in circles, e.g., there cannot exist a circle such as

$$x_1Rx_2, \qquad x_2Rx_3, \qquad x_3Rx_4, \qquad x_4Rx_5, \qquad x_5Rx_1.$$

This is because if we had such a circle, then by transitivity x_1Rx_1, contradicting part (i) of the foregoing theorem.

Of course "R" is not our favorite symbol for a linear ordering; our favorite is "$<$." For then we can write "$x < y$" to mean that the pair $\langle x, y\rangle$ is a member of the set $<$.

If $<$ is a linear ordering on A and if A is not too large, then we can draw a picture of the ordering. We represent the members of A by dots, placing the dot for x below the dot for y whenever $x < y$. Then we add vertical lines to connect the dots. The resulting picture has the points of A stretched out along a line, in the correct order. (The adjective "linear" reflects the possibility of drawing this picture.)

Figure 15 contains three such pictures. Part (a) is the picture for the usual order on $\{2, 3, 5\}$. Parts (b) and (c) portray the usual order on the natural numbers and on the integers, respectively. (Infinite pictures are more difficult to draw than finite pictures.)

In addition to the concept of linear ordering, there is the more general concept of a partial ordering. Partial orderings are discussed in the first section of Chapter 7. In fact you might want to read that section next, before going on to Chapter 4. At least look at Figs. 43 and 44 there, which contrast with Fig. 15.

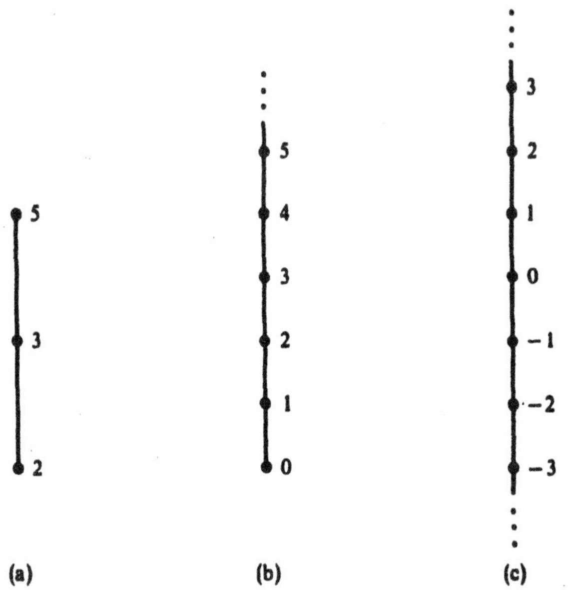

Fig. 15. Linear orderings look linear.

Exercises

43. Assume that R is a linear ordering on a set A. Show that R^{-1} is also a linear ordering on A.

44. Assume that $<$ is a linear ordering on a set A. Assume that $f: A \to A$ and that f has the property that whenever $x < y$, then $f(x) < f(y)$. Show that f is one-to-one and that whenever $f(x) < f(y)$, then $x < y$.

45. Assume that $<_A$ and $<_B$ are linear orderings on A and B, respectively. Define the binary relation $<_L$ on the Cartesian product $A \times B$ by:

$$\langle a_1, b_1\rangle <_L \langle a_2, b_2\rangle \quad \text{iff} \quad \text{either } a_1 <_A a_2 \text{ or } (a_1 = a_2 \mathbin{\&} b_1 <_B b_2)$$

Show that $<_L$ is a linear ordering on $A \times B$. (The relation $<_L$ is called *lexicographic* ordering, being the ordering used in making dictionaries.)

Review Exercises

46. Evaluate the following sets:
    (a) $\bigcap\bigcap\langle x, y\rangle$.
    (b) $\bigcap\bigcap\bigcap\{\langle x, y\rangle\}^{-1}$.

47. (a) Find all of the functions from $\{0, 1, 2\}$ into $\{3, 4\}$.
    (b) Find all of the functions from $\{0, 1, 2\}$ *onto* $\{3, 4, 5\}$.

48. Let T be the set $\{\varnothing, \{\varnothing\}\}$.
(a) Find all of the ordered pairs, if any, in $\mathscr{P}T$.
(b) Evaluate and simplify: $(\mathscr{P}T)^{-1} \circ (\mathscr{P}T \restriction \{\varnothing\})$

49. Find as many equivalence relations as you can on the set $\{0, 1, 2\}$.

50. (a) Find a linear ordering on $\{0, 1, 2, 3\}$ that contains the ordered pairs $\langle 0, 3\rangle$ and $\langle 2, 1\rangle$.
(b) Now find a different one meeting the same conditions.

51. Find as many linear orderings as possible on the set $\{0, 1, 2\}$ that contain the pair $\langle 2, 0\rangle$.

52. Suppose that $A \times B = C \times D$. Under what conditions can we conclude that $A = C$ and $B = D$?

53. Show that for any sets R and S we have $(R \cup S)^{-1} = R^{-1} \cup S^{-1}$, $(R \cap S)^{-1} = R^{-1} \cap S^{-1}$, and $(R - S)^{-1} = R^{-1} - S^{-1}$.

54. Prove that the following equations hold for any sets.
(a) $A \times (B \cap C) = (A \times B) \cap (A \times C)$.
(b) $A \times (B \cup C) = (A \times B) \cup (A \times C)$.
(c) $A \times (B - C) = (A \times B) - (A \times C)$.

55. Answer "yes" or "no." Where the answer is negative, supply a counterexample.
(a) Is it always true that $(A \times A) \cup (B \times C) = (A \cup B) \times (A \cup C)$?
(b) Is it always true that $(A \times A) \cap (B \times C) = (A \cap B) \times (A \cap C)$?

56. Answer "yes" or "no." Where the answer is negative, supply a counterexample.
(a) Is $\operatorname{dom}(R \cup S)$ always the same as $\operatorname{dom} R \cup \operatorname{dom} S$?
(b) Is $\operatorname{dom}(R \cap S)$ always the same as $\operatorname{dom} R \cap \operatorname{dom} S$?

57. Answer "yes" or "no." Where the answer is negative, supply a counterexample.
(a) Is $R \circ (S \cup T)$ always the same as $(R \circ S) \cup (R \circ T)$?
(b) Is $R \circ (S \cap T)$ always the same as $(R \circ S) \cap (R \circ T)$?

58. Give an example to show that $F[\![F^{-1}[\![S]\!]]\!]$ is not always the same as S.

59. Show that for any sets $Q \restriction (A \cap B) = (Q \restriction A) \cap (Q \restriction B)$ and $Q \restriction (A - B) = (Q \restriction A) - (Q \restriction B)$.

60. Prove that for any sets $(R \circ S) \restriction A = R \circ (S \restriction A)$.

CHAPTER 4

NATURAL NUMBERS

There are, in general, two ways of introducing new objects for mathematical study: the axiomatic approach and the constructive approach. The axiomatic approach is the one we have used for sets. The concept of set is one of our primitive notions, and we have adopted a list of axioms dealing with the primitive notions.

Now consider the matter of introducing the natural numbers[1]

$$0, \quad 1, \quad 2, \ldots$$

for further study. An axiomatic approach would consider "natural number" as a primitive notion and would adopt a list of axioms. Instead we will use the constructive approach for natural numbers. We will define natural numbers in terms of other available objects (sets, of course). In place of axioms for numbers we will be able to prove the necessary properties of numbers from known properties of sets.

[1] There is a curious point of terminology here. Is 0 a natural number? With surprising consistency, the present usage is for school books (through high-school level) to exclude 0 from the natural numbers, and for upper-division college-level books to include 0. Freshman and sophomore college books are in the transition zone. In this book we include 0 among the natural numbers.

Constructing the natural numbers in terms of sets is part of the process of "embedding mathematics in set theory." The process will be continued in Chapter 5 to obtain more exotic numbers, such as $\sqrt{2}$.

INDUCTIVE SETS

First we need to define natural numbers as suitable sets. Now numbers do not at first glance appear to be sets. Not that it is an easy matter to say what numbers *do* appear to be. They are abstract concepts, which are slippery things to handle. (See, for example, the section on "Two" in Chapter 5.) Nevertheless, we can construct specific sets that will serve perfectly well as numbers. In fact this can be done in a variety of ways. In 1908, Zermelo proposed to use

$$\varnothing, \quad \{\varnothing\}, \quad \{\{\varnothing\}\}, \ldots$$

as the natural numbers. Later von Neumann proposed an alternative, which has several advantages and has become standard. The guiding principle behind von Neumann's construction is to make each natural number be the set of all smaller natural numbers. Thus we define the first four natural numbers as follows:

$$\begin{aligned}
0 &= \varnothing, \\
1 &= \{\varnothing\} = \{\varnothing\}, \\
2 &= \{0, 1\} = \{\varnothing, \{\varnothing\}\}, \\
3 &= \{0, 1, 2\} = \{\varnothing, \{\varnothing\}, \{\varnothing, \{\varnothing\}\}\}.
\end{aligned}$$

We could continue in this way to define 17 or any chosen natural number. Notice, for example, that the set 3 has three members. It has been selected from the class of all three-member sets to represent the size of the sets in that class.

This construction of the numbers as sets involves some extraneous properties that we did not originally expect. For example,

$$0 \in 1 \in 2 \in 3 \in \cdots$$

and

$$0 \subseteq 1 \subseteq 2 \subseteq 3 \subseteq \cdots.$$

But these properties can be regarded as accidental side effects of the definition. They do no harm, and actually will be convenient at times.

Although we have defined the first four natural numbers, we do not yet have a definition of what it means in general for something to be a natural number. That is, we have not defined the set of all natural numbers. Such a definition cannot rely on linguistic devices such as three dots or phrases like "and so forth." First we define some preliminary concepts.

Definition For any set a, its *successor* a^+ is defined by

$$a^+ = a \cup \{a\}.$$

A set A is said to be *inductive* iff $\varnothing \in A$ and it is "closed under successor," i.e.,

$$(\forall a \in A)\, a^+ \in A.$$

In terms of the successor operation, the first few natural numbers can be characterized as

$$0 = \varnothing, \quad 1 = \varnothing^+, \quad 2 = \varnothing^{++}, \quad 3 = \varnothing^{+++}, \ldots.$$

These are all distinct, e.g., $\varnothing^+ \neq \varnothing^{+++}$ (Exercise 1). And although we have not yet given a formal definition of "infinite," we can see informally that any inductive set will be infinite.

We have as yet no axioms that provide for the existence of infinite sets. There are indeed infinitely many distinct sets whose existence we could establish. But there is no one set having infinitely many members that we can prove to exist. Consequently we cannot yet prove that any inductive set exists. We now correct that fault.

Infinity Axiom There exists an inductive set:

$$(\exists A)[\varnothing \in A \ \&\ (\forall a \in A)\, a^+ \in A].$$

Armed with this axiom, we can now define the concept of natural number.

Definition A *natural number* is a set that belongs to every inductive set.

We next prove that the collection of all natural numbers constitutes a set.

Theorem 4A There is a set whose members are exactly the natural numbers.

Proof Let A be an inductive set; by the infinity axiom it is possible to find such a set. By a subset axiom there is a set w such that for any x,

$$\begin{aligned} x \in w \quad &\Leftrightarrow \quad x \in A \ \&\ x \text{ belongs to every other inductive set} \\ &\Leftrightarrow \quad x \text{ belongs to every inductive set.} \end{aligned}$$

(This proof is essentially the same as the proof of Theorem 2B.) ⊣

The set of all natural numbers is denoted by a lowercase Greek omega:

$$\begin{aligned} x \in \omega \quad &\Leftrightarrow \quad x \text{ is a natural number} \\ &\Leftrightarrow \quad x \text{ belongs to every inductive set.} \end{aligned}$$

In terms of classes, we have

$$\omega = \bigcap \{A \mid A \text{ is inductive}\},$$

but the class of all inductive sets is not a set.

Theorem 4B ω is inductive, and is a subset of every other inductive set.

Proof First of all, $\varnothing \in \omega$ because $\varnothing$ belongs to every inductive set. And second,

$$\begin{aligned} a \in \omega \quad &\Rightarrow \quad a \text{ belongs to every inductive set} \\ &\Rightarrow \quad a^+ \text{ belongs to every inductive set} \\ &\Rightarrow \quad a^+ \in \omega. \end{aligned}$$

Hence ω is inductive. And clearly ω is included in every other inductive set. ⊣

Since ω is inductive, we know that 0 $(=\varnothing)$ is in ω. It then follows that 1 $(=0^+)$ is in ω, as are 2 $(=1^+)$ and 3 $(=2^+)$. Thus 0, 1, 2, and 3 are natural numbers. Unnecessary extraneous objects have been excluded from ω, since ω is the *smallest* inductive set. This fact can also be restated as follows.

Induction Principle for ω Any inductive subset of ω coincides with ω.

Suppose, for example, that we want to prove that for every natural number n, the statement __ n __ holds. We form the set

$$T = \{n \in \omega \mid \_\_ \, n \, \_\_\}$$

of natural numbers for which the desired conclusion is true. If we can show that T is inductive, then the proof is complete. Such a proof is said to be a proof *by induction*. The next theorem gives a very simple example of this method.

Theorem 4C Every natural number except 0 is the successor of some natural number.

Proof Let $T = \{n \in \omega \mid \text{either } n = 0 \text{ or } (\exists p \in \omega)\ n = p^+\}$. Then $0 \in T$. And if $k \in T$, then $k^+ \in T$. Hence by induction, $T = \omega$. ⊣

Exercise

See also the Review Exercises at the end of this chapter.

1. Show that $1 \neq 3$, i.e., that $\varnothing^{+} \neq \varnothing^{+++}$.

PEANO'S POSTULATES[2]

In 1889, Peano published a study giving an axiomatic approach to the natural numbers. He showed how the properties of natural numbers could be developed on the basis of a small number of axioms. Although he

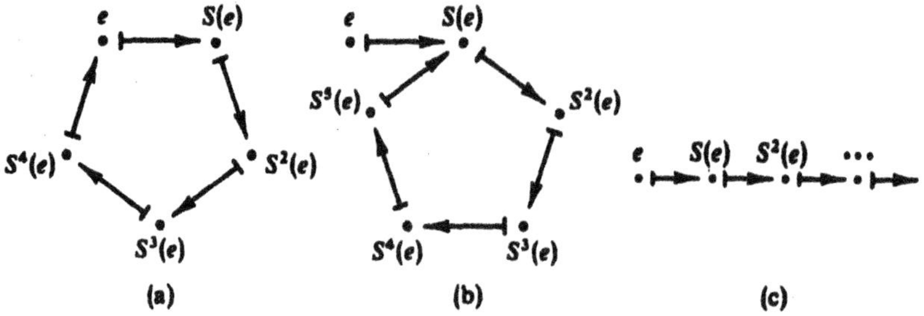

Fig. 16. Any Peano system must behave like (c).

attributed the formulation of the axioms to Dedekind, the axioms are generally known as "Peano's postulates." We will first show that the set ω we have constructed satisfies Peano's postulates, i.e., the "postulates" become provable when applied to ω. Later we will prove that anything satisfying Peano's postulates is, in a certain specific sense, "just like" ω.

To formulate these results more accurately, we must define the concept of a Peano system. First of all, if S is a function and A is a subset of dom S, then A is said to be *closed* under S iff whenever $x \in A$, then $S(x) \in A$. (This can equivalently be expressed as $S[A] \subseteq A$.) Define a *Peano system* to be a triple $\langle N, S, e\rangle$ consisting of a set N, a function $S: N \to N$, and a member $e \in N$ such that the following three conditions are met:

(i) $e \notin \operatorname{ran} S$.
(ii) S is one-to-one.
(iii) Any subset A of N that contains e and is closed under S equals N itself.

The condition "$e \notin \operatorname{ran} S$" rules out loops as in Fig. 16a—here the arrows indicate the action of S. And the requirement that S be one-to-one

[2] The material in this section on Peano systems is not essential to our later work. But the material on transitive sets is essential.

rules out the system of Fig. 16b. Consequently any Peano system must, in part, look like Fig. 16c. The last of the three conditions is the *Peano induction postulate*. Its function is to rule out any points other than the ones we expect. We expect the system to contain e, $S(e)$, $SS(e)$, $SSS(e)$, ... The Peano induction postulate replaces the three dots with a precise set-theoretic condition, stating that nothing smaller than N itself can contain e and be closed under S.

First we want to show that ω (with the successor operation and 0) is a Peano system. In particular, this will show that *some* Peano system exists. Let σ be the restriction of the successor operation to ω:

$$\sigma = \{\langle n, n^+\rangle \mid n \in \omega\}.$$

Theorem 4D $\langle \omega, \sigma, 0\rangle$ is a Peano system.

Proof Since ω is inductive, we have $0 \in \omega$ and $\sigma: \omega \to \omega$. The Peano induction postulate, as applied to $\langle \omega, \sigma, 0\rangle$, states that any subset A of ω containing 0 and closed under σ equals ω itself. This is just the induction principle for ω. Clearly $0 \notin \operatorname{ran} \sigma$, since $n^+ \neq \varnothing$. It remains only to show that σ is one-to-one. For that purpose (and others) we will use the concept of a transitive set.

Definition A set A is said to be a *transitive set* iff every member of a member of A is itself a member of A:

$$x \in a \in A \quad \Rightarrow \quad x \in A.$$

This condition can also be stated in any of the following (equivalent) ways:

$$\bigcup A \subseteq A,$$
$$a \in A \quad \Rightarrow \quad a \subseteq A,$$
$$A \subseteq \mathscr{P}A.$$

At this point we have violated a basic rule—we have defined "transitive" to mean two different things. In Chapter 3 we said that A is transitive if whenever xAy and yAz, then xAz. And now we define A to be a transitive set if a very different condition is met. But both usages of the word are well established. And so we will learn to live with the ambiguity. In practice, the context will make clear which sense of "transitive" is wanted. Furthermore when the concept of Chapter 3 is meant, we will refer to A as a transitive *relation* (luckily A will be a relation), reserving the phrase "transitive set" for the concept defined above.

Example The set $\{\varnothing, \{\{\varnothing\}\}\}$ is not a transitive set. This is because

$$\{\varnothing\} \in \{\{\varnothing\}\} \in \{\varnothing, \{\{\varnothing\}\}\},$$

but $\{\varnothing\} \notin \{\varnothing, \{\{\varnothing\}\}\}$. Also $\{0, 1, 5\}$ is not a transitive set, since $4 \in 5 \in \{0, 1, 5\}$ whereas $4 \notin \{0, 1, 5\}$.

Theorem 4E For a transitive set a,

$$\bigcup(a^+) = a.$$

Proof We proceed to calculate $\bigcup a^+$:

$$\begin{aligned} \bigcup a^+ &= \bigcup(a \cup \{a\}) \\ &= \bigcup a \cup \bigcup\{a\} \qquad \text{by Exercise 21 of Chapter 2} \\ &= \bigcup a \cup a \\ &= a. \end{aligned}$$

The last step is justified by the fact that $\bigcup a \subseteq a$ for a transitive set a. ⊣

Theorem 4F Every natural number is a transitive set.

Proof by induction We form the set of numbers for which the theorem is true; let $T = \{n \in \omega \mid n \text{ is a transitive set}\}$. It suffices to show that T is inductive, for then $T = \omega$.

Trivially $0 \in T$. If $k \in T$, then

$$\begin{aligned} \bigcup(k^+) &= k \qquad \text{by the preceding theorem} \\ &\subseteq k^+, \end{aligned}$$

whence $k^+ \in T$. Thus T is inductive. ⊣

We can now complete the proof of Theorem 4D; it remained for us to show that the successor operation on ω is one-to-one. If $m^+ = n^+$ for m and n in ω, then $\bigcup(m^+) = \bigcup(n^+)$. But since m and n are transitive sets, we have $\bigcup(m^+) = m$ and $\bigcup(n^+) = n$ by Theorem 4E. Hence $m = n$. ⊣

Theorem 4G The set ω is a transitive set.

This theorem can be stated as: Every natural number is itself a set of natural numbers. We will later strengthen this to: Every natural number is the set of all smaller natural numbers.

Proof by induction We want to show that $(\forall n \in \omega)\, n \subseteq \omega$. Form the set of n's for which this holds:

$$T = \{n \in \omega \mid n \subseteq \omega\}.$$

We must verify that T is inductive. Clearly $0 \in T$. If $k \in T$, then we have

$$k \subseteq \omega \quad \text{and} \quad \{k\} \subseteq \omega,$$

whereupon $k \cup \{k\} \subseteq \omega$, thus showing that $k^+ \in T$. And so T is inductive and therefore coincides with ω. ⊣

Exercises

2. Show that if a is a transitive set, then a^+ is also a transitive set.
3. (a) Show that if a is a transitive set, then $\mathscr{P}a$ is also a transitive set.
   (b) Show that if $\mathscr{P}a$ is a transitive set, then a is also a transitive set.
4. Show that if a is a transitive set, then $\bigcup a$ is also a transitive set.
5. Assume that every member of $\mathscr{A}$ is a transitive set.
   (a) Show that $\bigcup \mathscr{A}$ is a transitive set.
   (b) Show that $\bigcap \mathscr{A}$ is a transitive set (assuming that $\mathscr{A}$ is nonempty).
6. Prove the converse to Theorem 4E: If $\bigcup(a^+) = a$, then a is a transitive set.

RECURSION ON ω

Consider the following guessing game. Suppose I am thinking of a function $h: \omega \to A$. Possibly I am reluctant to tell you directly what the values of this function are. Instead I reveal (i) what $h(0)$ is, and (ii) a function $F: A \to A$ such that $h(n^+) = F(h(n))$ for all $n \in \omega$. This then gives away all the information; you can compute successively

$$\begin{aligned} &h(0), \\ &h(1) = F(h(0)), \\ &h(2) = F(h(1)), \end{aligned}$$

and so forth.

Now for a harder problem. Suppose we are given a set A, an element $a \in A$, and a function $F: A \to A$. How can we show that there exists a function $h: \omega \to A$ such that (i) $h(0) = a$, and (ii) $h(n^+) = F(h(n))$ for each $n \in \omega$. The preceding paragraph tells how to compute h *if it exists*. But we now want to prove that there exists a set h that is a function meeting the above conditions.

Recursion Theorem on ω Let A be a set, $a \in A$, and $F: A \to A$. Then there exists a unique function $h: \omega \to A$ such that

$$h(0) = a,$$

and for every n in ω,

$$h(n^+) = F(h(n)).$$

Proof[3] The idea is to let h be the union of many approximating functions. For the purposes of this proof, call a function v *acceptable* iff $\operatorname{dom} v \subseteq \omega$, $\operatorname{ran} v \subseteq A$, and the following conditions hold:

(i) If $0 \in \operatorname{dom} v$, then $v(0) = a$.
(ii) If $n^+ \in \operatorname{dom} v$ (where $n \in \omega$), then also $n \in \operatorname{dom} v$ and $v(n^+) = F(v(n))$.

Let $\mathscr{K}$ be the collection of all acceptable functions, and let $h = \bigcup \mathscr{K}$. Thus

$$(\text{☆}) \qquad \langle n, y\rangle \in h \quad \text{iff} \quad \langle n, y\rangle \text{ is a member of some acceptable } v$$
$$\text{iff} \quad v(n) = y \text{ for some acceptable } v.$$

We claim that this h meets the demands of the theorem. This claim can be broken down into four parts. The four parts involve showing that (1) h is a function, (2) h is acceptable, (3) $\operatorname{dom} h$ is all of ω, and (4) h is unique.

1. We first claim that h is a function. (Proving this will, in effect, amount to showing that two acceptable functions always agree with each other whenever both are defined.) Let S be the set of natural numbers at which there is no more than one candidate for $h(n)$:

$$S = \{n \in \omega \mid \text{for at most one } y, \langle n, y\rangle \in h\}.$$

We must check that S is inductive. If $\langle 0, y_1\rangle \in h$ and $\langle 0, y_2\rangle \in h$, then by (☆) there exist acceptable v_1 and v_2 such that $v_1(0) = y_1$ and $v_2(0) = y_2$. But by (i) it follows that $y_1 = a = y_2$. Thus $0 \in S$.

Next suppose that $k \in S$; we seek to show that $k^+ \in S$. Toward that end suppose that $\langle k^+, y_1\rangle \in h$ and $\langle k^+, y_2\rangle \in h$. Again there must exist acceptable v_1 and v_2 such that $v_1(k^+) = y_1$ and $v_2(k^+) = y_2$. By condition (ii) it follows that

$$y_1 = v_1(k^+) = F(v_1(k)) \qquad \text{and} \qquad y_2 = v_2(k^+) = F(v_2(k)).$$

But since $k \in S$, we have $v_1(k) = v_2(k)$. (This is because $\langle k, v_1(k)\rangle$ and $\langle k, v_2(k)\rangle$ are in h.) Therefore

$$y_1 = F(v_1(k)) = F(v_2(k)) = y_2.$$

This shows that $k^+ \in S$. Hence S is inductive and coincides with ω. Consequently h is a function.

[3] This proof is more involved than ones we have met up to now. In fact, you might want to postpone detailed study of it until after seeing some applications of the theorem. But it is an important proof, and the ideas in it will be seen again (in Chapters 7 and 9).

2. Next we claim that h itself is acceptable. We have just seen that h is a function, and it is clear from (☆) that $\operatorname{dom} h \subseteq \omega$ and $\operatorname{ran} h \subseteq A$.

First examine (i). If $0 \in \operatorname{dom} h$, then there must be some acceptable v with $v(0) = h(0)$. Since $v(0) = a$, we have $h(0) = a$.

Next examine (ii). Assume $n^+ \in \operatorname{dom} h$. Again there must be some acceptable v with $v(n^+) = h(n^+)$. Since v is acceptable we have $n \in \operatorname{dom} v$ (and $v(n) = h(n)$) and

$$h(n^+) = v(n^+) = F(v(n)) = F(h(n)).$$

Thus h satisfies (ii) and so is acceptable.

3. We now claim that $\operatorname{dom} h = \omega$. It suffices to show that $\operatorname{dom} h$ is inductive. The function $\{\langle 0, a\rangle\}$ is acceptable and hence $0 \in \operatorname{dom} h$. Suppose that $k \in \operatorname{dom} h$; we seek to show that $k^+ \in \operatorname{dom} h$. If this fails, then look at

$$v = h \cup \{\langle k^+, F(h(k))\rangle\}.$$

Then v is a function, $\operatorname{dom} v \subseteq \omega$, and $\operatorname{ran} v \subseteq A$. We will show that v is acceptable.

Condition (i) holds since $v(0) = h(0) = a$. For condition (ii) there are two cases. If $n^+ \in \operatorname{dom} v$ where $n^+ \neq k^+$, then $n^+ \in \operatorname{dom} h$ and $v(n^+) = h(n^+) = F(h(n)) = F(v(n))$. The other case occurs if $n^+ = k^+$. Since the successor operation is one-to-one, $n = k$. By assumption $k \in \operatorname{dom} h$. Thus

$$v(k^+) = F(h(k)) = F(v(k))$$

and (ii) holds. Hence v is acceptable. But then $v \subseteq h$, so that $k^+ \in \operatorname{dom} h$ after all. So $\operatorname{dom} h$ is inductive and therefore coincides with ω.

4. Finally we claim that h is unique. For let h_1 and h_2 both satisfy the conclusion of the theorem. Let S be the set on which h_1 and h_2 agree:

$$S = \{n \in \omega \mid h_1(n) = h_2(n)\}.$$

Then S is inductive; we leave the details of this to Exercise 7. Hence $S = \omega$ and $h_1 = h_2$. ⊣

Examples Let $\mathbb{Z}$ be the set of all integers, positive, negative, and zero:

$$\mathbb{Z} = \{\ldots, -1, 0, 1, 2, \ldots\}.$$

There is no function $h: \mathbb{Z} \to \mathbb{Z}$ such that for every $a \in \mathbb{Z}$,

$$h(a + 1) = h(a)^2 + 1.$$

(For notice that $h(a) > h(a - 1) > h(a - 2) > \cdots > 0$.) Recursion on ω relies on there being a starting point 0. $\mathbb{Z}$ has no analogous starting point.

For another example, let

$$F(a) = \begin{cases} a + 1 & \text{if } a < 0, \\ a & \text{if } a \geq 0. \end{cases}$$

Then there are infinitely many functions $h: \mathbb{Z} \to \mathbb{Z}$ such that $h(0) = 0$ and for every a in $\mathbb{Z}$, $h(a+1) = F(h(a))$. The graph of one such function is shown in Fig. 17.

Digression There is a classic erroneous proof of the recursion theorem that people have sometimes tried (even in print!) to give. The error is easier to analyze if we apply it not to ω but instead to an arbitrary Peano system $\langle N, S, e\rangle$. Given any a in A and any function $F: A \to A$, there is a unique function $h: N \to A$ such that $h(e) = a$ and $h(S(x)) = F(h(x))$ for each x in N. The erroneous proof of this statement runs as follows:

"We apply the Peano induction postulate to dom h. First of all, we are

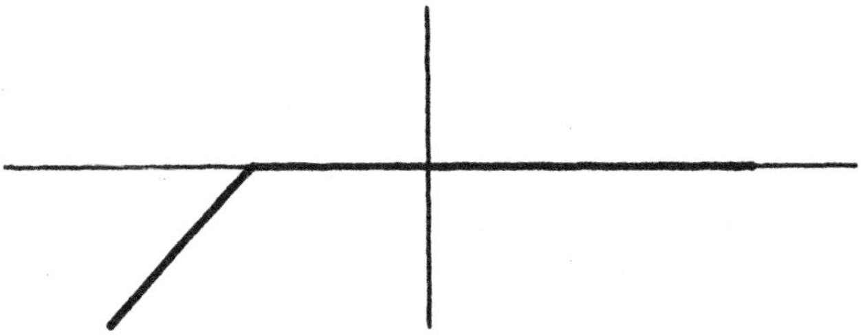

Fig. 17. $h(a+1)$ is $h(a)+1$ when $h(a) < 0$ and is $h(a)$ when $h(a) \geq 0$.

told that $h(e) = a$, and so h is defined at e, i.e., $e \in \text{dom } h$. And whenever $x \in \text{dom } h$ then immediately $h(S(x)) = F(h(x))$, so $S(x) \in \text{dom } h$ as well. Hence dom h is closed under S. It follows (by induction) that dom $h = N$, i.e., h is defined throughout N."

What is wrong? Well, for one thing, the proof talks about the function h *before* any such function is known to exist. One might think that a little rewording would get around this objection. But no, a closer examination of the proof shows that it does not utilize conditions (i) and (ii) in the definition of Peano systems. The recursion theorem is in general *false* for systems not meeting those conditions, such as the systems of Fig. 16a or Fig. 16b. So any correct proof of recursion absolutely must make use of conditions (i) and (ii), as well as using induction. (Our proof of recursion on ω uses these conditions in part 3.)

Our first application of the recursion theorem will be to show that any Peano system is "just like" $\langle\omega, \sigma, 0\rangle$. There are other Peano systems; for example, let N be the set $\{1, 2, 4, 8, \ldots\}$ of powers of 2, let $S(n) = 2n$, and let $e = 1$. Then $\langle N, S, e\rangle$ is a Peano system. The following theorem expresses the structural similarity between this Peano system and $\langle\omega, \sigma, 0\rangle$.

Theorem 4H Let $\langle N, S, e\rangle$ be a Peano system. Then $\langle \omega, \sigma, 0\rangle$ is isomorphic to $\langle N, S, e\rangle$, i.e., there is a function h mapping ω one-to-one onto N in a way that preserves the successor operation

$$h(\sigma(n)) = S(h(n))$$

and the zero element

$$h(0) = e.$$

Remark The equation $h(\sigma(n)) = S(h(n))$ (together with $h(0) = e$) implies that $h(1) = S(e)$, $h(2) = S(S(e))$, $h(3) = S(S(S(e)))$, etc. Thus the situation must be as shown in Fig. 18.

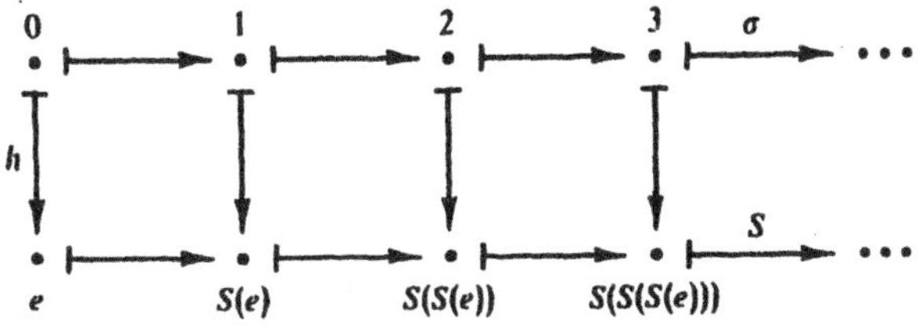

Fig. 18. Isomorphism of Peano systems.

Proof By the recursion theorem there is a unique function $h: \omega \to N$ such that $h(0) = e$ and for all $n \in \omega$, $h(n^+) = S(h(n))$. It remains to show that h is one-to-one and that $\operatorname{ran} h = N$.

To show that $\operatorname{ran} h = N$ we use the Peano induction postulate for $\langle N, S, e\rangle$. Clearly $e \in \operatorname{ran} h$. Also for any $x \in \operatorname{ran} h$ (say $x = h(n)$) we have $S(x) \in \operatorname{ran} h$ (since $S(x) = h(n^+)$). Hence by the Peano induction postulate applied to $\operatorname{ran} h$, we have $\operatorname{ran} h = N$.

To show that h is one-to-one we use induction in ω. Let

$$T = \{n \in \omega \mid \text{for every } m \text{ in } \omega \text{ different from } n,\ h(m) \neq h(n)\}.$$

First we claim that $0 \in T$. Any $m \in \omega$ different from 0 must be of the form p^+ (by Theorem 4C). And $h(p^+) = S(h(p)) \neq e$ since $e \notin \operatorname{ran} S$. Hence $h(0) = e \neq h(p^+)$, and consequently $0 \in T$.

Now assume that $k \in T$ and consider k^+. Suppose that $h(k^+) = h(m)$. Then $m \neq 0$ by the preceding paragraph, so $m = p^+$ for some p. Thus

$$S(h(k)) = h(k^+) = h(p^+) = S(h(p)).$$

Since S is one-to-one, this leaves the equation $h(k) = h(p)$. Since $k \in T$, we have $k = p$. Hence $k^+ = p^+ = m$. This shows that $k^+ \in T$. So T is inductive, and thus coincides with ω. Consequently h is one-to-one. ⊣

Theorems 4D and 4H relate the constructive approach to the natural numbers and the axiomatic approach. Theorem 4D shows that Peano's postulates are true of the number system we have constructed. And Theorem 4H shows that the number system we have constructed is, "to within isomorphism," the only system satisfying Peano's postulates.

Exercises

7. Complete part 4 of the proof of the recursion theorem on ω.

8. Let f be a one-to-one function from A into A, and assume that $c \in A - \operatorname{ran} f$. Define $h\colon \omega \to A$ by recursion:

$$h(0) = c,$$
$$h(n^+) = f(h(n)).$$

Show that h is one-to-one.

9. Let f be a function from B into B, and assume that $A \subseteq B$. We have two possible methods for constructing the "closure" C of A under f. First define C^* to be the intersection of the closed supersets of A:

$$C^* = \bigcap\{X \mid A \subseteq X \subseteq B \;\&\; f[\![X]\!] \subseteq X\}.$$

Alternatively, we could apply the recursion theorem to obtain the function h for which

$$h(0) = A,$$
$$h(n^+) = h(n) \cup f[\![h(n)]\!].$$

Clearly $h(0) \subseteq h(1) \subseteq \cdots$; define C_* to be $\bigcup \operatorname{ran} h$; in other words

$$C_* = \bigcup_{i \in \omega} h(i).$$

Show that $C^* = C_*$. [*Suggestion*: To show that $C^* \subseteq C_*$, show that $f[\![C_*]\!] \subseteq C_*$. To show that $C_* \subseteq C^*$, use induction to show that $h(n) \subseteq C^*$.]

10. In Exercise 9, assume that B is the set of real numbers, $f(x) = x^2$, and A is the closed interval $[\frac{1}{2}, 1]$. What is the set called C^* and C_*?

11. In Exercise 9, assume that B is the set of real numbers, $f(x) = x - 1$, and $A = \{0\}$. What is the set called C^* and C_*?

12. Formulate an analogue to Exercise 9 for a function $f\colon B \times B \to B$.

ARITHMETIC[4]

We can apply the recursion theorem to define addition and multiplication on ω. (Another way of obtaining these operations will be discussed in Chapter 6.) For example, suppose we want a function $A_5: \omega \to \omega$ such that $A_5(n)$ is the result of adding 5 to n. Then A_5 must satisfy the conditions

$$A_5(0) = 5,$$
$$A_5(n^+) = A_5(n)^+ \qquad \text{for } n \text{ in } \omega.$$

The recursion theorem assures us that a unique such function exists.

In general, for each $m \in \omega$ there exists (by the recursion theorem) a unique function $A_m: \omega \to \omega$ for which

$$A_m(0) = m,$$
$$A_m(n^+) = A_m(n)^+ \qquad \text{for } n \text{ in } \omega.$$

But we want one binary operation $+$, not all these little one-place functions.

Definition A *binary operation* on a set A is a function from $A \times A$ into A.

Definition *Addition* $(+)$ is the binary operation on ω such that for any m and n in ω,

$$m + n = A_m(n).$$

Thus when written as a relation,

$$+ = \{\langle\langle m, n\rangle, p\rangle \mid m \in \omega \;\&\; n \in \omega \;\&\; p = A_m(n)\}.$$

In conformity to everyday notation, we write $m + n$ instead of $+(m, n)$ or $+(\langle m, n\rangle)$.

Theorem 4I For natural numbers m and n,

(A1) $$m + 0 = m,$$
(A2) $$m + n^+ = (m + n)^+.$$

This theorem is an immediate consequence of the construction of A_m. Observe that (A1) and (A2) serve to characterize the binary operation $+$ in a recursive fashion. Our only reason for using the A_m's is that the recursion theorem applies directly to functions with domain ω, not domain $\omega \times \omega$. We can now forget the A_m's, and use $+$ and Theorem 4I instead.

We can now proceed to construct the multiplication operation in much the same way. We first apply the recursion theorem to obtain many functions $M_m: \omega \to \omega$ where $M_m(n)$ is the result of multiplying m by n. Specifically,

[4] Readers planning to omit Chapter 5 are permitted to skip this section also.

for each $m \in \omega$ there exists (by the recursion theorem) a unique function $M_m\colon \omega \to \omega$ for which

$$M_m(0) = 0,$$
$$M_m(n^+) = M_m(n) + m.$$

Definition *Multiplication* $(\cdot)$ is the binary operation on ω such that for any m and n in ω,

$$m \cdot n = M_m(n).$$

The theorem analogous to Theorem 4I is the following.

Theorem 4J For natural numbers m and n,

(M1) $$m \cdot 0 = 0,$$
(M2) $$m \cdot n^+ = m \cdot n + m.$$

We can now discard the M_m functions, and use $\cdot$ and Theorem 4J instead.

We could, in the same manner, define the exponentiation operation on ω. The equations that characterize exponentiation are

(E1) $$m^0 = 1,$$
(E2) $$m^{(n^+)} = m^n \cdot m.$$

Example $2 + 2 = 4$ (we would be alarmed if this failed), as the following calculation demonstrates:

$$\begin{aligned} 2 + 0 &= 2 && \text{by (A1)},\\ 2 + 1 &= 2 + 0^+ \\ &= (2 + 0)^+ && \text{by (A2)}\\ &= 2^+ \\ &= 3,\\ 2 + 2 &= 2 + 1^+ \\ &= (2 + 1)^+ && \text{by (A2)}\\ &= 3^+ \\ &= 4. \end{aligned}$$

Having now given set-theoretic definitions of the operations of arithmetic, we next verify that some of the common laws of arithmetic are provable within set theory. This verification is additional evidence, albeit at an elementary level, that mathematics can be embedded in set theory.

Theorem 4K The following identities hold for all natural numbers.

(1) Associative law for addition

$$m + (n + p) = (m + n) + p.$$

(2) Commutative law for addition

$$m + n = n + m.$$

(3) Distributive law

$$m \cdot (n + p) = m \cdot n + m \cdot p.$$

(4) Associative law for multiplication

$$m \cdot (n \cdot p) = (m \cdot n) \cdot p.$$

(5) Commutative law for multiplication

$$m \cdot n = n \cdot m.$$

Proof Each part is proved by induction. This exemplifies a general fact: When a function has been constructed by use of the recursion theorem, then general properties of the function must usually be proved by induction.

(1) The proof uses induction on p. That is, consider fixed natural numbers m and n, and define $A = \{p \in \omega \mid m + (n + p) = (m + n) + p\}$. We leave the verification that A is inductive for Exercise 15.

(2) It is necessary to prove two preliminary facts, each of which is proved by induction.

The first preliminary fact is that $0 + n = n$ for all $n \in \omega$. Let $A = \{n \in \omega \mid 0 + n = n\}$. Then $0 \in A$ by (A1). Suppose that $k \in A$. Then

$$\begin{aligned} 0 + k^+ &= (0 + k)^+ && \text{by (A2)} \\ &= k^+ && \text{since } k \in A, \end{aligned}$$

and hence $k^+ \in A$. So A is inductive.

The second preliminary fact is that $m^+ + n = (m + n)^+$ for m and n in ω. Consider any fixed $m \in \omega$ and let $B = \{n \in \omega \mid m^+ + n = (m + n)^+\}$. Again $0 \in B$ by (A1). Suppose that $k \in B$. Then

$$\begin{aligned} m^+ + k^+ &= (m^+ + k)^+ && \text{by (A2)} \\ &= (m + k)^{++} && \text{since } k \in B \\ &= (m + k^+)^+ && \text{by (A2),} \end{aligned}$$

showing that $k^+ \in B$. Hence B is inductive.

Finally we are ready to prove the commutative law. Consider any $n \in \omega$ and let $C = \{m \in \omega | m + n = n + m\}$. By the first preliminary fact, $0 + n = n = n + 0$, whence $0 \in C$. Suppose that $k \in C$. Then

$$\begin{aligned} k^+ + n &= (k + n)^+ && \text{by second fact} \\ &= (n + k)^+ && \text{since } k \in C \\ &= n + k^+ && \text{by (A2),} \end{aligned}$$

so that $k^+ \in C$. Hence C is inductive.

(3) Consider fixed m and n in ω and let

$$A = \{p \in \omega \mid m \cdot (n + p) = m \cdot n + m \cdot p\}.$$

To check that $0 \in A$, observe that

$$\begin{aligned} m \cdot (n + 0) &= m \cdot n && \text{by (A1)} \\ &= m \cdot n + 0 && \text{by (A1)} \\ &= m \cdot n + m \cdot 0 && \text{by (M1).} \end{aligned}$$

Now suppose that $k \in A$. Then

$$\begin{aligned} m \cdot (n + k^+) &= m \cdot (n + k)^+ && \text{by (A2)} \\ &= m \cdot (n + k) + m && \text{by (M2)} \\ &= (m \cdot n + m \cdot k) + m && \text{since } k \in A \\ &= m \cdot n + (m \cdot k + m) && \text{by part (1)} \\ &= m \cdot n + m \cdot k^+ && \text{by (M2),} \end{aligned}$$

which shows that $k^+ \in A$. Hence A is inductive.

The reader has no doubt observed that each inductive argument here is quite straightforward. And each is, for that matter, much like the next.

(4) Consider fixed m and n in ω and let

$$A = \{p \in \omega \mid m \cdot (n \cdot p) = (m \cdot n) \cdot p\}.$$

To check that $0 \in A$ we note that $m \cdot (n \cdot 0) = m \cdot 0 = 0$ by (M1), and $(m \cdot n) \cdot 0 = 0$ as well. Now suppose that $k \in A$. Then

$$\begin{aligned} m \cdot (n \cdot k^+) &= m \cdot (n \cdot k + n) && \text{by (M2)} \\ &= m \cdot (n \cdot k) + m \cdot n && \text{by part (3)} \\ &= (m \cdot n) \cdot k + m \cdot n && \text{since } k \in A \\ &= (m \cdot n) \cdot k^+ && \text{by (M2),} \end{aligned}$$

which shows that $k^+ \in A$. Hence A is inductive.

(5) The proof here follows the outline of part (2). There are the analogous two preliminary facts to be proved: $0 \cdot n = 0$ and $m^+ \cdot n = m \cdot n + n$. The details of the three inductive arguments are left for Exercise 16. ⊣

Exercises

13. Let m and n be natural numbers such that $m \cdot n = 0$. Show that either $m = 0$ or $n = 0$.

14. Call a natural number *even* if it has the form $2 \cdot m$ for some m. Call it *odd* if it has the form $(2 \cdot p) + 1$ for some p. Show that each natural number is either even or odd, but never both.

15. Complete the proof of part (1) of Theorem 4K.

16. Complete the proof of part (5) of Theorem 4K.

17. Prove that $m^{n+p} = m^n \cdot m^p$.

ORDERING ON ω

We have defined natural numbers in such a way that, for example, $4 \in 7$. This may have appeared to be a spurious side effect of our definition, but we now want to turn it to our advantage. We have the following strikingly simple definition of order on ω: For natural numbers m and n, define m to be *less than* n iff $m \in n$. We could introduce a special symbol "<" for this:

$$m < n \quad \text{iff} \quad m \in n.$$

But the special symbol seems unnecessary; we can just use "$\in$". But it will be necessary to keep in mind the dual role of this symbol, which denotes both membership and ordering. In place of an $\leq$ symbol, we define

$$m \underline{\in} n \qquad \text{iff} \qquad \text{either} \quad m \in n \quad \text{or} \quad m = n.$$

Observe that

$$p \in k^+ \quad \Leftrightarrow \quad p \underline{\in} k,$$

a fact we will use in later calculations.

We are now entitled to state the following fact: Any natural number is just the set of all smaller natural numbers. That is, for any n in ω,

$$x \text{ is a member of } n \quad \Leftrightarrow \quad x \in \omega \;\&\; x \text{ is less than } n.$$

To verify this, note that we can restate it as

$$x \in n \quad \Leftrightarrow \quad x \in \omega \;\&\; x \in n,$$

which is true because ω is a transitive set, and thus $x \in n \in \omega \Rightarrow x \in \omega$.

We should show that we do indeed have a linear ordering relation on ω, in the sense defined in Chapter 3. The relation in question is the set of ordered pairs $\in_\omega$ defined by

$$\in_\omega = \{\langle m, n\rangle \in \omega \times \omega \mid m \in n\}.$$

We will prove that this is a linear ordering relation on ω, i.e., that it is a transitive relation that satisfies trichotomy on ω.

Because each natural number is a transitive set, we have for m, n, p in ω:

$$m \in n \;\&\; n \in p \quad \Rightarrow \quad m \in p.$$

That is, our ordering relation on ω is a transitive relation. It is somewhat harder to show from our definitions that of any two distinct natural numbers, one is larger than the other. For that result, we will need the following lemma.

Lemma 4L (a) For any natural numbers m and n,

$$m \in n \quad \text{iff} \quad m^+ \in n^+.$$

(b) No natural number is a member of itself.

Proof (a) First assume that $m^+ \in n^+$. Then we have $m \in m^+ \subseteq n$. Hence (by the transitivity of n) we obtain $m \in n$.

To prove the converse we use induction on n. That is, form

$$T = \{n \in \omega \mid (\forall m \in n)\, m^+ \in n^+\}.$$

Then $0 \in T$ vacuously. Consider any $k \in T$. In order to show that $k^+ \in T$, we must show that whenever $m \in k^+$, then $m^+ \in k^{++}$. Given $m \in k^+$, we have either $m = k$ (in which case $m^+ = k^+ \in k^{++}$) or $m \in k$. In the latter case (since $k \in T$), $m^+ \in k^+ \subseteq k^{++}$. So in either case we get $m^+ \in k^{++}$ and thus $k^+ \in T$. Hence T is inductive and coincides with ω.

Part (b) follows easily from (a). Let

$$T = \{n \in \omega \mid n \notin n\}.$$

Then $0 \in T$ since nothing is a member of 0. And by part (a), $k \notin k \Rightarrow k^+ \notin k^+$. Hence T is inductive and coincides with ω. ⊣

(In Chapter 7 we will come to the regularity axiom, which implies among other things that *no* set is a member of itself. But for natural numbers we can get along without the regularity axiom.)

We next use the lemma to prove that for two distinct natural numbers, one is always a member of the other. (It is the smaller one that is a member of the larger one.)

Trichotomy Law for ω For any natural numbers m and n, exactly one of the three conditions

$$m \in n, \qquad m = n, \qquad n \in m$$

holds.

Proof First note that *at most* one can hold. If $m \in n$ and $m = n$, then $m \in m$, in violation of Lemma 4L(b). Also if $m \in n \in m$, then because m is a transitive set we again have $m \in m$.

It remains to show that *at least* one holds. For that we use induction; let

$$T = \{n \in \omega \mid (\forall m \in \omega)(m \in n \text{ or } m = n \text{ or } n \in m)\}.$$

In order to show that $0 \in T$, we want to show that $0 \mathrel{\underline{\in}} m$ for all m. This we do by induction on m. (An induction within an induction!) Clearly $0 \mathrel{\underline{\in}} 0$, and if $0 \mathrel{\underline{\in}} k$, then $0 \in k^+$. Hence $0 \in T$.

Now assume that $k \in T$ and consider k^+. For any m in ω we have (since $k \in T$) either $m \mathrel{\underline{\in}} k$ (in which case $m \in k^+$) or $k \in m$. In the latter case $k^+ \in m^+$ by Lemma 4L(a), and so $k^+ \mathrel{\underline{\in}} m$. Thus in every case, either $m \in k^+$ or $k^+ = m$ or $k^+ \in m$. And so $k^+ \in T$, T is inductive, and we are done. ⊣

A set A is said to be a *proper subset* of B ($A \subset B$) iff it is a subset of B that is unequal to B.

$$A \subset B \quad \Leftrightarrow \quad A \subseteq B \;\&\; A \neq B.$$

Ordering on ω is given not only by the membership relation, but also by the proper subset relation:

Corollary 4M For any natural numbers m and n,

$$m \in n \quad \text{iff} \quad m \subset n$$

and

$$m \mathrel{\underline{\in}} n \quad \text{iff} \quad m \subseteq n.$$

Proof Since n is a transitive set,

$$m \in n \quad \Rightarrow \quad m \subseteq n,$$

and the inclusion is proper by Lemma 4L(b). Conversely assume that $m \subset n$. Then $m \neq n$, and $n \notin m$ lest $n \in n$. So by trichotomy $m \in n$ and we are done. ⊣

The above proof uses trichotomy in a typical way: To show that $m \in n$, it suffices to eliminate the other two alternatives.

The following theorem gives the order-preserving properties of addition and multiplication. The theorem will be used in Chapter 5 (but not elsewhere).

Theorem 4N For any natural numbers m, n, and p,

$$m \in n \quad \Leftrightarrow \quad m + p \in n + p.$$

If, in addition, $p \neq 0$, then

$$m \in n \quad \Leftrightarrow \quad m \cdot p \in n \cdot p.$$

Proof First consider addition. For the "⇒" half we use induction on p. Consider fixed $m \in n \in \omega$ and let

$$A = \{p \in \omega \mid m + p \in n + p\}.$$

Clearly $0 \in A$, and

$$\begin{aligned} k \in A \quad &\Rightarrow \quad m + k \in n + k \\ &\Rightarrow \quad (m + k)^+ \in (n + k)^+ \qquad \text{by Lemma 4L(a)} \\ &\Rightarrow \quad m + k^+ \in n + k^+ \qquad \text{by (A2)} \\ &\Rightarrow \quad k^+ \in A. \end{aligned}$$

Hence A is inductive and so $A = \omega$.

For the "⇐" half we use the trichotomy law and the "⇒" half. If $m + p \in n + p$, then we cannot have $m = n$ (lest $n + p \in n + p$) nor $n \in m$ (lest $n + p \in m + p \in n + p$). The only alternative is $m \in n$.

For multiplication the procedure is similar. For the "⇒" direction, consider fixed $m \in n \in \omega$ and let

$$B = \{q \in \omega \mid m \cdot q^+ \in n \cdot q^+\}.$$

(Recall that for a natural number $p \neq 0$ there is some $q \in \omega$ with $q^+ = p$.) It is easy to see that $0 \in B$, since $m \cdot 0^+ = m \cdot 0 + m = m$. Suppose that $k \in B$; we need to show that $m \cdot k^{++} \in n \cdot k^{++}$. Thus

$$\begin{aligned} m \cdot k^{++} &= m \cdot k^+ + m \\ &\in n \cdot k^+ + m \end{aligned}$$

by applying the first part of the theorem to the fact that $m \cdot k^+ \in n \cdot k^+$. And by again applying the first part of the theorem (this time to the fact that $m \in n$),

$$\begin{aligned} n \cdot k^+ + m &\in n \cdot k^+ + n \\ &= n \cdot k^{++}. \end{aligned}$$

Hence $k^+ \in B$, B is inductive, and $B = \omega$.

The "⇐" half then follows exactly as for addition. ⊣

Corollary 4P The following cancellation laws hold for m, n, and p in ω:

$$m + p = n + p \quad \Rightarrow \quad m = n,$$

$$m \cdot p = n \cdot p \;\&\; p \neq 0 \quad \Rightarrow \quad m = n.$$

Proof Apply trichotomy and the preceding theorem. ⊣

Well Ordering of ω Let A be a nonempty subset of ω. Then there is some $m \in A$ such that $m \in n$ for all $n \in A$.

Note Such an m is said to be *least* in A. Thus the theorem asserts that any nonempty subset of ω has a least element. The least element is always unique, for if m_1 and m_2 are both least in A, then $m_1 \in m_2$ and $m_2 \in m_1$. Consequently $m_1 = m_2$.

Proof Assume that A is a subset of ω without a least element; we will show that $A = \varnothing$. We could attempt to do this by showing that the complement $\omega - A$ was inductive. But in order to show that $k^+ \in \omega - A$, it is not enough to know merely that $k \in \omega - A$, we must know that all numbers smaller than k are in $\omega - A$ as well. Given this additional information, we can argue that $k^+ \in \omega - A$ lest it be a least element for A.

To write down what is approximately this argument, let

$$B = \{m \in \omega \mid \text{no number less than } m \text{ belongs to } A\}.$$

We claim that B is inductive. $0 \in B$ vacuously. Suppose that $k \in B$. Then if n is less than k^+, either n is less than k (in which case $n \notin A$ since $k \in B$) or $n = k$ (in which case $n \notin A$ lest, by trichotomy, it be least in A). In either case, n is outside of A. Hence $k^+ \in B$ and so B is inductive. It clearly follows that $A = \varnothing$; for example, $7 \notin A$ because $8 \in B$. ⊣

Corollary 4Q There is no function $f: \omega \to \omega$ such that $f(n^+) \in f(n)$ for every natural number n.

Proof The range of f would be a nonempty subset of ω without a least element, contradicting the well ordering of ω. ⊣

Our proof of the well ordering of ω suggests that it might be useful to have a second induction principle.

Strong Induction Principle for ω Let A be a subset of ω, and assume that for every n in ω,

$$\text{if every number less than } n \text{ is in } A, \text{ then } n \in A.$$

Then $A = \omega$.

Proof Suppose, to the contrary, that $A \neq \omega$. Then $\omega - A \neq \varnothing$, and by the well ordering it has a least number m. Since m is least in $\omega - A$, all numbers less than m are in A. But then by the hypothesis of the theorem $m \in A$, contradicting the fact that $m \in \omega - A$. ⊣

The well-ordering principle provides an alternative to proofs by induction. Suppose we want to show that for every natural number, a certain statement holds. Instead of forming the set of numbers for which the statement is true, consider the set of numbers for which it is *false*, i.e., the set C of counterexamples. To show that $C = \varnothing$, it suffices to show that C cannot have a least element.

Exercises

18. Simplify: $\in_\omega^{-1}[\![\{7, 8\}]\!]$.

19. Prove that if m is a natural number and d is a nonzero number, then there exist numbers q and r such that $m = (d \cdot q) + r$ and r is less than d.

20. Let A be a nonempty subset of ω such that $\bigcup A = A$. Show that $A = \omega$.

21. Show that no natural number is a subset of any of its elements.

22. Show that for any natural numbers m and p we have $m \in m + p^+$.

23. Assume that m and n are natural numbers with m less than n. Show that there is some p in ω for which $m + p^+ = n$. (It follows from this and the preceding exercise that m is less than n iff $(\exists p \in \omega)\ m + p^+ = n$.)

24. Assume that $m + n = p + q$. Show that

$$m \in p \quad \Leftrightarrow \quad q \in n.$$

25. Assume that $n \in m$ and $q \in p$. Show that

$$(m \cdot q) + (n \cdot p) \in (m \cdot p) + (n \cdot q).$$

[*Suggestion:* Use Exercise 23.]

26. Assume that $n \in \omega$ and $f : n^+ \to \omega$. Show that ran f has a largest element.

27. Assume that A is a set, G is a function, and f_1 and f_2 map ω into A. Further assume that for each n in ω both $f_1 \upharpoonright n$ and $f_2 \upharpoonright n$ belong to dom G and

$$f_1(n) = G(f_1 \upharpoonright n)\ \&\ f_2(n) = G(f_2 \upharpoonright n).$$

Show that $f_1 = f_2$.

28. Rewrite the proof of Theorem 4G using, in place of induction, the well ordering of ω.

Review Exercises

29. Write an expression for the set 4 using only symbols $\varnothing$, {, }, and commas.

30. What is $\bigcup 4$? What is $\bigcap 4$?

31. What is $\bigcup\bigcup 7$?

32. (a) Let $A = \{1\}$. Calculate A^+ and $\bigcup(A^+)$.
(b) What is $\bigcup(\{2\}^+)$?

33. Which of the following sets are transitive? (For each set S that is not transitive, specify a member of $\bigcup S$ not belonging to S.)

(a) $\{0, 1, \{1\}\}$.

(b) $\{1\}$.

(c) $\langle 0, 1\rangle$.

34. Find suitable a, b, etc. making each of the following sets transitive.

(a) $\{\{\{\varnothing\}\}, a, b\}$.

(b) $\{\{\{\{\varnothing\}\}\}, c, d, e\}$.

35. Let S be the set $\langle 1, 0\rangle$.

(a) Find a transitive set T_1 for which $S \subseteq T_1$.

(b) Find a transitive set T_2 for which $S \in T_2$.

36. There is a function $h: \omega \to \omega$ for which $h(0) = 3$ and $h(n^+) = 2 \cdot h(n)$. What is $h(4)$?

37. We will say that a set S *has n elements* (where $n \in \omega$) iff there is a one-to-one function from n onto S. Assume that A has m elements and B has n elements.

(a) Show that if A and B are disjoint, then $A \cup B$ has $m + n$ elements.

(b) Show that $A \times B$ has $m \cdot n$ elements.

38. Assume that h is the function from ω into ω for which $h(0) = 1$ and $h(n^+) = h(n) + 3$. Give an explicit (not recursive) expression for $h(n)$.

39. Assume that h is the function from ω into ω for which $h(0) = 1$ and $h(n^+) = h(n) + (2 \cdot n) + 1$. Give an explicit (not recursive) expression for $h(n)$.

40. Assume that h is the function from ω into ω defined by $h(n) = 5 \cdot n + 2$. Express $h(n^+)$ in terms of $h(n)$ as simply as possible.

CHAPTER 5

CONSTRUCTION OF THE REAL NUMBERS[1]

In Chapter 4 we gave a set-theoretic construction of the set ω of natural numbers. In the present chapter we will continue to show how mathematics can be embedded in set theory, by giving a set-theoretic construction of the real numbers. (The operative phrase is "can be," not "is" or "must be." We will return to this point in the section on "Two.")

INTEGERS

First we want to extend our set ω of natural numbers to a set $\mathbb{Z}$ of integers (both positive and negative). Here "extend" is to be loosely interpreted, since ω will not actually be a subset of $\mathbb{Z}$. But $\mathbb{Z}$ will include an "isomorphic copy" of ω (Fig. 19).

A negative integer can be named by using two natural numbers and a subtraction symbol: $2 - 3$, $5 - 10$, etc. We need some sets to stand behind these names.

As a first guess, we could try taking the integer -1 to be the pair $\langle 2, 3 \rangle$ of natural numbers used to name -1 in the preceding paragraph. And

[1] Other chapters do not depend on Chapter 5.

similarly we could try letting the integer -5 be the pair $\langle 5, 10\rangle$ of natural numbers. But this first guess fails, because -1 has a multiplicity of names: $2-3=0-1$ but $\langle 2, 3\rangle \neq \langle 0, 1\rangle$.

As a second guess, we can define an equivalence relation $\sim$ such that $\langle 2, 3\rangle \sim \langle 0, 1\rangle$. (Imposing such an equivalence relation is sometimes described as "identifying" $\langle 2, 3\rangle$ and $\langle 0, 1\rangle$.) Then we will have the one equivalence class

$$[\langle 2, 3\rangle] = [\langle 0, 1\rangle],$$

and we can take -1 to be this equivalence class. Then for the set $\mathbf{Z}$ of all integers, we can take the set of all equivalence classes:

$$\mathbf{Z} = (\omega \times \omega)/\sim.$$

This is in fact what we do. Call a pair of natural numbers a *difference*; then an integer will be an equivalence class of differences. Consider two differences $\langle m, n\rangle$ and $\langle p, q\rangle$. When should we call them equivalent?

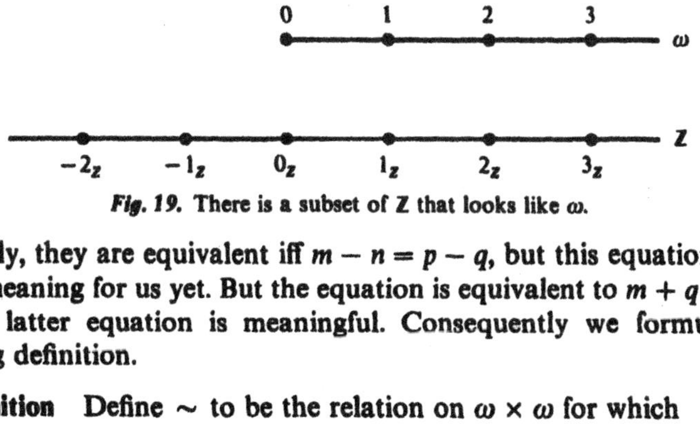

Fig. 19. There is a subset of $\mathbf{Z}$ that looks like ω.

Informally, they are equivalent iff $m-n=p-q$, but this equation has no official meaning for us yet. But the equation is equivalent to $m+q=p+n$, and the latter equation is meaningful. Consequently we formulate the following definition.

Definition Define $\sim$ to be the relation on $\omega \times \omega$ for which

$$\langle m, n\rangle \sim \langle p, q\rangle \quad \text{iff} \quad m+q=p+n.$$

Thus $\sim$ is a set of ordered pairs whose domain and range are also sets of ordered pairs. In more explicit (but less readable) form, the above definition can be stated:

$$\sim \; = \{\langle\langle m, n\rangle, \langle p, q\rangle\rangle \mid m+q=p+n \text{ and all are in } \omega\}.$$

Theorem 5ZA The relation $\sim$ is an equivalence relation on $\omega \times \omega$.

Proof We leave it to you to check that $\sim$ is reflexive on $\omega \times \omega$ and is symmetric. To show transitivity, suppose that $\langle m, n\rangle \sim \langle p, q\rangle$ and $\langle p, q\rangle \sim \langle r, s\rangle$. Then (by the definition of $\sim$)

$$m+q+p+s=p+n+r+q.$$

By use of the cancellation law (Corollary 4P), we obtain $m + s = r + n$, and thus $\langle m, n\rangle \sim \langle r, s\rangle$. ⊣

Definition The set $\mathbf{Z}$ of *integers* is the set $(\omega \times \omega)/\sim$ of all equivalence classes of differences.

For example, the integer $2_{\mathbf{Z}}$ is the equivalence class

$$[\langle 2, 0\rangle] = \{\langle 2, 0\rangle, \langle 3, 1\rangle, \langle 4, 2\rangle, \ldots\},$$

and the integer $-3_{\mathbf{Z}}$ is the equivalence class

$$[\langle 0, 3\rangle] = \{\langle 0, 3\rangle, \langle 1, 4\rangle, \langle 2, 5\rangle, \ldots\}.$$

These equivalence classes can be pictured as 45° lines in the Cartesian product $\omega \times \omega$ (Fig. 20).

Next we want to endow $\mathbf{Z}$ with a suitable addition operation. Informally, we can add differences:

$$(m - n) + (p - q) = (m + p) - (n + q).$$

This indicates that the correct addition function $+_{\mathbf{Z}}$ for integers will satisfy the equation

$$[\langle m, n\rangle] +_{\mathbf{Z}} [\langle p, q\rangle] = [\langle m + p, n + q\rangle].$$

This equation will serve to define $+_{\mathbf{Z}}$, once we have verified that it makes sense. The situation here is of the sort discussed in Theorem 3Q. We want to specify the value of the operation $+_{\mathbf{Z}}$ at a pair of equivalence classes by (1) selecting representatives $\langle m, n\rangle$ and $\langle p, q\rangle$ from the classes, (2)

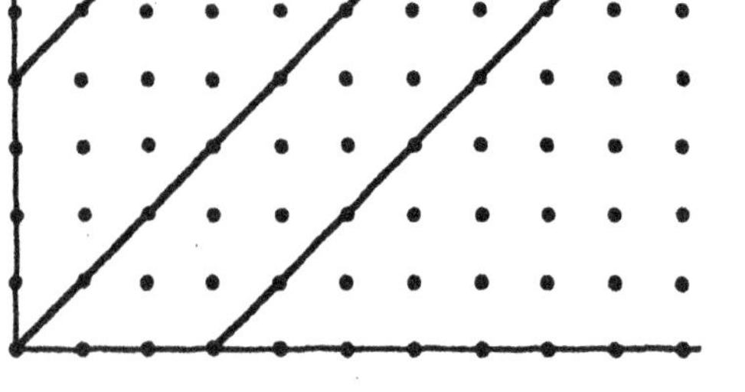

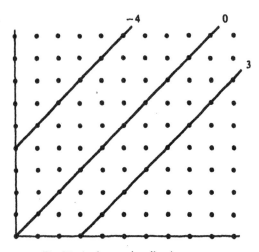

Fig. 20. An integer is a line in $\omega \times \omega$.

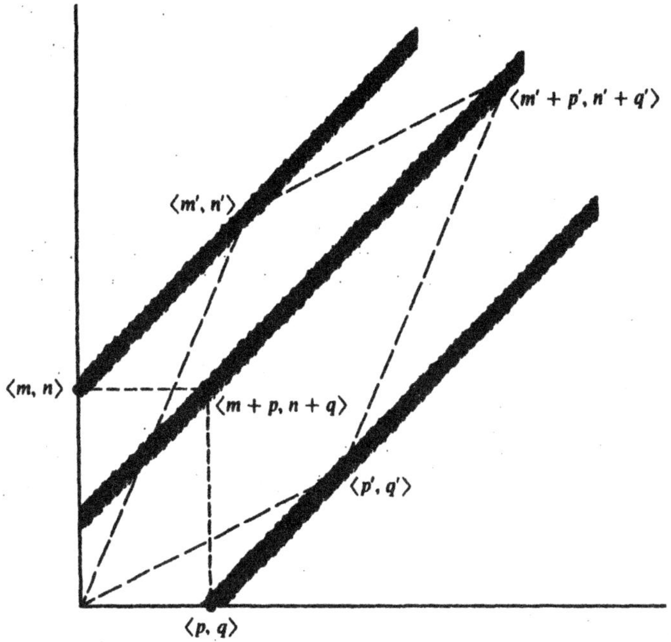

Fig. 21. Addition of lines is well defined.

operating on the representatives (by vector addition in this case), and then (3) forming the equivalence class of the result of the vector addition. For $+_Z$ to be well defined, we must verify that choice of other representatives $\langle m', n' \rangle$ and $\langle p', q' \rangle$ from the given classes would yield the same equivalence class for the sum (Fig. 21).

Lemma 5ZB If $\langle m, n \rangle \sim \langle m', n' \rangle$ and $\langle p, q \rangle \sim \langle p', q' \rangle$, then

$$\langle m + p, n + q \rangle \sim \langle m' + p', n' + q' \rangle.$$

Proof We are given, by hypothesis, the two equations

$$m + n' = m' + n \qquad \text{and} \qquad p + q' = p' + q.$$

We want to obtain the equation

$$m + p + n' + q' = m' + p' + n + q.$$

But this results from just adding together the two given equations. ⊣

This lemma justifies the definition of $+_Z$. In the terminology of Theorem 3Q, it says that the function F of vector addition

$$F(\langle m, n \rangle, \langle p, q \rangle) = \langle m + p, n + q \rangle$$

is compatible with $\sim$. Hence there is a well-defined function $\hat{F}$ on the quotient set; $\hat{F}$ is just our operation $+_Z$. It satisfies the equation

$$[\langle m, n\rangle] +_Z [\langle p, q\rangle] = [\langle m + p, n + q\rangle].$$

In other words, for integers a and b our addition formula is

$$a +_Z b = [\langle m + p, n + q\rangle],$$

where $\langle m, n\rangle$ is chosen from a and $\langle p, q\rangle$ is chosen from b. Theorem 3Q assures us that the equivalence class on the right is independent of how these choices are made.

Example We can calculate $2_Z +_Z (-3_Z)$. Since $2_Z = [\langle 2, 0\rangle]$ and $-3_Z = [\langle 0, 3\rangle]$, we have

$$\begin{aligned} 2_Z +_Z (-3_Z) &= [\langle 2, 0\rangle] +_Z [\langle 0, 3\rangle] \\ &= [\langle 2 + 0, 0 + 3\rangle] \\ &= [\langle 2, 3\rangle] \\ &= -1_Z. \end{aligned}$$

The familiar properties of addition, such as commutativity and associativity, now follow easily from the corresponding properties of addition of natural numbers.

Theorem 5ZC The operation $+_Z$ is commutative and associative:

$$a +_Z b = b +_Z a,$$
$$(a +_Z b) +_Z c = a +_Z (b +_Z c).$$

Proof The integer a must be of the form $[\langle m, n\rangle]$ for some natural numbers m and n; similarly b is $[\langle p, q\rangle]$. Then:

$$\begin{aligned} a +_Z b &= [\langle m, n\rangle] +_Z [\langle p, q\rangle] \\ &= [\langle m + p, n + q\rangle] && \text{by definition of } +_Z \\ &= [\langle p + m, q + n\rangle] && \text{by commutativity of } + \text{ on } \omega \\ &= [\langle p, q\rangle] +_Z [\langle m, n\rangle] \\ &= b +_Z a. \end{aligned}$$

The calculation for associativity is similar (Exercise 4). ⊣

Let $0_Z = [\langle 0, 0\rangle]$. Then it is straightforward to verify that $a +_Z 0_Z = a$ for any integer a, i.e., 0_Z is an identity element for addition. And the new feature that $\mathbf{Z}$ has (and the feature for which the extension from ω to $\mathbf{Z}$ was made), is the existence of additive inverses.

Theorem 5ZD (a) 0_Z is an identity element for $+_Z$:

$$a +_Z 0_Z = a$$

for any a in $\mathbb{Z}$.

(b) Additive inverses exist: For any integer a, there is an integer b such that

$$a +_Z b = 0_Z.$$

Proof (b) Given an integer a, it must be of the form $[\langle m, n\rangle]$. Take b to be $[\langle n, m\rangle]$. Then $a +_Z b = [\langle m + n, n + m\rangle] = [\langle 0, 0\rangle] = 0_Z$. ⊣

Theorems 5ZC and 5ZD together say that $\mathbb{Z}$ with the operation $+_Z$ and the identity element 0_Z is an *Abelian group*. The concept of an Abelian group is central to abstract algebra, but in this book the concept will receive only passing attention.

Inverses are unique. That is, if $a +_Z b = 0_Z$ and $a +_Z b' = 0_Z$, then $b = b'$. To prove this, observe that

$$b = b +_Z (a +_Z b') = (b +_Z a) +_Z b' = b'.$$

(This proof works in any Abelian group.) The inverse of a is denoted as $-a$. Then as the proof to Theorem 5ZD shows,

$$-[\langle m, n\rangle] = [\langle n, m\rangle].$$

Inverses provide us with a subtraction operation, which we define by the equation

$$b - a = b +_Z (-a).$$

We can also endow the set $\mathbb{Z}$ with a multiplication operation, which we obtain in much the same way as we obtained the addition operation. First we look at the informal calculation with differences

$$(m - n) \cdot (p - q) = (mp + nq) - (mq + np),$$

which tells us that the desired operation $\cdot_Z$ will satisfy the equation

$$[\langle m, n\rangle] \cdot_Z [\langle p, q\rangle] = [\langle mp + nq, mq + np\rangle].$$

(Here we write, as usual, mp in place of $m \cdot p$.) Again we must verify that the above equation characterizes a well-defined operation on equivalence classes. That is, we must verify that the operation on differences

$$G(\langle m, n\rangle, \langle p, q\rangle) = \langle mp + nq, mq + np\rangle$$

is compatible with $\sim$. This verification is accomplished by the following lemma.

Lemma 5ZE If $\langle m, n\rangle \sim \langle m', n'\rangle$ and $\langle p, q\rangle \sim \langle p', q'\rangle$, then

$$\langle mp + nq, mq + np\rangle \sim \langle m'p' + n'q', m'q' + n'p'\rangle.$$

Proof We are given the two equations

$$m + n' = m' + n, \tag{1}$$

$$p + q' = p' + q, \tag{2}$$

and we want to obtain the equation

$$mp + nq + m'q' + n'p' = m'p' + n'q' + mq + np.$$

The idea is take multiples of (1) and (2) that contain the terms we need. First multiply Eq. (1) by p; this gives us an mp on the left and an np on the right. Second, multiply the reverse of Eq. (1) by q; this gives us an nq on the left and an mq on the right. Third, multiply Eq. (2) by m'. Fourth, multiply the reverse of Eq. (2) by n'. Now add the four equations we have obtained from (1) and (2). All the unwanted terms cancel, and we are left with the desired equation. It works. ⊣

As for addition, we can prove the basic properties of multiplication from the corresponding properties of multiplication of natural numbers.

Theorem 5ZF The multiplication operation $\cdot_Z$ is commutative, associative, and distributive over $+_Z$:

$$a \cdot_Z b = b \cdot_Z a$$

$$(a \cdot_Z b) \cdot_Z c = a \cdot_Z (b \cdot_Z c)$$

$$a \cdot_Z (b +_Z c) = (a \cdot_Z b) +_Z (a \cdot_Z c)$$

Proof Say that $a = [\langle m, n\rangle]$ and $b = [\langle p, q\rangle]$. For the commutative law, we have

$$a \cdot_Z b = [\langle mp + nq, mq + np\rangle],$$

whereas

$$b \cdot_Z a = [\langle pm + qn, pn + qm\rangle].$$

The equality of these two follows at once from the commutativity of addition and multiplication in ω.

The other parts of the theorem are proved by the same method. Say that $c = [\langle r, s\rangle]$. Then $(a \cdot_Z b) \cdot_Z c$ is

$$[\langle (mp + nq)r + (mq + np)s, (mp + nq)s + (mq + np)r\rangle],$$

where $a \cdot_Z (b \cdot_Z c)$ is

$$[\langle m(pr + qs) + n(ps + qr), m(ps + qr) + n(pr + qs)\rangle].$$

The equality of these follows from laws of arithmetic in ω (Theorem 4K).

As for the distributive law, when we expand $a \cdot_Z (b +_Z c)$, we obtain

$$[\langle m(p + r) + n(q + s), m(q + s) + n(p + r)\rangle],$$

whereas when we expand $a \cdot_Z b +_Z a \cdot_Z c$ we obtain

$$[\langle mp + nq + mr + ns, mq + np + ms + nr\rangle].$$

Again equality is clear from laws of arithmetic in ω. ⊣

The remaining properties of multiplication that we will need constitute the next theorem. Let 1_Z be the integer $[\langle 1, 0\rangle]$.

Theorem 5ZG (a) The integer 1_Z is a multiplicative identity element:

$$a \cdot_Z 1_Z = a$$

for any integer a.

(b) $0_Z \neq 1_Z$.

(c) Whenever $a \cdot_Z b = 0_Z$, then either $a = 0_Z$ or $b = 0_Z$.

Part (c) is sometimes stated: There are no "zero divisors" in $\mathbb{Z}$.

Proof Part (a) is a trivial calculation.

For part (b) it is necessary to check that $\langle 0, 0\rangle \not\sim \langle 1, 0\rangle$. This reduces to checking that $0 \neq 1$ in ω, which is true.

For part (c), assume that $a \neq 0_Z$ and $b \neq 0_Z$; it will suffice to prove that $a \cdot_Z b \neq 0_Z$. We know that for some m, n, p, and q:

$$a = [\langle m, n\rangle], \qquad b = [\langle p, q\rangle],$$

$$a \cdot_Z b = [\langle mp + nq, mq + np\rangle].$$

Since $a \neq [\langle 0, 0\rangle]$, we have $m \neq n$. So either $m \in n$ or $n \in m$. Similarly, either $p \in q$ or $q \in p$. This leads to a total of four cases, but in each case we have

$$mp + nq \neq mq + np$$

by Exercise 25 of Chapter 4. Hence $a \cdot_Z b \neq [\langle 0, 0\rangle]$. ⊣

In algebraic terminology, we can say that $\mathbb{Z}$ together with $+_Z, \cdot_Z, 0_Z$, and 1_Z forms an *integral domain*. This means that:

(i) $\mathbb{Z}$ with $+_Z$ and 0_Z forms as Abelian group (Theorems 5ZC and 5ZD).

(ii) Multiplication is commutative and associative, and is distributive over addition (Theorem 5ZF).

(iii) 1_Z is a multiplicative identity (different from 0_Z), and no zero divisors exist (Theorem 5ZG).

There is a summary of these algebraic concepts near the end of this chapter. The value of the concepts stems from the large array of structures that satisfy the various conditions. In this book, however, we are concerned with only the most standard cases.

Example The calculation

$$[\langle 0, 1\rangle] \cdot_Z [\langle m, n\rangle] = [\langle n, m\rangle]$$

shows that $-1_Z \cdot_Z a = -a$.

Next we develop an ordering relation $<_Z$ on the integers. The informal calculation

$$m - n < p - q \quad \text{iff} \quad m + q < p + n$$

indicates that ordering $<_Z$ on $\mathbb{Z}$ should be defined by

$$[\langle m, n\rangle] <_Z [\langle p, q\rangle] \quad \text{iff} \quad m + q \in p + n.$$

As usual, it is necessary to check that this condition yields a well-defined relation on the integers. That is, we want to define

$$a <_Z b \quad \text{iff} \quad m + q \in p + n,$$

where m, n, p, and q are chosen so that $a = [\langle m, n\rangle]$ and $b = [\langle p, q\rangle]$. But that choice can be made in infinitely many ways; we must verify that we have the same outcome each time. The following lemma does just this.

Lemma 5ZH If $\langle m, n\rangle \sim \langle m', n'\rangle$ and $\langle p, q\rangle \sim \langle p', q'\rangle$, then

$$m + q \in p + n \quad \text{iff} \quad m' + q' \in p' + n'.$$

Proof The hypotheses give us the equations

$$m + n' = m' + n \qquad \text{and} \qquad p + q' = p' + q.$$

In order to utilize these equations in the inequality $m + q \in p + n$, we add n' and q' to each side of this inequality:

$$\begin{aligned} m + q \in p + n &\Leftrightarrow m + q + n' + q' \in p + n + n' + q' \\ &\Leftrightarrow m' + n + q + q' \in p' + q + n + n' \\ &\Leftrightarrow m' + q' \in p' + n'. \end{aligned}$$

Here the first and third steps use Theorem 4N, while the middle step uses the given equations. ⊣

Theorem 5ZI The relation $<_Z$ is a linear ordering relation on the set of integers.

Proof We must show that $<_Z$ is a transitive relation that satisfies trichotomy on $\mathbb{Z}$.

To prove transitivity, consider integers $a = [\langle m, n\rangle]$, $b = [\langle p, q\rangle]$, and $c = [\langle r, s\rangle]$. Then

$$\begin{aligned} a <_Z b \ \&\ b <_Z c &\Rightarrow m + q \in p + n \ \&\ p + s \in r + q \\ &\Rightarrow m + q + s \in p + n + s \ \&\ p + s + n \in r + q + n \\ &\Rightarrow m + q + s \in r + q + n \\ &\Rightarrow m + s \in r + n \qquad \text{by Theorem 4N} \\ &\Rightarrow a <_Z c. \end{aligned}$$

Proving trichotomy is easy. To say that exactly one of

$$a <_Z b, \qquad a = b, \qquad b <_Z a$$

holds is to say that exactly one of

$$m + q \in p + n, \qquad m + q = p + n, \qquad p + n \in m + q$$

holds. Thus the result follows from trichotomy in ω. ⊣

An integer b is called *positive* iff $0_Z <_Z b$. It is easy to check that

$$b <_Z 0_Z \quad \text{iff} \quad 0_Z <_Z -b.$$

Thus a consequence of trichotomy is the fact that for an integer b, exactly one of the three alternatives

$$b \text{ is positive}, \qquad b \text{ is zero}, \qquad -b \text{ is positive}$$

holds.

The next theorem shows that addition preserves order, as does multiplication by a positive integer. (The corresponding theorem for ω was Theorem 4N.)

Theorem 5ZJ The following are valid for any integers a, b, and c:

(a) $a <_Z b \Leftrightarrow a +_Z c <_Z b +_Z c$.

(b) If $0_Z <_Z c$, then

$$a <_Z b \quad \Leftrightarrow \quad a \cdot_Z c <_Z b \cdot_Z c.$$

Proof Assume that a, b, and c are $[\langle m, n\rangle]$, $[\langle p, q\rangle]$, and $[\langle r, s\rangle]$, respectively. The result to be proved in part (a) then translates to the following statement about natural numbers:

$$m + q \in p + n \quad \Leftrightarrow \quad m + r + q + s \in p + r + n + s.$$

This is an immediate consequence of the fact that addition in ω preserves order (Theorem 4N).

Part (b) is similar in spirit. As in Theorem 4N, it suffices to prove one direction:

$$0_Z <_Z c \;\&\; a <_Z b \quad\Rightarrow\quad a \cdot_Z c <_Z b \cdot_Z c.$$

This translates to:

$$s \in r \;\&\; m + q \in p + n \quad\Rightarrow\quad mr + ns + ps + qr \in pr + qs + ms + nr.$$

This is not as bad as it looks. If we let $k = m + q$ and $l = p + n$, then it becomes

$$s \in r \;\&\; k \in l \quad\Rightarrow\quad kr + ls \in ks + lr.$$

This is just Exercise 25 of Chapter 4. ⊣

Corollary 5ZK For any integers a, b, and c the cancellation laws hold:

$$a +_Z c = b +_Z c \quad\Rightarrow\quad a = b,$$

$$a \cdot_Z c = b \cdot_Z c \;\&\; c \neq 0_Z \quad\Rightarrow\quad a = b.$$

Proof This follows from the preceding theorem in the same way that the cancellation laws in ω (Corollary 4P) followed from the order-preserving properties (Theorem 4N). ⊣

Although ω is not actually a subset of $\mathbf{Z}$, nonetheless $\mathbf{Z}$ has a subset that is "just like" ω. To make this precise, define the function $E\colon \omega \to \mathbf{Z}$ by

$$E(n) = [\langle n, 0\rangle].$$

For example, $E(0) = 0_Z$ and $E(1) = 1_Z$.

The following theorem, in algebraic terminology, says that E is an "isomorphic embedding" of the system $\langle \omega, +, \cdot, \in_\omega\rangle$ into the system $\langle \mathbf{Z}, +_Z, \cdot_Z, <_Z\rangle$. That is, E is a one-to-one function that preserves addition, multiplication, and order.

Theorem 5ZL E maps ω one-to-one into $\mathbf{Z}$, and satisfies the following properties for any natural numbers m and n:

(a) $E(m + n) = E(m) +_Z E(n)$.
(b) $E(mn) = E(m) \cdot_Z E(n)$.
(c) $m \in n$ iff $E(m) <_Z E(n)$.

Proof To show that E is one-to-one we calculate

$$\begin{aligned} E(m) = E(n) &\Rightarrow [\langle m, 0\rangle] = [\langle n, 0\rangle] \\ &\Rightarrow \langle m, 0\rangle \sim \langle n, 0\rangle \\ &\Rightarrow m = n. \end{aligned}$$

Parts (a), (b), and (c) are proved by routine calculations (Exercise 8). ⊣

Finally we can give a precise counterpart to our motivating guideline that the difference $\langle m, n\rangle$ should name $m - n$. For any m and n,

$$[\langle m, n\rangle] = E(m) - E(m)$$

as is verified by evaluating the right side of this equation (Exercise 9).

Henceforth we will streamline our notation by omitting the subscript "Z" on $+_Z$, $\cdot_Z$, $<_Z$, 0_Z, 1_Z, etc. Furthermore $a \cdot b$ will usually be written as just ab.

Exercises

1. Is there a function $F: \mathbb{Z} \to \mathbb{Z}$ satisfying the equation
$$F([\langle m, n\rangle]) = [\langle m + n, n\rangle]?$$
2. Is there a function $G: \mathbb{Z} \to \mathbb{Z}$ satisfying the equation
$$G([\langle m, n\rangle]) = [\langle m, m\rangle]?$$
3. Is there a function $H: \mathbb{Z} \to \mathbb{Z}$ satisfying the equation
$$H([\langle m, n\rangle]) = [\langle n, m\rangle]?$$
4. Prove that $+_Z$ is associative. (This is part of Theorem 5ZC.)
5. Give a formula for subtraction of integers:
$$[\langle m, n\rangle] - [\langle p, q\rangle] = ?$$
6. Show that $a \cdot_Z 0_Z = 0_Z$ for every integer a.
7. Show that
$$a \cdot_Z (-b) = (-a) \cdot_Z b = -(a \cdot_Z b)$$
for all integers a and b.
8. Prove parts (a), (b), and (c) of Theorem 5ZL.
9. Show that
$$[\langle m, n\rangle] = E(m) - E(n)$$
for all natural numbers m and n.

RATIONAL NUMBERS

We can extend our set $\mathbb{Z}$ of integers to the set $\mathbb{Q}$ of rational numbers in much the same way as we extended ω to $\mathbb{Z}$. In fact, the extension from $\mathbb{Z}$ to $\mathbb{Q}$ is to multiplication what the extension from ω to $\mathbb{Z}$ is to addition. In the integers we get additive inverses, i.e., solutions to the equation $a + x = 0$.

In the rationals we will get multiplicative inverses, i.e., solutions to the equation $r \cdot_Q x = 1_Q$ (for nonzero r).

We can name a rational number by using two integers and a symbol for division:

$$1/2, \quad -3/4, \quad 6/12.$$

But as before, each number has a multiplicity of names, e.g., $1/2 = 6/12$. So the name "1/2" must be identified with the name "6/12."

By a *fraction* we mean an ordered pair of integers, the second component of which (called the *denominator*) is nonzero. For example, $\langle 1, 2\rangle$ and $\langle 6, 12\rangle$ are fractions; we want a suitable equivalence relation $\sim$ for which $\langle 1, 2\rangle \sim \langle 6, 12\rangle$. Since $a/b = c/d$ iff $a \cdot d = c \cdot b$, we choose to define $\sim$ as follows. Let $\mathbb{Z}'$ be the set $\mathbb{Z} - \{0\}$ of nonzero integers. Then $\mathbb{Z} \times \mathbb{Z}'$ is the set of all fractions.

Definition Define $\sim$ to be the binary relation on $\mathbb{Z} \times \mathbb{Z}'$ for which

$$\langle a, b\rangle \sim \langle c, d\rangle \quad \text{iff} \quad a \cdot d = c \cdot b.$$

The set $\mathbb{Q}$ of *rational numbers* is the set $(\mathbb{Z} \times \mathbb{Z}')/\sim$ of all equivalence classes of fractions.

We use the same symbol "$\sim$" that has been used previously for other equivalence relations, but as we discuss only one equivalence relation at a time, no confusion should result.

For example, $\langle 1, 2\rangle \sim \langle 6, 12\rangle$ since $1 \cdot 12 = 6 \cdot 2$. The equivalence class $[\langle 1, 2\rangle]$ is the rational number "one-half." The rationals zero and one are

$$0_Q = [\langle 0, 1\rangle] \qquad \text{and} \qquad 1_Q = [\langle 1, 1\rangle].$$

These are distinct, because $\langle 0, 1\rangle \not\sim \langle 1, 1\rangle$. Of course we must check that $\sim$ is indeed an equivalence relation.

Theorem 5QA The relation $\sim$ is an equivalence relation on $\mathbb{Z} \times \mathbb{Z}'$.

Proof You should verify that the relation is reflexive on $\mathbb{Z} \times \mathbb{Z}'$ and is symmetric. As for transitivity, suppose that

$$\langle a, b\rangle \sim \langle c, d\rangle \qquad \text{and} \qquad \langle c, d\rangle \sim \langle e, f\rangle.$$

Then

$$ad = cb \qquad \text{and} \qquad cf = ed.$$

Multiply the first equation by f and the second by b to get

$$adf = cbf \qquad \text{and} \qquad cfb = edb.$$

From this we conclude that $adf = edb$ and hence (by canceling the nonzero d) $af = eb$. This tells us that $\langle a, b\rangle \sim \langle e, f\rangle$. ⊣

We can picture the equivalence classes as nonhorizontal lines (in the "plane" $\mathbb{Z} \times \mathbb{Z}'$) through the origin (Fig. 22). The fraction $\langle 1, 2\rangle$ lies on the line with slope 2; in general, $[\langle a, b\rangle]$ is the line with slope b/a.

We arrive at addition and multiplication operations for $\mathbb{Q}$ by the same methods used for $\mathbb{Z}$. For addition, the informal calculation

$$\frac{a}{b} + \frac{c}{d} = \frac{ad + cb}{bd}$$

indicates that $+_Q$ should be defined by the equation

$$[\langle a, b\rangle] +_Q [\langle c, d\rangle] = [\langle ad + cb, bd\rangle].$$

Note that $bd \neq 0$ since $b \neq 0$ and $d \neq 0$. Hence $\langle ad + cb, bd\rangle$ *is* a fraction. As usual, we must check that there is a well-defined function $+_Q$ on equivalence classes that satisfies the above equation. The following lemma, together with Theorem 3Q, does just that.

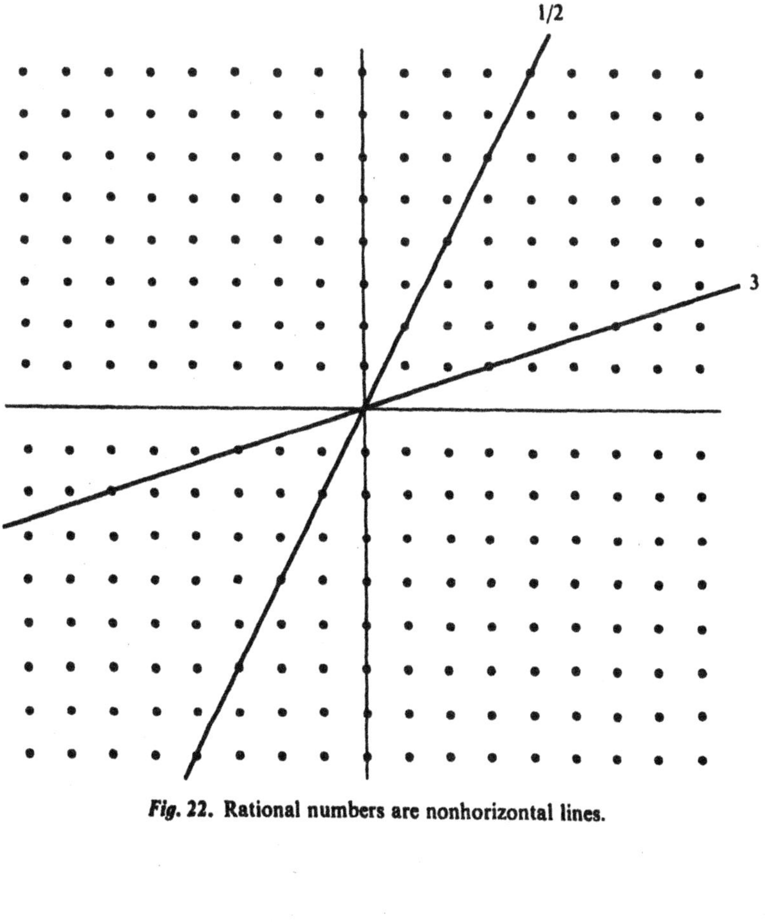

Fig. **22.** Rational numbers are nonhorizontal lines.

Lemma 5QB If $\langle a, b\rangle \sim \langle a', b'\rangle$ and $\langle c, d\rangle \sim \langle c', d'\rangle$, then

$$\langle ad + cb, bd\rangle \sim \langle a'd' + c'b', b'd'\rangle.$$

Proof We are given the equations

$$ab' = a'b \qquad \text{and} \qquad cd' = c'd.$$

We want the equation

$$(ad + cb)b'd' = (a'd' + c'b')bd,$$

which, when expanded (with the factors in alphabetic order), becomes

$$ab'dd' + bb'cd' = a'bdd' + bb'c'd.$$

This clearly is obtainable from the given equations. ⊣

Example Just to be on the safe side, we will check that $2 + 2 = 4$ in $\mathbb{Q}$. Let $2_Q = [\langle 2, 1\rangle]$ and $4_Q = [\langle 4, 1\rangle]$. Then

$$\begin{aligned} 2_Q +_Q 2_Q &= [\langle 2, 1\rangle] +_Q [\langle 2, 1\rangle] \\ &= [\langle 2 + 2, 1\rangle] \\ &= [\langle 4, 1\rangle] = 4_Q, \end{aligned}$$

where we use the fact that $2 + 2 = 4$ in $\mathbb{Z}$.

The rationals with $+_Q$ and 0_Q also form an Abelian group:

Theorem 5QC (a) Addition $+_Q$ is associative and commutative:

$$(q +_Q r) +_Q s = q +_Q (r +_Q s),$$
$$r +_Q s = s +_Q r.$$

(b) 0_Q is an identity element for $+_Q$:

$$r +_Q 0_Q = r$$

for any r in $\mathbb{Q}$.

(c) Additive inverses exist: For any r in $\mathbb{Q}$ there is an s in $\mathbb{Q}$ such that $r +_Q s = 0_Q$.

Proof First we verify commutativity. On the one hand,

$$[\langle a, b\rangle] +_Q [\langle c, d\rangle] = [\langle ad + cb, bd\rangle],$$

and on the other

$$[\langle c, d\rangle] +_Q [\langle a, b\rangle] = [\langle cb + ad, db\rangle].$$

But the right sides of these two equations are equal, by known commutative laws for arithmetic in $\mathbb{Z}$.

The verification of associativity is similar. Consider three rational numbers $[\langle a, b\rangle]$, $[\langle c, d\rangle]$, and $[\langle e, f\rangle]$. Then one grouping for the sum is

$$
\begin{aligned}
([\langle a, b\rangle] +_Q [\langle c, d\rangle]) +_Q [\langle e, f\rangle] &= [\langle ad + cb, bd\rangle] +_Q [\langle e, f\rangle] \\
&= [\langle (ad + cb)f + e(bd), (bd)f\rangle] \\
&= [\langle adf + cbf + ebd, bdf\rangle].
\end{aligned}
$$

The same expansion for the other grouping is

$$
\begin{aligned}
[\langle a, b\rangle] +_Q ([\langle c, d\rangle] +_Q [\langle e, f\rangle]) &= [\langle a, b\rangle] +_Q [\langle cf + ed, df\rangle] \\
&= [\langle a(df) + (cf + ed)b, b(df)\rangle] \\
&= [\langle adf + cfb + edb, bdf\rangle],
\end{aligned}
$$

which agrees with the first calculation.

Part (b) is a routine calculation. We know that $r = [\langle a, b\rangle]$ for some integers a and b. Then

$$
\begin{aligned}
r +_Q 0_Q &= [\langle a, b\rangle] +_Q [\langle 0, 1\rangle] \\
&= [\langle a \cdot 1 + 0 \cdot b, b \cdot 1\rangle] \\
&= [\langle a, b\rangle] = r.
\end{aligned}
$$

Finally for part (c) we select (with r as above) $s = [\langle -a, b\rangle]$. Then it is easy to calculate that

$$
\begin{aligned}
r +_Q s &= [\langle a, b\rangle] +_Q [\langle -a, b\rangle] \\
&= [\langle ab + (-a)b, bb\rangle] \\
&= [\langle 0, bb\rangle] = 0_Q,
\end{aligned}
$$

since $\langle 0, bb\rangle \sim \langle 0, 1\rangle$. ⊣

As in any Abelian group, the inverse of r is unique; we denote it as $-r$. The above proof shows that $-[\langle a, b\rangle] = [\langle -a, b\rangle]$.

For rational numbers, multiplication is simpler than addition. The informal calculation

$$
\frac{a}{b} \cdot \frac{c}{d} = \frac{ac}{bd}
$$

indicates that $\cdot_Q$ should be defined by the equation

$$
[\langle a, b\rangle] \cdot_Q [\langle c, d\rangle] = [\langle ac, bd\rangle].
$$

(Notice the close analogy with $+_Z$.) This multiplication function is well defined, as the following lemma verifies.

Lemma 5QD If $\langle a, b\rangle \sim \langle a', b'\rangle$ and $\langle c, d\rangle \sim \langle c', d'\rangle$, then

$$\langle ac, bd\rangle \sim \langle a'c', b'd'\rangle.$$

Proof The proof is exactly as in Lemma 5ZB, but with addition replaced by multiplication. ⊣

Example Recall that $1_Q = [\langle 1, 1\rangle]$. We can now check that 1_Q is a multiplicative identity element, i.e., that $r \cdot_Q 1_Q = r$. We know that $r = [\langle a, b\rangle]$ for some a and b. Thus

$$\begin{aligned} r \cdot_Q 1_Q &= [\langle a, b\rangle] \cdot_Q [\langle 1, 1\rangle] \\ &= [\langle a \cdot 1, b \cdot 1\rangle] \\ &= [\langle a, b\rangle] \\ &= r. \end{aligned}$$

You should also verify that $r \cdot_Q 0_Q = 0_Q$.

Theorem 5QE Multiplication of rationals is associative, commutative, and distributive over addition:

$$\begin{aligned} (p \cdot_Q q) \cdot_Q r &= p \cdot_Q (q \cdot_Q r), \\ q \cdot_Q r &= r \cdot_Q q, \\ p \cdot_Q (q +_Q r) &= (p \cdot_Q q) +_Q (p \cdot_Q r). \end{aligned}$$

Proof The verification of associativity and commutativity is directly analogous to verification of the same properties for $+_Z$.

We will proceed to prove the distributive law. We know that we can write $p = [\langle a, b\rangle]$, $q = [\langle c, d\rangle]$, and $r = [\langle e, f\rangle]$ for some integers a, b, c, d, e, and f. Then

$$\begin{aligned} p \cdot_Q (r +_Q s) &= [\langle a, b\rangle] \cdot_Q ([\langle c, d\rangle] +_Q [\langle e, f\rangle]) \\ &= [\langle a, b\rangle] \cdot_Q [\langle cf + ed, df\rangle] \\ &= [\langle acf + aed, bdf\rangle]. \end{aligned}$$

On the other side of the expected equation we have

$$\begin{aligned} (p \cdot_Q r) +_Q (p \cdot_Q s) &= ([\langle a, b\rangle] \cdot_Q [\langle c, d\rangle]) +_Q ([\langle a, b\rangle] \cdot_Q [\langle e, f\rangle]) \\ &= [\langle ac, bd\rangle] +_Q [\langle ae, bf\rangle] \\ &= [\langle acbf + aebd, bdbf\rangle]. \end{aligned}$$

This agrees with the first calculation because $\langle i, j\rangle \sim \langle bi, bj\rangle$. ⊣

The new property the rationals have (and that integers lack) is the existence of multiplicative inverses.

Theorem 5QF For every nonzero r in $\mathbb{Q}$ there is a nonzero q in $\mathbb{Q}$ such that $r \cdot_Q q = 1_Q$.

Proof The given r must be of the form $[\langle a, b\rangle]$, where $a \neq 0$, lest $r = 0_Q$. Let $q = [\langle b, a\rangle]$. Then $q \neq 0_Q$ and $r \cdot_Q q = [\langle ab, ab\rangle] = 1_Q$. ⊣

We can use the existence of multiplicative inverses to show that there are no zero divisors in $\mathbb{Q}$:

Corollary 5QG If r and s are nonzero rational numbers, then $r \cdot_Q s$ is also nonzero.

Proof The preceding theorem provides us with rationals r' and s' for which $r \cdot_Q r' = s \cdot_Q s' = 1_Q$. Hence

$$(r \cdot_Q s) \cdot_Q (r' \cdot_Q s') = 1_Q$$

by using commutative and associative laws. But this implies that $r \cdot_Q s \neq 0_Q$, because $0_Q \cdot_Q (r' \cdot_Q s') = 0_Q \neq 1_Q$. ⊣

We can restate this corollary by saying that the set of nonzero rational numbers is *closed* under multiplication; i.e., the product of numbers in this set is again in this set.

As a result of the foregoing theorems, we can assert that the nonzero rationals with *multiplication* form an Abelian group. That is, multiplication gives us a binary operation on the nonzero rationals that is associative and commutative, we have an identity element 1_Q, and we have multiplicative inverses. As in any Abelian group, the inverse of r is unique; we denote it as r^{-1}. The proof of Theorem 5QF shows that

$$[\langle a, b\rangle]^{-1} = [\langle b, a\rangle].$$

Inverses provide us with a division operation. For a nonzero rational r we can define

$$s \div r = s \cdot_Q r^{-1}.$$

Then we have

$$\begin{aligned} [\langle c, d\rangle] \div [\langle a, b\rangle] &= [\langle c, d\rangle] \cdot_Q [\langle b, a\rangle] \\ &= [\langle cb, da\rangle], \end{aligned}$$

a version of the "invert and multiply" rule for division of fractions.

The algebraic concept exemplified by the rational numbers is the concept of a *field*. To say that $\langle \mathbb{Q}, +_Q, \cdot_Q, 0_Q, 1_Q\rangle$ is a field means that it is an integral domain with the further property that multiplicative inverses exist. (Other examples of fields are provided by the real numbers and by the complex numbers.) The method we have used to extend from $\mathbb{Z}$ to $\mathbb{Q}$ can be applied to extend any integral domain to a field.

Next we want to define the ordering relation for the rational numbers. The informal calculation

$$\frac{a}{b} < \frac{c}{d} \quad \text{iff} \quad ad < cb$$

is correct if b and d are positive. There is no guarantee that denominators are always positive. But because

$$[\langle a, b\rangle] = [\langle -a, -b\rangle],$$

every rational number can be represented by some fraction with a positive denominator. (Recall that for nonzero b, either b or $-b$ is positive.) The above informal calculation then suggests that we define $<_Q$ so that

$$[\langle a, b\rangle] <_Q [\langle c, d\rangle] \quad \text{iff} \quad ad < cb$$

whenever b and d are positive. As with $<_Z$, we must verify that this condition yields a well-defined relation. The following lemma accomplishes the verification.

Lemma 5QH Assume that $\langle a, b\rangle \sim \langle a', b'\rangle$ and $\langle c, d\rangle \sim \langle c', d'\rangle$. Further assume that b, b', d, and d' are all positive. Then

$$ad < cb \quad \text{iff} \quad a'd' < c'b'.$$

Proof The proof is the same as the proof of Lemma 5ZH, but with multiplication of integers in place of addition of natural numbers. ⊣

This lemma guarantees that when we test to see whether or not $r <_Q s$, it does not matter which fractions with positive denominators we choose from r and s.

Example To check that $0_Q <_Q 1_Q$, we choose fractions $\langle 0, 1\rangle \in 0_Q$ and $\langle 1, 1\rangle \in 1_Q$. Then since $0 \cdot 1 < 1 \cdot 1$, we do indeed have $0_Q <_Q 1_Q$. But we could also have chosen fractions $\langle 0, 4\rangle \in 0_Q$ and $\langle 3, 3\rangle \in 1_Q$. Then since $0 \cdot 3 < 3 \cdot 4$, we again find, in consistency with the first calculation, that $0_Q <_Q 1_Q$.

Theorem 5QI The relation $<_Q$ is a linear ordering on $\mathbb{Q}$.

Proof The proof is the same as the proof of Theorem 5ZI, with multiplication in place of addition. For example, to prove trichotomy, we consider rational numbers r and s. For suitable integers we can write

$$r = [\langle a, b\rangle] \qquad \text{and} \qquad s = [\langle c, d\rangle],$$

where b and d are positive. Then trichotomy for $\mathbb{Z}$ tells us that exactly one of

$$ad < cb, \qquad ad = cb, \qquad cb < ad$$

holds, whence exactly one of

$$r <_Q s, \qquad r = s, \qquad s <_Q r$$

holds. ⊣

One can check that $r <_Q 0_Q$ iff $0_Q <_Q -r$ (Exercise 12). Call q *positive* iff $0_Q <_Q q$. Then as a consequence of trichotomy, we have the fact for any rational number r, exactly one of the three alternatives

$$r \text{ is positive}, \qquad r \text{ is zero}, \qquad -r \text{ is positive}$$

holds. We can define the *absolute value* $|r|$ of r by

$$|r| = \begin{cases} -r & \text{if} \quad -r \text{ is positive}, \\ r & \text{otherwise}. \end{cases}$$

Then $0_Q \leq_Q |r|$ for every r.

Next we prove that order is preserved by addition and by multiplication by a positive factor.

Theorem 5QJ Let r, s, and t be rational numbers.

(a) $r <_Q s$ iff $r +_Q t <_Q s +_Q t$.
(b) If t is positive, then

$$r <_Q s \quad \text{iff} \quad r \cdot_Q t <_Q s \cdot_Q t.$$

Proof Part (b) has the same proof as part (a) of Theorem 5ZJ, but with multiplication in place of addition. To prove part (a), we first write r, s, and t in the form

$$r = [\langle a, b\rangle], \qquad s = [\langle c, d\rangle], \qquad t = [\langle e, f\rangle],$$

where b, d, and f are positive. Since t is a positive rational, e is also a positive integer. Then

$$\begin{aligned} r +_Q t <_Q s +_Q t \quad &\Leftrightarrow \quad [\langle af + eb, bf\rangle] <_Q [\langle cf + ed, df\rangle] \\ &\Leftrightarrow \quad (af + eb)df < (cf + ed)bf \\ &\Leftrightarrow \quad adff + bdef < bcff + bdef \\ &\Leftrightarrow \quad ad < bc \qquad \text{by Theorem 5ZJ} \\ &\Leftrightarrow \quad r <_Q s \end{aligned}$$

as desired. ⊣

We have already said that the rational numbers form a field; the two preceding theorems state that $\langle \mathbb{Q}, +_Q, \cdot_Q, 0_Q, 1_Q, <_Q\rangle$ is an *ordered field*.

Theorem 5QK The following cancellation laws hold for any rational numbers.

(a) If $r +_Q t = s +_Q t$, then $r = s$.
(b) If $r \cdot_Q t = s \cdot_Q t$ and t is nonzero, then $r = s$.

Proof We can prove this as a corollary of the preceding theorem, following our past pattern. But there is now a simpler option open to us. In part (a) we add $-t$ to both sides of the given equation, and in part (b) we multiply both sides of the given equation by t^{-1}. (This proof works in any Abelian group.) ⊣

Finally, we want to show that, although $\mathbb{Z}$ is not a subset of $\mathbb{Q}$, nevertheless $\mathbb{Q}$ has a subset that is "just like" $\mathbb{Z}$. Define the embedding function $E: \mathbb{Z} \to \mathbb{Q}$ by

$$E(a) = [\langle a, 1\rangle].$$

This functions gives us an isomorphic embedding in the sense that the following theorem holds.

Theorem 5QL E is a one-to-one function from $\mathbb{Z}$ into $\mathbb{Q}$ satisfying the following conditions:

(a) $E(a + b) = E(a) +_Q E(b)$.
(b) $E(ab) = E(a) \cdot_Q E(b)$.
(c) $E(0) = 0_Q$ and $E(1) = 1_Q$.
(d) $a < b$ iff $E(a) <_Q E(b)$.

Proof Each part of the theorem can be proved by direct calculation. First we check that E is one-to-one:

$$\begin{aligned} E(a) = E(b) &\Rightarrow [\langle a, 1\rangle] = [\langle b, 1\rangle] \\ &\Rightarrow \langle a, 1\rangle \sim \langle b, 1\rangle \\ &\Rightarrow a = b. \end{aligned}$$

Parts (a), (b), and (d) are proved by the following calculations:

$$\begin{aligned} E(a) +_Q E(b) &= [\langle a, 1\rangle] +_Q [\langle b, 1\rangle] \\ &= [\langle a + b, 1\rangle] \\ &= E(a + b), \end{aligned}$$

$$\begin{aligned} E(a) \cdot_Q E(b) &= [\langle a, 1\rangle] \cdot_Q [\langle b, 1\rangle] \\ &= [\langle ab, 1\rangle] \\ &= E(ab), \end{aligned}$$

$$\begin{aligned} E(a) <_Q E(b) &\Leftrightarrow [\langle a, 1\rangle] <_Q [\langle b, 1\rangle] \\ &\Leftrightarrow a \cdot 1 < b \cdot 1 \\ &\Leftrightarrow a < b. \end{aligned}$$

Finally part (c) is a restatement of the definitions of 0_Q and 1_Q. ⊣

We also obtain the following relation between fractions and division:

$$[\langle a, b\rangle] = E(a) \div E(b).$$

Since $b \neq 0$, we have $E(b) \neq 0_Q$, and so the indicated division is possible.

Henceforth we will simplify the notation by omitting the subscript "Q" on $+_Q$, $\cdot_Q$, 0_Q, and so forth. Also the product $r \cdot s$ will usually be written as just rs.

Exercises

10. Show that $r \cdot_Q 0_Q = 0_Q$ for every rational number r.

11. Give a direct proof (not using Theorem 5QF) that if $r \cdot_Q s = 0_Q$, then either $r = 0_Q$ or $s = 0_Q$.

12. Show that

$$r <_Q 0_Q \quad \text{iff} \quad 0_Q <_Q -r.$$

13. Give a new proof of the cancellation law for $+_Z$ (Corollary 5ZK(a)), using Theorem 5ZD instead of Theorem 5ZJ.

14. Show that the ordering of rationals is dense, i.e., between any two rationals there is a third one:

$$p <_Q s \Rightarrow (\exists r)(p <_Q r <_Q s).$$

REAL NUMBERS

The last number system that we will consider involves the set $\mathbb{R}$ of all real numbers. The ancient Pythagoreans discovered, to their dismay, that there was a *need* to go beyond the rational numbers. They found that there simply was no rational number to measure the length of the hypotenuse of a right triangle whose other two sides had unit length.

In our previous extensions of number systems, we relied on the facts that an integer could be named by a pair of natural numbers, and a rational number could be named by a pair of integers. But we *cannot* hope to name real number by a pair of rationals, because, as we will prove in Chapter 6, there are too many real numbers and not enough pairs of rationals. Hence we must look at new techniques in searching for a way to name real numbers.

Actually there are several methods that can be used successfully to construct the real numbers. One approach is to utilize decimal expansions, so that a real number is determined by an integer and an infinite sequence of digits (a function from ω into 10). This approach may be found in Claude Burrill's book, *Foundations of Real Numbers*, McGraw-Hill, 1967.

A more common method of constructing a suitable set $\mathbb{R}$ is to utilize the fact that a real number can be named by giving a sequence of rationals (a function from ω into $\mathbb{Q}$) converging to it. So one can take the set of all convergent sequences and then divide out by an equivalence relation (where two sequences are equivalent iff they converge to the same limit). But there is one hitch: The concepts of "convergent" and "equivalent" must be defined without reference to the real number to which the sequence is converging. This can be done by a technique named after Cauchy.

Define a *Cauchy sequence* to be a function $s: \omega \to \mathbb{Q}$ such that $|s_m - s_n|$ is arbitrarily small for all sufficiently large m and n; i.e.,

$$(\forall \text{ positive } \varepsilon \text{ in } \mathbb{Q})(\exists k \in \omega)(\forall m > k)(\forall n > k)\ |s_m - s_n| < \varepsilon.$$

(Here we write s_n in place of $s(n)$, as usual for ω-sequences.) The concept of a Cauchy sequence is useful here because of the theorem of calculus asserting that a sequence is convergent iff it is a Cauchy sequence.

Let C be the set of all Cauchy sequences. For r and s in C, we define r and s to be *equivalent* $(r \sim s)$ iff $|r_n - s_n|$ is arbitrarily small for all sufficiently large n; i.e.,

$$(\forall \text{ positive } \varepsilon \text{ in } \mathbb{Q})(\exists k \in \omega)(\forall n > k)\ |r_n - s_n| < \varepsilon.$$

Then the quotient set $C/\sim$ is a suitable candidate for $\mathbb{R}$. (This approach to constructing $\mathbb{R}$ is due to Cantor.)

An alternative construction of $\mathbb{R}$ uses so-called *Dedekind cuts*. This is the method we follow henceforth in this section. The Cauchy sequence construction and the Dedekind cut construction each have their own advantages. The Dedekind cut construction of $\mathbb{R}$ has an initial advantage of simplicity, in that it provides a simple definition of $\mathbb{R}$ and its ordering. But multiplication of Dedekind cuts is awkward, and verification of the properties of multiplication is a tedious business. The Cauchy sequence construction of $\mathbb{R}$ also has the advantage of generality, since it can be used with an arbitrary metric space in place of $\mathbb{Q}$.

With these considerations in mind, we choose the following strategy. We will present the Dedekind cut construction, and will prove that least upper bounds exist in $\mathbb{R}$. (This is the property that distinguishes $\mathbb{R}$ from the other ordered fields.) Although we will define addition and multiplication of real numbers, we will not give complete verification of the algebraic properties. The Cauchy sequence construction may be found, among other places, in Norman Hamilton and Joseph Landin's book, *Set Theory and the Structure of Arithmetic*, Allyn and Bacon, 1961.

The idea behind Dedekind cuts is that a real number x can be named by giving an infinite set of rationals, namely all the rationals less than x. We will in effect define x to be the set of all rationals smaller than x. To avoid circularity in the definition, we must be able to characterize the sets of rationals obtainable in this way. The following definition does the job.

Definition A *Dedekind cut* is a subset x of $\mathbb{Q}$ such that:

(a) $\varnothing \neq x \neq \mathbb{Q}$.

(b) x is "closed downward," i.e.,

$$q \in x \,\&\, r < q \quad \Rightarrow \quad r \in x.$$

(c) x has no largest member.

We then define the set $\mathbb{R}$ of real numbers to be the set of all Dedekind cuts. Note that there is no equivalence relation here; a real (i.e., a real number) is a cut, not an equivalence class of cuts.

The ordering on $\mathbb{R}$ is particularly simple. For x and y in $\mathbb{R}$, define

$$x <_R y \quad \text{iff} \quad x \subset y.$$

In other words, $<_R$ is the relation of being a proper subset:

$$<_R = \{\langle x, y\rangle \in \mathbb{R} \times \mathbb{R} \mid x \subset y\}.$$

Theorem 5RA The relation $<_R$ is a linear ordering on $\mathbb{R}$.

Proof The relation is clearly transitive; we must show that it satisfies trichotomy on $\mathbb{R}$. So consider any x and y in $\mathbb{R}$. Obviously *at most* one of the alternatives,

$$x \subset y, \qquad x = y, \qquad y \subset x,$$

can hold, but we must prove that at least one holds. Suppose that the first two fail, i.e., that $x \nsubseteq y$. We must prove that $y \subset x$.

Since $x \nsubseteq y$ there is some rational r in the relative complement $x - y$ (see Fig. 23). Consider any $q \in y$. If $r \leq q$, then since y is closed downward,

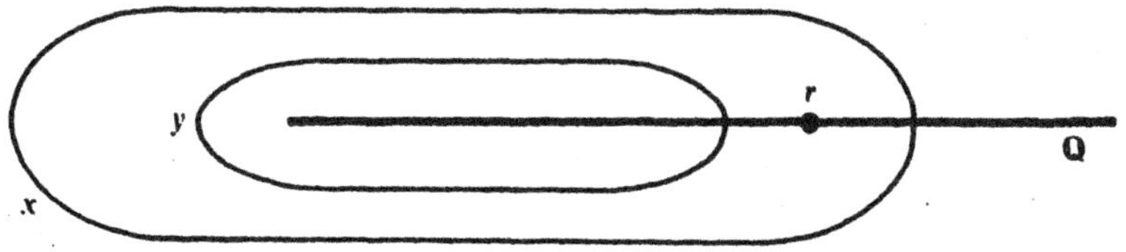

Fig. 23. The proof of Theorem 5RA.

we would have $r \in y$. But $r \notin y$, so we must have $q < r$. Since x is closed downward, it follows that $q \in x$. Since q was arbitrary (and $x \neq y$), we have $y \subset x$. ⊣

Now consider a set A of reals; a real number x is said to be an *upper bound* of A iff $y \leq_R x$ for every y in A. The number x itself might or might not belong to A. The set A is *bounded* (i.e., *bounded above*) iff there exists some upper bound of A. A *least upper bound* of A is an upper bound that is less than any other upper bound.

First consider an example not in $\mathbb{R}$, but in $\mathbb{Q}$. The set

$$\{r \in \mathbb{Q} \mid r \cdot r < 2\}$$

of rationals whose square is less than 2 is a bounded set of rationals that has no least upper bound in $\mathbb{Q}$. (We are stating this, not proving it, but it follows from the fact that $\sqrt{2}$ is irrational.) The following theorem shows that examples of this sort cannot be found in $\mathbb{R}$.

Theorem 5RB Any bounded nonempty subset of $\mathbb{R}$ has a least upper bound in $\mathbb{R}$.

Proof Let A be the set of real numbers in question. We will show that the least upper bound is just $\bigcup A$.

Simply by the definition of $\bigcup A$, we have $x \subseteq \bigcup A$ for all $x \in A$. Furthermore let z be any upper bound for A, so that $x \subseteq z$ for all $x \in A$. It then follows that $\bigcup A \subseteq z$; compare Exercise 5 of Chapter 2. The argument so far is not tied to $\mathbb{R}$; we have only shown that $\bigcup A$ is the least upper bound of the set A with respect to ordering by inclusion.

What remains to be shown is that $\bigcup A \in \mathbb{R}$. Because A is nonempty, it is easy to see that $\bigcup A \neq \varnothing$. Also $\bigcup A \neq \mathbb{Q}$ because $\bigcup A \subseteq z$ where z is an upper bound for A. You can easily verify (Exercise 15) that $\bigcup A$ is closed downward and has no largest element. ⊣

The foregoing theorem is important in mathematical analysis. For example, it is needed in order to prove that a continuous function on a closed interval assumes a maximum. And this in turn is used to prove the mean value theorem of calculus.

The addition operation for $\mathbb{R}$ is easily defined from addition of rationals. For reals x and y, define:

$$x +_R y = \{q + r \mid q \in x \;\&\; r \in y\}.$$

Lemma 5RC For real numbers x and y, the sum $x +_R y$ is also in $\mathbb{R}$.

Proof Clearly $x +_R y$ is a nonempty subset of $\mathbb{Q}$. To show that $x +_R y \neq \mathbb{Q}$, choose some q' in $\mathbb{Q} - x$ and r' in $\mathbb{Q} - y$. Then

$$\begin{aligned} q \in x \,\&\, r \in y \;&\Rightarrow\; q < q' \,\&\, r < r' \\ &\Rightarrow\; q + r < q' + r' \end{aligned}$$

so that any member $q + r$ of $x +_R y$ is strictly less than $q' + r'$. Hence $q' + r' \notin x +_R y$.

To show that $x +_R y$ is closed downward, consider any

$$p < q + r \in x +_R y$$

(where $q \in x$ and $r \in y$). Then adding $-q$ to both sides of the inequality, we have $p - q < r$. Since y is closed downward, we have $p - q \in y$. Thus we can write p as the sum

$$p = q + (p - q)$$

of q from x and $p - q$ from y; this is what we need to have $p \in x +_R y$. (*Note*: Here "$p - q$" refers to subtraction of rationals, $p + (-q)$. Earlier in this proof "$\mathbb{Q} - x$" referred to the relative complement of x in $\mathbb{Q}$. If this sort of thing happened often, we would use a different symbol "$\mathbb{Q} \setminus x$" for complements. But in fact the opportunities for confusion will be rare.)

We leave it to you to verify that $x +_R y$ has no largest member (Exercise 16). ⊣

Theorem 5RD Addition of real numbers is associative and commutative:

$$(x +_R y) +_R z = x +_R (y +_R z),$$
$$x +_R y = y +_R x.$$

Proof Since addition of rationals is commutative, it is clear from the definition of $+_R$ that it is commutative as well. As for associativity, we have

$$\begin{aligned} (x +_R y) +_R z &= \{s + r \mid s \in x +_R y \,\&\, r \in z\} \\ &= \{(p + q) + r \mid p \in x \,\&\, q \in y \,\&\, r \in z\}, \end{aligned}$$

and a similar calculation applies to the other grouping. Thus associativity of $+_R$ follows from associativity of addition of rationals. ⊣

The zero element of $\mathbb{R}$ is defined to be the set of negative rational numbers:

$$0_R = \{r \in \mathbb{Q} \mid r < 0\}.$$

Theorem 5RE (a) 0_R is a real number.
(b) For any x in $\mathbb{R}$, we have $x +_R 0_R = x$.

Proof (a) It is easy to see that $\varnothing \neq 0_R \neq \mathbb{Q}$; for example, $-1 \in 0_R$ and $1 \notin 0_R$. And it is clear that 0_R is closed downward. The fact that 0_R has no largest member follows immediately from the density of the rationals (Exercise 14).

For part (b), we must prove that

$$\{r + s \mid r \in x \ \&\ s < 0\} = x.$$

The "$\subseteq$" inclusion holds because x is closed downward. To prove the "$\supseteq$" half, consider any p in x. Since x has no largest member, there is some r with $p < r \in x$. Let $s = p - r$. Then $s < 0$ and $p = r + s \in 0_R$. Hence both inclusions hold. ⊣

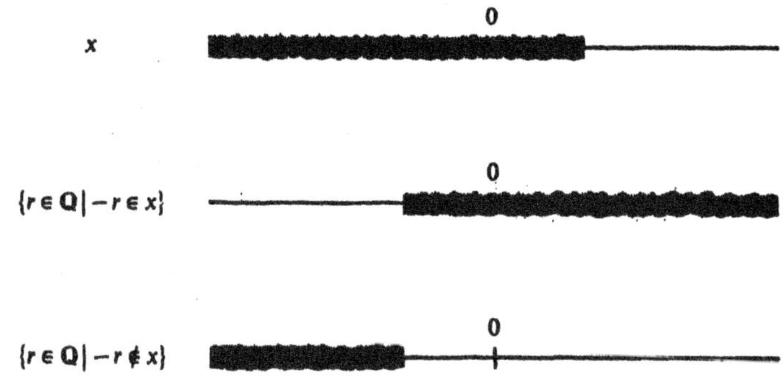

Fig. 24. Sometimes $-x$ is $\{r \in \mathbb{Q} \mid -r \notin x\}$, but not always.

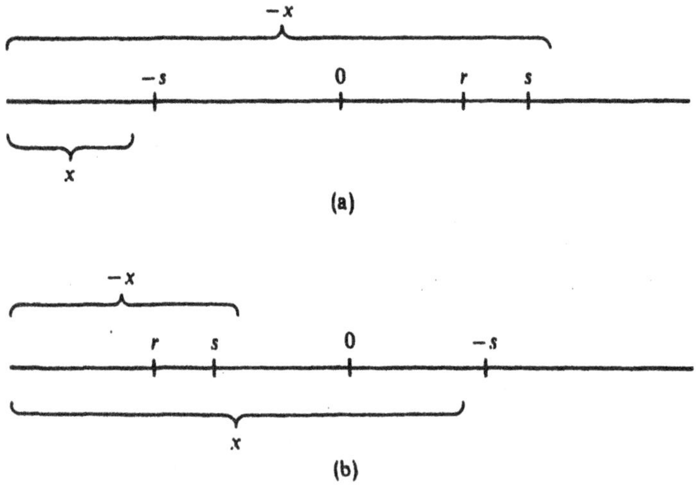

Fig. 25. In (a) x is negative; in (b) x is positive.

Before we can conclude that the real numbers form an Abelian group with $+_R$ and 0_R, we must prove that additive inverses exist. First we need to say just what set $-x$ should be (where $x \in \mathbb{R}$). We think of the real number $-x$ as the set of all smaller rational numbers. If we draw a picture as in Fig. 24, we might be tempted to think that $-x$ is the complement of $\{r \in \mathbb{Q} \mid -r \in x\}$, or in other words, that $-x$ should be $\{r \in \mathbb{Q} \mid -r \notin x\}$. This choice is not quite right, because it may have a largest element. Instead we define

$$-x = \{r \in \mathbb{Q} \mid (\exists s > r)\, -s \notin x\}.$$

In Fig. 25, $-x$ is illustrated as a subset of the rational number line.

Theorem 5RF For every x in $\mathbb{R}$:

(a) $-x \in \mathbb{R}$,
(b) $x +_R (-x) = 0_R$.

Proof To prove that $-x$ is a real number, we first must show that $\varnothing \neq -x \neq \mathbb{Q}$. There is some rational t with $t \notin x$; let $r = -t - 1$. Then $r \in -x$ because $r < -t$ and $-(-t) \notin x$. Hence $-x \neq \varnothing$. To show that $-x \neq \mathbb{Q}$, take any $p \in x$. We claim that $-p \notin -x$. This holds because if $s > -p$, then $-s < p \in x$, whence $-s \in x$. Hence $-p \notin -x$ and so $-x \neq \mathbb{Q}$.

It is easier to show that $-x$ is closed downward. Suppose that $q < r \in -x$. Then $(\exists s > r)\, -s \notin x$. Consequently $(\exists s > q)\, -s \notin x$, since the same s can be used. Hence $q \in -x$.

It remains to show that $-x$ has no largest element. Consider then any element r in $-x$. We know that for some $s > r$ we have $-s \notin x$. Because the rationals are densely ordered (Exercise 14), there is some p with $s > p > r$. Then $p \in -x$, and p is larger than r. This completes the proof that $-x$ is a real number.

Now we turn to part (b). By definition

$$x +_R (-x) = \{q + r \mid q \in x \;\&\; (\exists s > r)\, -s \notin x\}.$$

For any such member $q + r$ of this set, we have $r < s$ and $q < -s$ (lest $-s \leq q \in x$). Hence by the order-preserving property of addition,

$$q + r < (-s) + s = 0.$$

This shows that $x +_R (-x) \subseteq 0_R$.

To establish the other inclusion, consider any p in 0_R. Then $p < 0$, and so $-p$ is positive. By Exercise 19, there is some $q \in x$ for which $q + (-p \div 2) \notin x$. Let $s = (p \div 2) - q$, so that $-s \notin x$. Then p is the sum of q (which is in x) and $p - q$ (which is less than s, where $-s \notin x$). This makes p a member of $x +_R (-x)$. Thus we have both inclusions. ⊣

We have now shown that $\langle \mathbb{R}, +_R, 0_R \rangle$ is an Abelian group. As in any Abelian group, the cancellation law holds:

Corollary 5RG For any real numbers,

$$x +_R z = y +_R z \Rightarrow x = y.$$

Proof Simply add $-z$ to both sides of the given equation. ⊣

Next we prove that addition preserves order.

Theorem 5RH For any real numbers,

$$x <_R y \Leftrightarrow x +_R z <_R y +_R z.$$

Proof It is easy to prove that

$$(1) \qquad x \leq_R y \Rightarrow x +_R z \leq_R y +_R z,$$

because this amounts to the statement that if $x \subseteq y$, then

$$\{q + s \mid q \in x \ \&\ s \in z\} \subseteq \{r + s \mid r \in y \ \&\ s \in z\},$$

which is obvious. And by Corollary 5RG we have

$$(2) \qquad x \neq y \Rightarrow x +_R z \neq y +_R z,$$

which together with (1) gives the "$\Rightarrow$" half of the theorem. The "$\Leftarrow$" half then follows by trichotomy (as in the proof to Theorem 4N). ⊣

We can define the absolute value $|x|$ of a real number x to be the larger of x and $-x$. Since our ordering is inclusion, the larger of the two is just their union. Thus our definition becomes

$$|x| = x \cup -x.$$

Then by Exercise 20, $|x|$ is always nonnegative, i.e., $0_R \leq_R |x|$.

Consider now the definition of multiplication. For the product of x and y we cannot use $\{rs \mid r \in x \ \&\ s \in y\}$ (in analogy to the definition of $x +_R y$), because both x and y contain negative rationals of large magnitude. Instead we use the following variation on the above idea.

Definition (a) If x and y are nonnegative real numbers, then

$$x \cdot_R y = 0_R \cup \{rs \mid 0 \leq r \in x \ \&\ 0 \leq s \in y\}.$$

(b) If x and y are both negative real numbers, then

$$x \cdot_R y = |x| \cdot_R |y|.$$

(c) If one of the real numbers x and y is negative and one is nonnegative, then

$$x \cdot_R y = -(|x| \cdot_R |y|).$$

The facts we want to know about multiplication are gathered into the following theorem. Let $1_R = \{r \in \mathbb{Q} \mid r < 1\}$. Clearly $0_R <_R 1_R$. We will not give a proof for this theorem; a proof can be found in Appendix F of *Number Systems and the Foundations of Analysis* by Elliott Mendelson (Academic Press, 1973).

Theorem 5RI For any real numbers, the following hold:

(a) $x \cdot_R y$ is a real number.

(b) Multiplication is associative, commutative, and distributive over addition.

(c) $0_R \neq 1_R$ and $x \cdot_R 1_R = x$.

(d) For nonzero x there is a nonzero real number y with $x \cdot_R y = 1_R$.

(e) Multiplication by a positive number preserves order: If $0_R <_R z$, then

$$x <_R y \quad \Leftrightarrow \quad x \cdot_R z <_R y \cdot_R z.$$

The foregoing theorems show that, like the rationals, the reals (with $+_R$, $\cdot_R$, 0_R, 1_R, and $<_R$) form an ordered field. But unlike the rationals, the reals have the least-upper-bound property (Theorem 5RB). An ordered field is said to be *complete* iff it has the least-upper-bound property. It can be shown that the reals, in a sense, yield the *only* complete ordered field. That is, any other complete ordered field is "just like" (or more precisely, is isomorphic to) the ordered field of real numbers. For an exact statement of this theorem and for its proof, see any of the books we have referred to in this section, or p. 110 of Andrew Gleason's book, *Fundamentals of Abstract Analysis*, Addison-Wesley, 1966.

The correct embedding function E from $\mathbb{Q}$ into $\mathbb{R}$ assigns to each rational number r the corresponding real number

$$E(r) = \{q \in \mathbb{Q} \mid q < r\},$$

consisting of all smaller rationals.

Theorem 5RJ E is a one-to-one function from $\mathbb{Q}$ into $\mathbb{R}$ satisfying the following conditions:

(a) $E(r + s) = E(r) +_R E(s)$.

(b) $E(rs) = E(r) \cdot_R E(s)$.

(c) $E(0) = 0_R$ and $E(1) = 1_R$.

(d) $r < s$ iff $E(r) <_R E(s)$.

Proof First of all, we must show that $E(r)$ is a real number. Obviously $E(r)$ is a set of rationals, and it is closed downward. Furthermore $\varnothing \neq E(r) \neq \mathbb{Q}$ because $r - 1 \in E(r)$ and $r \notin E(r)$. $E(r)$ has no largest element, because if $q \in E(r)$, then by Exercise 14 there is a larger element p with $q < p < r$. Hence $E(r)$ is indeed a real number.

To show that E is one-to-one, suppose that $r \neq s$. Then one is less than the other; we may suppose that $r < s$. Then $r \in E(s)$ whereas $s \notin E(s)$. Hence $E(r) \neq E(s)$.

Next let us prove part (d), because it is easy. If $r < s$, then clearly $E(r) \subseteq E(s)$. The inclusion is proper since E is one-to-one. Thus

$$r < s \quad \Rightarrow \quad E(r) \subset E(s).$$

The converse follows from trichotomy. If $E(r) \subset E(s)$, then we cannot have $r = s$ nor $s < r$ (lest $E(s) \subset E(r)$), so we must have $r < s$.

For part (a), we have

$$\begin{aligned} E(r) +_R E(s) &= \{p + q \mid p \in E(r) \,\&\, q \in E(s)\} \\ &= \{p + q \mid p < r \,\&\, q < s\}. \end{aligned}$$

We must show that this is the same as $E(r + s)$, i.e., that

$$\{p + q \mid p < r \,\&\, q < s\} = \{t \mid t < r + s\}.$$

The "$\subseteq$" inclusion holds because by Theorem 5QJ,

$$p + q < r + q < r + s.$$

To establish the "$\supseteq$" inclusion, suppose that $t < r + s$. Let $\varepsilon = (r + s - t) \div 2$; then $\varepsilon > 0$. Define $p = r - \varepsilon$ and $q = s - \varepsilon$. Then $p < r$ and $q < s$ and $p + q = t$. Thus we can represent t as a sum in the desired form. Hence both inclusions hold.

Finally, we omit the (awkward) proof of part (b), and part (c) is only a restatement of definitions. ⊣

Exercises

15. In Theorem 5RB, show that $\bigcup A$ is closed downward and has no largest element.

16. In Lemma 5RC, show that $x +_R y$ has no largest element.

17. Assume that a is a positive integer. Show that for any integer b there is some k in ω with

$$b < a \cdot E(k).$$

18. Assume that p is a positive rational number. Show that for any rational number r there is some k in ω with

$$r < p \cdot E(E(k)).$$

(Here k is in ω, $E(k)$ is the corresponding integer, and $E(E(k))$ is the corresponding rational.)

19. Assume that p is a positive rational number. Show that for any real number x there is some rational q in x such that

$$p + q \notin x.$$

20. Show that for any real number x, we have $0_R \leq_R |x|$.

21. Show that if $x <_R y$, then there is a rational number r with

$$x <_R E(r) <_R y.$$

22. Assume that $x \in \mathbb{R}$. How do we know that $|x| \in \mathbb{R}$?

SUMMARIES

In this chapter we have given one way of constructing the real numbers as particular sets. Along the way, some concepts from abstract algebra have naturally arisen. For convenient reference, we have collected in the present section certain definitions that have played a key role in this chapter.

Integers Let m, n, p, and q be natural numbers.

$$\begin{aligned}
[\langle m, n\rangle] \sim [\langle p, q\rangle] \quad &\Leftrightarrow \quad m + q = p + n, \\
[\langle m, n\rangle] +_Z [\langle p, q\rangle] &= [\langle m + p, n + q\rangle], \\
-[\langle m, n\rangle] &= [\langle n, m\rangle], \\
[\langle m, n\rangle] \cdot_Z [\langle p, q\rangle] &= [\langle mp + nq, mp + np\rangle], \\
[\langle m, n\rangle] <_Z [\langle p, q\rangle] \quad &\Leftrightarrow \quad m + q \in p + n, \\
E(n) &= [\langle n, 0\rangle].
\end{aligned}$$

Rational numbers Let a, b, c, and d be integers with $bd \neq 0$.

$$\begin{aligned}
\langle a, b\rangle \sim \langle c, d\rangle \quad &\Leftrightarrow \quad ad = cb, \\
[\langle a, b\rangle] +_Q [\langle c, d\rangle] &= [\langle ad + cb, bd\rangle], \\
-[\langle a, b\rangle] &= [\langle -a, b\rangle], \\
[\langle a, b\rangle] \cdot_Q [\langle c, d\rangle] &= [\langle ac, bd\rangle], \\
[\langle a, b\rangle] <_Q [\langle c, d\rangle] \quad &\Leftrightarrow \quad ad < cb, \quad \text{when } b \text{ and } d \text{ are positive}, \\
E(a) &= [\langle a, 1\rangle].
\end{aligned}$$

Real numbers. A real number is a set x such that $\varnothing \subset x \subset \mathbb{Q}$, x is closed downward, and x has no largest member.

$$\begin{aligned}
x <_R y \quad &\Leftrightarrow \quad x \subset y, \\
x +_R y &= \{q + r \mid q \in x \;\&\; r \in y\}, \\
-x &= \{r \in \mathbb{Q} \mid (\exists s > r)\, -s \notin x\}, \\
|x| &= x \cup -x, \\
|x| \cdot_R |y| &= 0_R \cup \{rs \mid 0 \leq r \in |x| \;\&\; 0 \leq s \in |y|\}, \\
E(r) &= \{q \in \mathbb{Q} \mid q < r\}.
\end{aligned}$$

Next we turn to the definitions from abstract algebra that are relevant to the number systems in this chapter.

An *Abelian group* (in additive notation) is a triple[2] $\langle A, +, 0\rangle$ consisting of a set A, a binary operation $+$ on A, and an element ("zero") of A, such that the following conditions are met:

1. $+$ is associative and commutative.
2. 0 is an identity element, i.e., $x + 0 = x$.
3. Inverses exist, i.e., $\forall x\ \exists y\ x + y = 0$.

An Abelian group (in multiplicative notation) is a triple $\langle A, \cdot, 1\rangle$ consisting of a set A, a binary operation $\cdot$ on A, and an element 1 of A, such that the following conditions are met:

1. $\cdot$ is associative and commutative.
2. 1 is an identity element, i.e., $x \cdot 1 = x$.
3. Inverses exist, i.e., $\forall x\ \exists y\ x \cdot y = 1$.

This is, of course, the same as the preceding definition.

A *group* has the same definition, except that we do not require that the binary operation be commutative. All of the groups that we have considered have, in fact, been Abelian groups. But some of our results (e.g., the uniqueness of inverses) are correct in any group, Abelian or not.

A *commutative ring with identity* is a quintuple $\langle D, +, \cdot, 0, 1\rangle$ consisting of a set D, binary operations $+$ and $\cdot$ on D, and distinguished elements 0 and 1 of D, such that the following conditions are met:

1. $\langle D, +, 0\rangle$ is an Abelian group.
2. The operation $\cdot$ is associative and commutative, and is distributive over addition.
3. 1 is a multiplicative identity ($x \cdot 1 = x$) and $0 \neq 1$.

An *integral domain* is a commutative ring with identity with the additional property that there are no zero divisors:

4. If $x \neq 0$ and $y \neq 0$, then also $x \cdot y \neq 0$.

A *field* is a commutative ring with identity in which multiplicative inverses exist:

4′. If x is a nonzero element of D, then $x \cdot y = 1$ for some y.

Any field is also an integral domain, because condition 4′ implies condition 4 (see the proof to Corollary 5QG).

[2] It is also possible to define a group to be a pair $\langle A, +\rangle$, since the zero element turns out to be uniquely determined. We have formulated these definitions to match the exposition in this chapter.

An *ordered field* is a sextuple $\langle D, +, \cdot, 0, 1, <\rangle$ such that the following conditions are met:

1. $\langle D, +, \cdot, 0, 1\rangle$ is a field.
2. $<$ is a linear ordering on D.
3. Order is preserved by addition and by multiplication by a positive element:

$$x < y \iff x + z < y + z.$$

If $0 < z$, then

$$x < y \iff x \cdot z < y \cdot z.$$

We can define *ordered integral domain* or even *ordered commutative ring with identity* by adjusting the first condition. A *complete ordered field* is an ordered field in which for every bounded nonempty subset of D there is a least upper bound.

The constructions in this chapter can be viewed as providing an existence proof for such fields. The conditions for a complete ordered field are not impossible to meet, for we have constructed a field meeting them.

TWO

What is a two? What are numbers? These are awkward questions; yet when we discuss numbers one might naively expect us to know what it is we are talking about.

In the Real World, we do not encounter (directly) abstract objects such as numbers. Instead we find physical objects: a number of similar apples, a ruler, partially filled containers (Fig. 26).

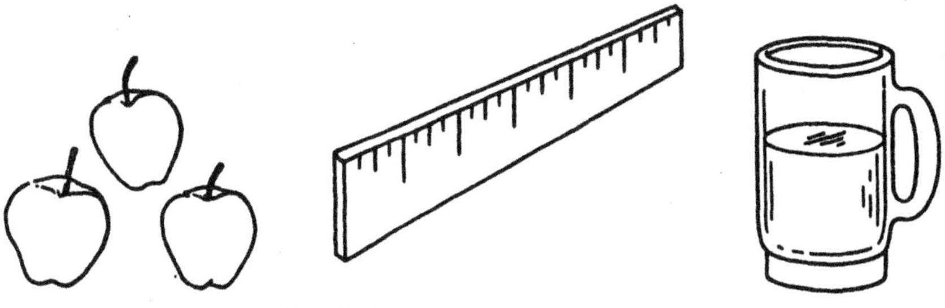

Fig. 26. A picture of the Real World.

Somehow we manage to abstract from this physical environment the concept of numbers. Not in any precise sense, of course, but we feel inwardly that we know what numbers are, or at least some numbers like 2 and 3. And we have various mental images that we use when thinking about numbers (Fig. 27).

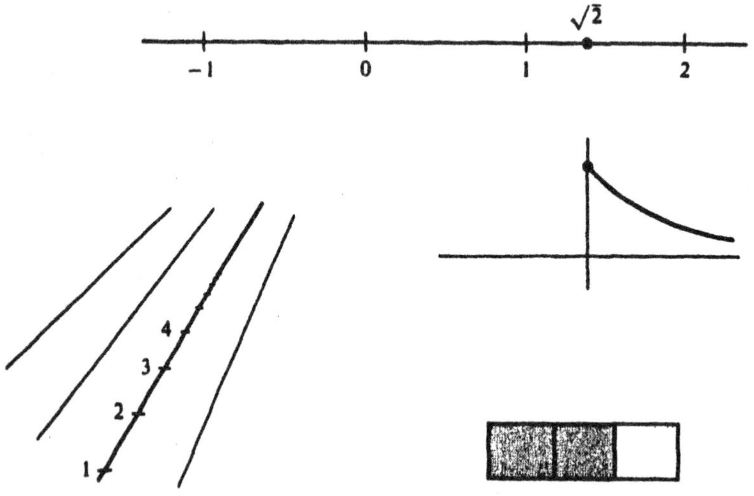

Fig. 27. Pictures of mental images.

Then at some point we acquire a mathematical education. This teaches us many formal manipulations of symbols (Fig. 28). It all seems fairly reasonable, but it can all be done without paying very much attention to what the symbols stand for. In fact computers can be programmed to carry out the manipulations without any understanding at all.

Since numbers are abstract objects (as contrasted with physical objects) it might be helpful to consider first other abstractions. Take honesty, for example. Honesty is a *property* possessed by those people whose utterances are true sentences, who do not fudge on their income tax, and so forth. By way of imitation, we can try characterizing two as the property that is true of exactly those sets that, for some distinct x and y, have as members x and y and nothing else. A slight variation on this proposal would be to eliminate sets and characterize two as the property that is true of exactly those properties that, for some distinct x and y, are true of x and y and nothing else. (This proposal is due to Frege.)

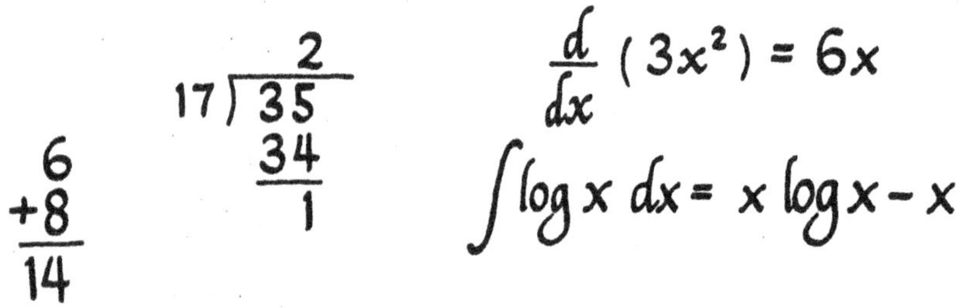

Fig. 28. Pictures of manipulated symbols.

If we define numbers in terms of properties, someone might ask us what a property is. And no matter how we define numbers, the procedure of defining one concept in terms of another cannot go on forever, producing an infinite regress of definitions. Eventually the procedure must be founded on a commonly agreed upon basis. Properties might form such a basis. For a mathematician, sets form a more workable basis.

Actually there is a close connection between properties and sets. Define the *extension* of a property to be the set of all objects of which that property is true. If a couple of properties have the same extension (i.e., if they are true of exactly the same objects), are they in fact one and the same property? Is the property of being a prime number less than 10 the same as the property of being a solution to

$$x^4 - 17x^3 + 101x^2 - 247x + 210 = 0?$$

If you answer "yes," then one says that you are thinking of properties *extensionally*, whereas if you answer "no," then you are thinking of properties *intensionally*. Both alternatives are legitimate, as long as the choice of alternatives is made clear. Either way, sets are the extensions of properties. (All sets are obtainable in this way; the set x is the extension of the property of belonging to x.)

We can recast Frege's proposal in terms of sets as follows: Two is the set having as members exactly those sets that, for some distinct x and y, have as members x and y and nothing else. But in Zermelo–Fraenkel set theory, there is no such set. (For the number one, you should check Exercise 8 of Chapter 2.) Our response to this predicament was to select artificially one particular set $\{\varnothing, \{\varnothing\}\}$ as a paradigm. Now $\{\varnothing, \{\varnothing\}\}$ is very different from the property that is true of exactly those sets that, for some distinct x and y, have as members x and y and nothing else, but it serves as an adequate substitute. Then rapidly one thing led to another, until we had the complete ordered field of real numbers. And as we have mentioned, one complete ordered field is very much like any other.

But let us back up a little. In mathematics there are two ways to introduce new objects:

(i) The new objects might be defined in terms of other already known objects.

(ii) The new objects can be introduced as primitive notions and axioms can be adopted to describe the notions. (This is not so much a way of answering foundational questions about the objects as it is a way of circumventing them.)

In constructing real numbers as certain sets we have selected the first path. The axiomatic approach would regard the definition of a complete ordered field as *axioms* concerning the real number system (as a primitive notion). On the other hand, for sets themselves we have followed (with stripes) the axiomatic method.

But what about the Real World and those mental images? And the manipulated symbols? We want not just any old concept of number, but a concept that accurately reflects our experiences with apples and rulers and containers, and accurately mirrors our mental images. This is not a precise criterion, since it demands that a precise mathematical concept be compared with informal and intuitive ideas. And consequently the question whether our concept is indeed accurate must be evaluated on informal grounds. Throughout this chapter our formulation of definitions has been motivated ultimately by our intuitive ideas. Is there any way we could have gone wrong?

Yes, we could have gone wrong. In seeking a number system applicable to problems dealing with physical objects and physical space, we might have been guided by erroneous ideas. There is always the possibility that lines in the Real World do not really resemble $\mathbb{R}$. For example, over very short distances, space might be somehow quantized instead of being continuous. (Experimental evidence forced us to accept the fact that matter is quantized; experimental evidence has not *yet* forced us to accept similar ideas about space.) Or over very large distances, space might not be Euclidean (a possibility familiar to science-fiction buffs). In such events, mathematical theorems about $\mathbb{R}$, while still true of $\mathbb{R}$, would be less interesting, as they might be inapplicable to certain problems in the Real World.

Mathematical concepts are useful in solving problems from the Real World to the extent that the concepts accurately reflect the essential features of those problems. The process of solving a problem mathematically has three parts (Fig. 29). We begin with a Real World problem. Then we need to model the original problem by a mathematical problem. This typically requires simplifying or idealizing some aspects of the original problem. (For example, we might decide to ignore air resistance or friction.) The middle step in the process consists of finding a mathematical solution to the mathematical problem. The final step is to interpret the mathematical solution in terms of the original problem. The middle step in this process is called "pure mathematics," and the entire process is called "applied mathematics."

We have, for example, all been given problems such as: If Johnny has six pennies and steals eight more, how many does he have? We first convert this to the mathematical problem: $6 + 8 = ?$ Then by pure mathematics (addition, in this case) we obtain 14 as the solution. Finally, we decide that Johnny has fourteen pennies.

The mathematical modeling of a Real World problem is not always this

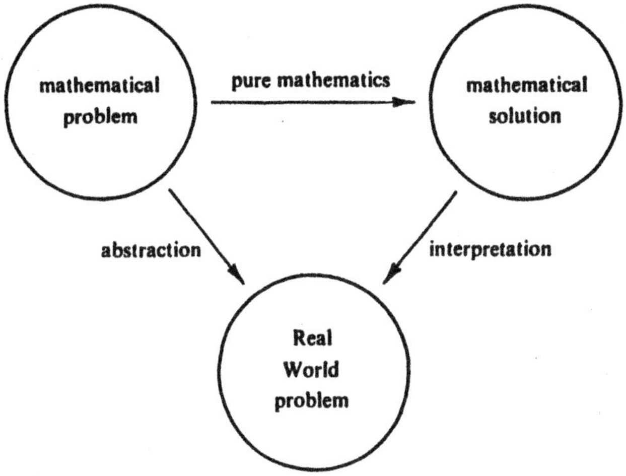

Fig. 29. Applied mathematics.

straightforward. When we try to interpret our mathematical solution in terms of the original problem, we might discover that it just does not fit. If we start with six blobs of water and add eight more blobs, we may end up with only four or five rather large puddles. This outcome does not shake our faith in arithmetic at all. It does show that we need to revise the model and try again (perhaps by measuring volume instead of counting blobs). From the vast array of mathematical concepts we must select those (if any!) that accurately model the essential feature of the problem to be solved.

CHAPTER 6

CARDINAL NUMBERS AND THE AXIOM OF CHOICE

EQUINUMEROSITY

We want to discuss the *size* of sets. Given two sets A and B, we want to consider such questions as:

(a) Do A and B have the same size?
(b) Does A have more elements than B?

Now for finite sets, this is not very complicated. We can just count the elements in the two sets, and compare the resulting numbers. And if one of the sets is finite and the other is infinite, it seems conservative enough to say that the infinite set definitely has more elements than does the finite set.

But now consider the case of two infinite sets. Our first need is for a definition: What exactly should "A has the same size as B" mean when A and B are infinite? After we select a reasonable definition, we can then ask, for example, whether any two infinite sets have the same size. (We have not yet officially defined "finite" or "infinite," but we will soon be in a position to define these terms in a precise way.)

An Analogy In order to find a solution to the above problem, we can first consider an analogous problem, but one on a very simple level.

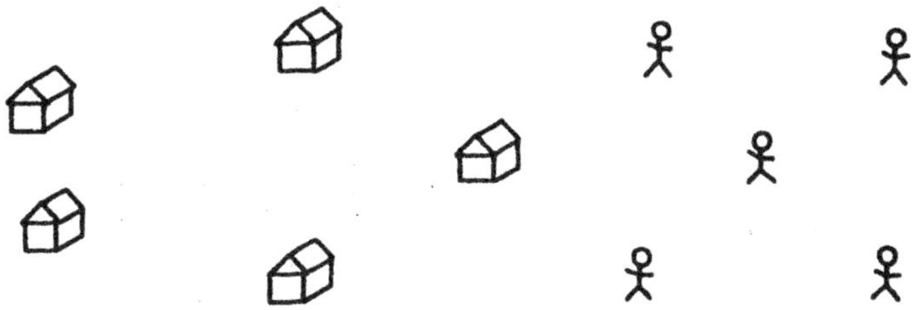

Fig. 30. Are there exactly as many houses as people?

Imagine that your mathematical education is just beginning—that you are on your way to nursery school. You are apprehensive about going, because you have heard that they have mathematics lessons and you cannot count past three. Sure enough, on the very first day they show you Fig. 30 and ask, "Are there exactly as many houses as people?" Your heart sinks. There are too many houses and too many people for you to count. This is just the predicament described earlier, where we had sets A and B that, being infinite, had too many elements to count.

But wait! All is not lost. You take your crayon and start pairing people with houses (Fig. 31). You soon discover that there are indeed exactly as many houses as people. And you did not have to count past three. You get a gold star and go home happy. We adopt the same solution.

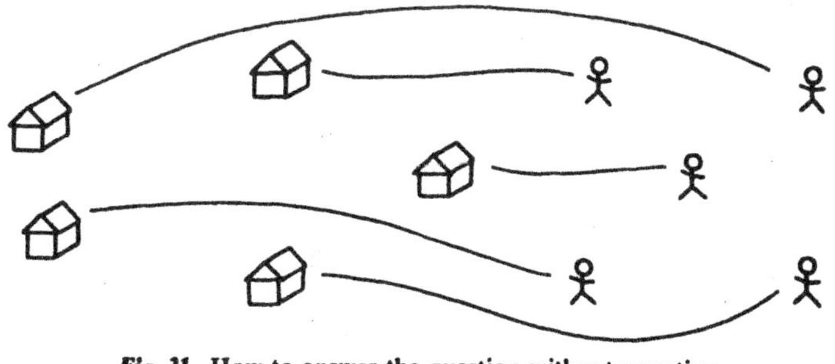

Fig. 31. How to answer the question without counting.

Definition A set A is *equinumerous* to a set B (written $A \approx B$) iff there is a one-to-one function from A onto B.

A one-to-one function from A onto B is called a *one-to-one correspondence* between A and B. For example, in Fig. 30 the set of houses is equinumerous to the set of people. A one-to-one correspondence between the sets is exhibited in Fig. 31.

Example The set $\omega \times \omega$ is equinumerous to ω. There is a function J mapping $\omega \times \omega$ one-to-one onto ω, shown in Fig. 32, where the value of

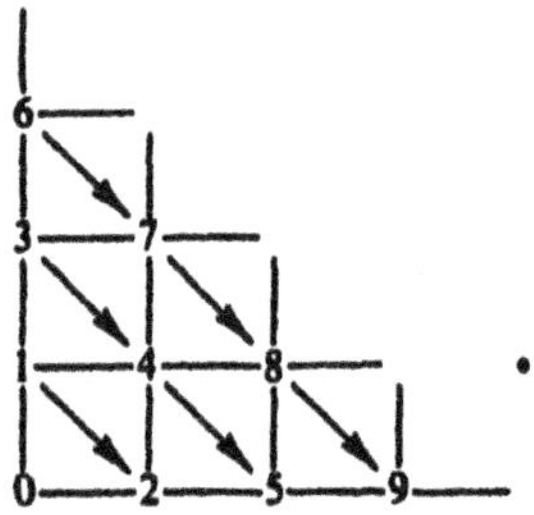

Fig. 32. $\omega \times \omega \approx \omega$.

$J(m, n)$ is written at the point with coordinates $\langle m, n\rangle$. In fact we can give a polynomial expression for J:

$$J(m, n) = \tfrac{1}{2}[(m + n)^2 + 3m + n],$$

as you are asked to verify in Exercise 2.

Example The set of natural numbers is equinumerous to the set $\mathbb{Q}$ of rational numbers, i.e., $\omega \approx \mathbb{Q}$. The method to be used here is like the one used in the preceding example. We arrange $\mathbb{Q}$ in an orderly pattern, then thread a path through the pattern, pairing natural numbers with the rationals as we go. The pattern is shown in Fig. 33. We define $f\colon \omega \to \mathbb{Q}$, where $f(n)$ is the rational next to the bracketed numeral for n in Fig. 33. To ensure that f is one-to-one, we skip rationals met for the second (or third or later) time. Thus $f(4) = -1$, and we skip $-2/2$, $-3/3$, and so forth.

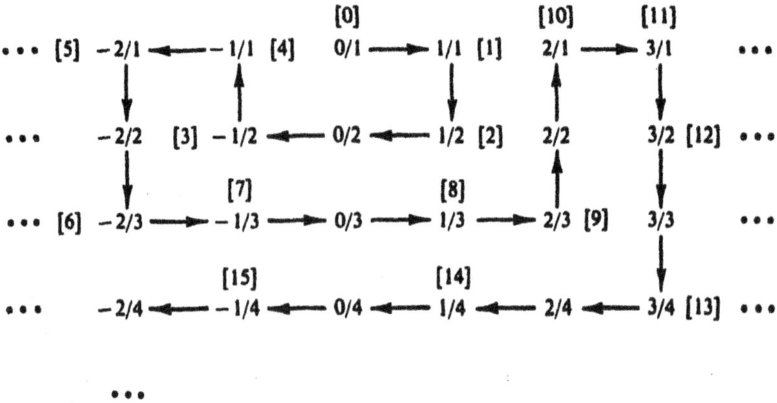

Fig. 33. $\omega \approx \mathbb{Q}$.

Example The open unit interval

$$(0, 1) = \{x \in \mathbb{R} \mid 0 < x < 1\}$$

is equinumerous to the set $\mathbb{R}$ of all real numbers. A geometric construction of a one-to-one correspondence is shown in Fig. 34. Here (0, 1) has been bent into a semicircle with center P. Each point in (0, 1) is paired with its projection (from P) on the real line.

There is also an analytical proof that $(0, 1) \approx \mathbb{R}$. Let $f(x) = \tan \pi(2x - 1)/2$. Then f maps (0, 1) one-to-one (and continuously) onto $\mathbb{R}$.

As the above example shows, it is quite possible for an infinite set, such as $\mathbb{R}$, to be equinumerous to a proper subset of itself, such as (0, 1). (For finite sets this never happens, as we will prove shortly.) Galileo remarked in 1638 that ω was equinumerous to the set $\{0, 1, 4, 9, \ldots\}$ of squares of natural numbers, and found this to be a curious fact. The

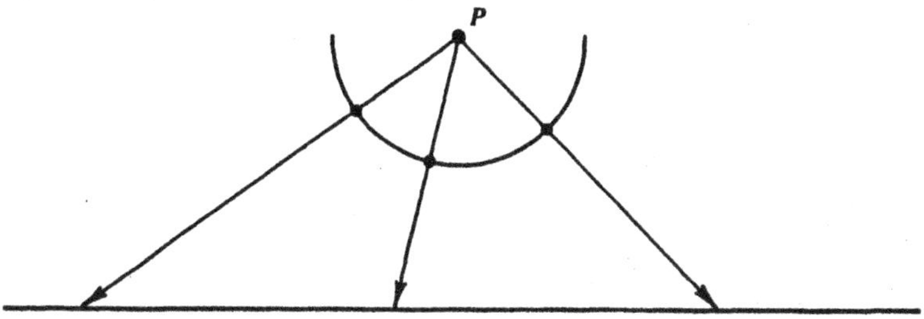

Fig. 34. $(0, 1) \approx \mathbb{R}$.

squares are in some sense a small part of the natural numbers, e.g., the fraction of the natural numbers less than n that are squares converges to 0 as n tends to infinity. But when viewed simply as two abstract sets, the set of natural numbers and the set of squares have the same size. Similarly the set of even integers is equinumerous to the set of all integers. If we focus attention on the way in which even integers are placed among the others, then we are tempted to say that there are only half as many even integers as there are integers altogether. But if we instead view the two sets as two different abstract sets, then they have the same size.

Example For any set A, we have $\mathscr{P}A \approx {}^A2$. To prove this, we define a one-to-one function H from $\mathscr{P}A$ onto ${}^A2$ as follows: For any subset B of A, $H(B)$ is the characteristic function of B, i.e., the function f_B from A into 2 for which

$$f_B(x) = \begin{cases} 1 & \text{if } x \in B, \\ 0 & \text{if } x \in A - B. \end{cases}$$

Then any function $g \in {}^A2$ is in ran H, since

$$g = H(\{x \in A \mid g(x) = 1\}).$$

The following theorem shows that equinumerosity has the property of being reflexive (on the class of all sets), symmetric, and transitive. But it cannot be represented by an equivalence relation, because it concerns *all* sets. In von Neumann–Bernays set theory, one can form the class

$$\mathsf{E} = \{\langle A, B\rangle \mid A \approx B\}.$$

Then E is an "equivalence relation on V," in the sense that it is a class of ordered pairs that is reflexive on V, symmetric, and transitive. But E is not a set, lest its field V be a set. In Zermelo–Fraenkel set theory, we have only the "equivalence concept" of equinumerosity.

Theorem 6A For any sets A, B, and C:

(a) $A \approx A$.
(b) If $A \approx B$, then $B \approx A$.
(c) If $A \approx B$ and $B \approx C$, then $A \approx C$.

Proof See Exercise 5. ⊣

In light of the examples presented up to now, you might well ask whether any two infinite sets are equinumerous. Such is not the case; some infinite sets are much larger than others.

Theorem 6B (Cantor 1873) (a) The set ω is not equinumerous to the set $\mathbb{R}$ of real numbers.

(b) No set is equinumerous to its power set.

Proof We will show that for any function $f : \omega \to \mathbb{R}$, there is a real number z not belonging to $\operatorname{ran} f$. Imagine a list of the successive values of f, expressed as infinite decimals:

$$\begin{aligned} f(0) &= 236.001\ldots, \\ f(1) &= -7.777\ldots, \\ f(2) &= 3.1415\ldots, \\ &\vdots \end{aligned}$$

(In Chapter 5 we did not go into the matter of decimal expansions, but you are surely familiar with them.) We will proceed to construct the real z. The integer part is 0, and the $(n + 1)$st decimal place of z is 7 unless the $(n + 1)$st decimal place of $f(n)$ is 7, in which case the $(n + 1)$st decimal place of z is 6. For example, in the case shown,

$$z = 0.767\ldots.$$

Then z is a real number not in $\operatorname{ran} f$, as it differs from $f(n)$ in the $(n + 1)$st decimal place.

The proof of (b) is similar. Let $g : A \to \mathscr{P}A$; we will construct a subset B of A that is not in $\operatorname{ran} g$. Specifically, let

$$B = \{x \in A \mid x \notin g(x)\}.$$

Then $B \subseteq A$, but for each $x \in A$,

$$x \in B \quad \Leftrightarrow \quad x \notin g(x).$$

Hence $B \neq g(x)$. ⊣

The set $\mathbb{R}$ happens to be equinumerous to $\mathscr{P}\omega$, as we will soon be able to prove. A larger set is $\mathscr{P}\mathbb{R}$, and $\mathscr{P}\mathscr{P}\mathbb{R}$ is larger still.

Before continuing our consideration of infinite sets, we will study the other alternative: the sets that are "small" at least to the extent of being finite.

Exercises

1. Show that the equation

$$f(m, n) = 2^m(2n + 1) - 1$$

defines a one-to-one correspondence between $\omega \times \omega$ and ω.

2. Show that in Fig. 32 we have:

$$\begin{aligned} J(m, n) &= [1 + 2 + \cdots + (m + n)] + m \\ &= \tfrac{1}{2}[(m + n)^2 + 3m + n]. \end{aligned}$$

3. Find a one-to-one correspondence between the open unit interval (0, 1) and $\mathbb{R}$ that takes rationals to rationals and irrationals to irrationals.

4. Construct a one-to-one correspondence between the closed unit interval

$$[0, 1] = \{x \in \mathbb{R} \mid 0 \leq x \leq 1\}$$

and the open unit interval (0, 1).

5. Prove Theorem 6A.

FINITE SETS

Although we have long been using the words "finite" and "infinite" in an informal way, we have not yet given them precise definitions. Now is the time.

Definition A set is *finite* iff it is equinumerous to some natural number. Otherwise it is *infinite*.

Here we rely on the fact that in our construction of ω, each natural number is the set of all smaller natural numbers. For example, any natural number is itself a finite set.

We want to check that each finite set S is equinumerous to a *unique* number n. The number n can then be used as a count of the elements in S.

We first need the following theorem, which implies that if n objects are placed into fewer than n pigeonholes, then some pigeonhole receives more than one object. Recall that a set A is a *proper subset* of B iff $A \subseteq B$ and $A \neq B$.

Pigeonhole Principle No natural number is equinumerous to a proper subset of itself.

Proof Assume that f is a one-to-one function from the set n into the set n. We will show that ran f is all of the set n (and not a proper subset of n). This suffices to prove the theorem.

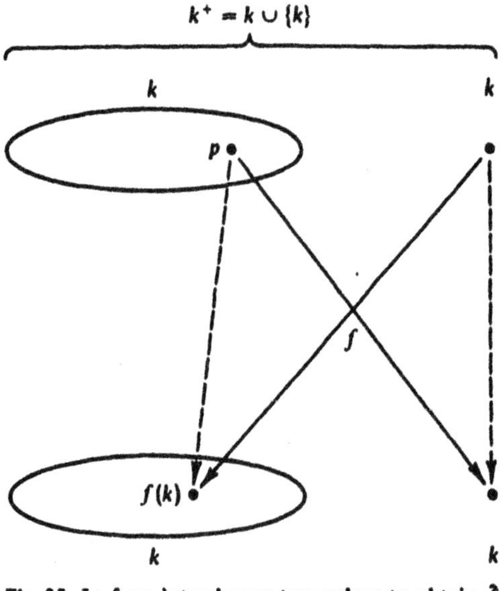

Fig. 35. In f we interchange two values to obtain $\hat{f}$.

We use induction on n. Define:

$$T = \{n \in \omega \mid \text{any one-to-one function from } n \text{ into } n \text{ has range } n\}.$$

Then $0 \in T$; the only function from the set 0 into the set 0 is $\varnothing$ and its range is the set 0. Suppose that $k \in T$ and that f is a one-to-one function from the set k^+ into the set k^+. We must show that the range of f is all of the set k^+; this will imply that $k^+ \in T$. Note that the restriction $f \upharpoonright k$ of f to the set k maps the set k one-to-one into the set k^+.

Case I Possibly the set k is closed under f. Then $f \upharpoonright k$ maps the set k into the set k. Then because $k \in T$ we may conclude that ran $(f \upharpoonright k)$ is all of the set k. Since f is one-to-one, the only possible value for $f(k)$ is the number k. Hence ran f is $k \cup \{k\}$, which is the set k^+.

Case II Otherwise $f(p) = k$ for some number p less than k. In this case we interchange two values of the function. Define $\hat{f}$ by

$$\hat{f}(p) = f(k),$$
$$\hat{f}(k) = f(p) = k,$$
$$\hat{f}(x) = f(x) \qquad \text{for other } x \in k^+$$

(see Fig. 35). Then $\hat{f}$ maps the set k^+ one-to-one into the set k^+, and the set k is closed under $\hat{f}$. So by Case I, ran $\hat{f} = k^+$. But ran $\hat{f}$ = ran f.

Thus in either case, ran $f = k^+$. So T is inductive and equals ω. ⊣

Corollary 6C No finite set is equinumerous to a proper subset of itself.

Proof This is the same as the pigeonhole principle, but for an arbitrary finite set A instead of a natural number. Since A is equinumerous to a natural number n, we can use the one-to-one correspondence g between A and n to "transfer" the pigeonhole principle to the set A.

Suppose that, contrary to our hopes, there is a one-to-one correspondence f between A and some proper subset of A. Consider the composition $g \circ f \circ g^{-1}$, illustrated in Fig. 36. This composition maps n into n, and it is one-to-one by Exercise 17 of Chapter 3. Furthermore its range C is a proper subset of n. (Consider any a in A − ran f; then $g(a) \in n - C$.) Thus n is equinumerous to C, in contradiction to the pigeonhole principle. ⊣

The foregoing proof uses an argument that is useful elsewhere as well. We have sets A and n that are "alike" in that $A \approx n$, but different in that they have different members. Think of the members of A as being red, the members of n as being blue. Then the function $g: A \to n$ paints red

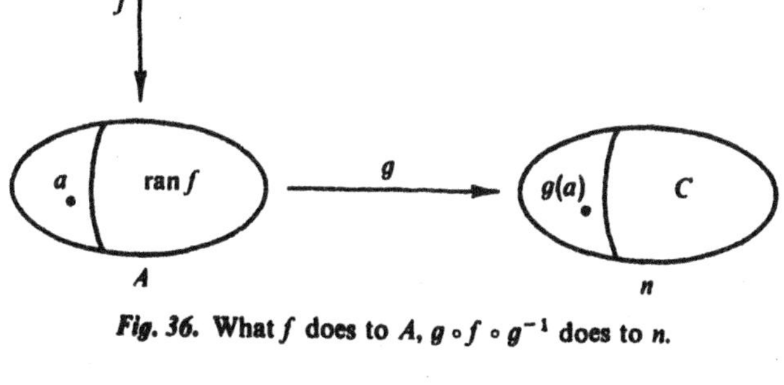

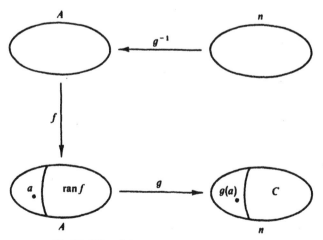

Fig. 36. What f does to A, $g \circ f \circ g^{-1}$ does to n.

things blue, and the function $g^{-1}: n \to A$ paints blue things red. The composition $g \circ f \circ g^{-1}$ paints things red, applies f, and then restores the blue color.

Corollary 6D (a) Any set equinumerous to a proper subset of itself is infinite.

(b) The set ω is infinite.

Proof The preceding corollary proves part (a). Part (b) follows at once from part (a), since the function σ whose value at each number n is n^+ maps ω one-to-one onto $\omega - \{0\}$. ⊣

Corollary 6E Any finite set is equinumerous to a *unique* natural number.

Proof Assume that $A \approx m$ and $A \approx n$ for natural numbers m and n. Then $m \approx n$. By trichotomy and Corollary 4M, either $m = n$ or one is a proper subset of the other. But the latter alternative is impossible since $m \approx n$. Hence $m = n$. ⊣

For a finite set A, the unique $n \in \omega$ such that $A \approx n$ is called the *cardinal number* of A, abbreviated card A. For example, card $n = n$ for $n \in \omega$. And if a, b, c, and d are all distinct objects, then card$\{a, b, c, d\} = 4$. This is because $\{a, b, c, d\} \approx 4$; selecting a one-to-one correspondence is the process called "counting." Observe that for any finite sets A and B, we have $A \approx$ card A and

$$\text{card } A = \text{card } B \quad \text{iff} \quad A \approx B.$$

What about infinite sets? The number card A measures the size of a finite set A. We want "numbers" that similarly measure the size of infinite sets. Just what sets these "numbers" are is not too crucial, any more than it was crucial just what set the number 2 was. The essential demand is that we define card A for arbitrary A in such a way that

$$\text{card } A = \text{card } B \quad \text{iff} \quad A \approx B.$$

Now it turns out that there is no way of defining card A that is really simple. We therefore postpone until Chapter 7 the actual definition of the set card A. The information we need for the present chapter is embodied in the following promise.

Promise For any set A we will define a set card A in such a way that:

(a) For any sets A and B,

$$\text{card } A = \text{card } B \quad \text{iff} \quad A \approx B.$$

(b) For a finite set A, card A is the natural number n for which $A \approx n$.

(In making good on this promise, we will use in Chapter 7 additional axioms, namely the replacement axioms and the axiom of choice. If you plan to omit Chapter 7, then regard card A as an additional primitive notion and the promise as an additional axiom.)

We define a cardinal number to be something that is card A for some set A. By part (b) of the promise, any natural number n is also a cardinal number, since $n = \text{card } n$. But card ω is not a natural number (card $\omega \neq n = \text{card } n$, since ω is not equinumerous to n). Just what set card ω is will not be revealed until Chapter 7. Meanwhile we give it the name that Cantor gave it:

$$\text{card } \omega = \aleph_0 .$$

The symbol $\aleph$ is aleph, the first letter of the Hebrew alphabet.

In general, for a cardinal number κ, there will be a great many sets A of cardinality κ, i.e., sets with card $A = \kappa$. (The one exception to this occurs when $\kappa = 0$.) In fact, for any nonzero cardinal κ, the class

$$\mathsf{K}_\kappa = \{X \mid \text{card } X = \kappa\}$$

of sets of cardinality κ is too large to be a set (Exercise 6). But all of the sets of cardinality κ look, from a great distance, very much alike—the elements of two such sets may differ but the number of elements is always κ. In particular, if one set X of cardinality κ is finite, then all of them are; in this case κ is a *finite cardinal.* And if not, then κ is an *infinite cardinal.* Thus the finite cardinals are exactly the natural numbers. $\aleph_0$ is an infinite cardinal, as are card $\mathbb{R}$, card $\mathscr{P}\omega$, card $\mathscr{P}\mathscr{P}\omega$, etc.

Before leaving this section on finite sets, we will verify a fact that, on an informal level, appears inevitable: Any subset of a finite set is finite. (Later we will find another proof of this.)

Lemma 6F If C is a proper subset of a natural number n, then $C \approx m$ for some m less than n.

Proof We use induction. Let

$$T = \{n \in \omega \mid \text{any proper subset of } n \text{ is equinumerous to a member of } n\}.$$

Then $0 \in T$ vacuously, 0 having no proper subsets. Assume that $k \in T$ and consider a proper subset C of k^+.

Case I $C = k$. Then $C \approx k \in k^+$.

Case II C is a proper subset of k. Then since $k \in T$, we have $C \approx m$ for $m \in k \in k^+$.

Case III Otherwise $k \in C$. Then $C = (C \cap k) \cup \{k\}$ and $C \cap k$ is a proper subset of k. Because $k \in T$, there is $m \in k$ with $C \cap k \approx m$. Say f is a one-to-one correspondence between $C \cap k$ and m; then $f \cup \{\langle k, m\rangle\}$ is a one-to-one correspondence between C and m^+. Since $m \in k$, we have $m^+ \in k^+$. Hence $C \approx m^+ \in k^+$ and $k^+ \in T$.

Thus T is inductive and coincides with ω. ⊣

Corollary 6G Any subset of a finite set is finite.

Proof Consider $A \subseteq B$ and let f be a one-to-one correspondence between B and some n in ω. Then $A \approx f[\![A]\!] \subseteq n$ and by the lemma $f[\![A]\!] \approx m$ for some $m \in n$. Hence $A \approx m \in n \in \omega$. ⊣

Exercises

6. Let κ be a nonzero cardinal number. Show that there does not exist a set to which every set of cardinality κ belongs.

7. Assume that A is finite and $f: A \rightarrow A$. Show that f is one-to-one iff $\operatorname{ran} f = A$.

8. Prove that the union of two finite sets is finite (Corollary 6K), without any use of arithmetic.

9. Prove that the Cartesian product of two finite sets is finite (Corollary 6K), without any use of arithmetic.

CARDINAL ARITHMETIC

The operations of addition, multiplication, and exponentiation are well known to be useful for finite cardinals. The operations can be useful for arbitrary cardinals as well. To extend the concept of addition from the finite to the infinite case, we need a characterization of addition that is correct in the finite case, and is meaningful (and plausible) in the infinite case. In Chapter 4 we obtained addition on ω by use of the recursion theorem. That approach is unsuitable here, so we seek another approach.

The answer to our search lies in the way addition is actually explained in the elementary schools. First-graders are not told about the recursion theorem. Instead, if they want to add 2 and 3, they select two sets K and L with card $K = 2$ and card $L = 3$. Sets of fingers are handy; sets of apples are preferred by textbooks. Then they look at card$(K \cup L)$. If they had the good sense to select K and L to be disjoint, then card$(K \cup L) = 5$.

The same idea is embodied in the following definition of addition. In the same vein, we can include multiplication and exponentiation.

Definition Let κ and λ be any cardinal numbers.

(a) $\kappa + \lambda = \text{card}(K \cup L)$, where K and L are any disjoint sets of cardinality κ and λ, respectively.
(b) $\kappa \cdot \lambda = \text{card}(K \times L)$, where K and L are any sets of cardinality κ and λ, respectively.
(c) $\kappa^\lambda = \text{card}\,{}^LK$, where K and L are any sets of cardinality κ and λ, respectively.

In every case it is necessary to prove that the operation is well defined; that is, that the outcome is independent of the particular sets K and L selected to represent the cardinals. Also for the case of finite cardinals, we should check that the above definitions are not in open conflict with operations on ω defined in Chapter 4.

First consider addition. To add two cardinals κ and λ, the definition demands that we first select disjoint sets K and L with $\text{card } K = \kappa$ and $\text{card } L = \lambda$. This is possible; if our first choices for K and L fail to be disjoint, we can switch to $K \times \{0\}$ and $L \times \{1\}$. For then, since

$$K \approx K \times \{0\} \qquad \text{and} \qquad L \approx L \times \{1\},$$

we have $\text{card}(K \times \{0\}) = \kappa$ and $\text{card}(L \times \{1\}) = \lambda$. And the sets $K \times \{0\}$ and $L \times \{1\}$ *are* disjoint.

Having selected disjoint representatives K and L for κ and λ, we form their union $K \cup L$. Then by definition

$$\kappa + \lambda = \text{card}(K \cup L).$$

We must verify that this sum is independent of the particular disjoint sets K and L selected. This verification is accomplished by part (a) of the following theorem. For suppose K_1, L_1 and K_2, L_2 are two selections of disjoint sets of cardinality κ, λ. Since $\text{card } K_1 = \text{card } K_2 = \kappa$, we have $K_1 \approx K_2$; similarly $L_1 \approx L_2$. The following theorem then yields $K_1 \cup L_1 \approx K_2 \cup L_2$, whence

$$\text{card}(K_1 \cup L_1) = \text{card}(K_2 \cup L_2).$$

Thus we have an unambiguous sum $\kappa + \lambda$. Parts (b) and (c) of the theorem perform the same service for multiplication and exponentiation.

Theorem 6H Assume that $K_1 \approx K_2$ and $L_1 \approx L_2$.

(a) If $K_1 \cap L_1 = K_2 \cap L_2 = \varnothing$, then $K_1 \cup L_1 \approx K_2 \cup L_2$.
(b) $K_1 \times L_1 \approx K_2 \times L_2$.
(c) ${}^{(L_1)}K_1 \approx {}^{(L_2)}K_2$.

Proof Since $K_1 \approx K_2$, there is a one-to-one function f from K_1 onto K_2; since $L_1 \approx L_2$, there is a one-to-one function g from L_1 onto L_2. Then for (a), the function h, defined by

$$h(x) = \begin{cases} f(x) & \text{if } x \in K_1, \\ g(x) & \text{if } x \in L_1, \end{cases}$$

maps $K_1 \cup L_1$ one-to-one onto $K_2 \cup L_2$. (We need $K_1 \cap L_1 = \varnothing$ to be sure h is a function; we need $K_2 \cap L_2 = \varnothing$ to be sure it is one-to-one.)

For (b), the function h, defined by

$$h(\langle x, y\rangle) = \langle f(x), g(y)\rangle$$

(for $x \in K_1$ and $y \in L_1$), maps $K_1 \times L_1$ one-to-one onto $K_2 \times L_2$.

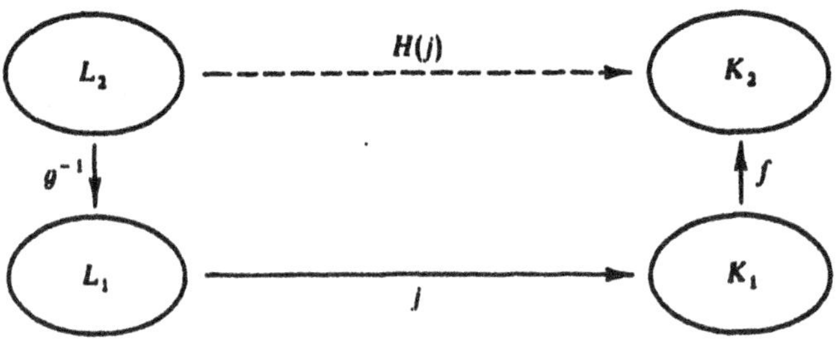

Fig. 37. $H(j) = f \circ j \circ g^{-1}$.

Finally for (c), the function H, defined by

$$H(j) = f \circ j \circ g^{-1},$$

maps ${}^{(L_1)}K_1$ one-to-one onto ${}^{(L_2)}K_2$ (see Fig. 37). For clearly $H(j)$ is a function from L_2 to K_2. To see that H is one-to-one, suppose that j and j' are different functions from L_1 into K_1. Then $j(t) \neq j'(t)$ for some $t \in L_1$. Then compute $H(j)$ and $H(j')$ at $g(t)$:

$$H(j)(g(t)) = f(j(t)) \neq f(j'(t)) = H(j')(g(t))$$

since f is one-to-one. Hence the function $H(j)$ is different from the function $H(j')$, i.e., H is one-to-one. And to see that ran H is all of ${}^{(L_2)}K_2$, consider any function d from L_2 into K_2. Then $d = H(j)$, where $j = f^{-1} \circ d \circ g$. ⊣

Remark The proof of part (c) is our first proof dealing with cardinal exponentiation. Observe that it involves constructing a function H whose arguments and values are themselves functions. In such cases it is imperative to use a rationally chosen notation, e.g., an uppercase letter for H and lowercase letters for the arguments of H, as in the expression "$H(j)$." Here the function H must be one-to-one. But the function j is in general not one-to-one, nor is $H(j)$. It may help to think of H as a "super function" that assigns certain lower-level functions to other lower-level functions. The super function in this proof is one-to-one, but the lower-level functions are not necessarily one-to-one.

Examples 1. $2 + 2 = 4$. Proving this amounts to finding disjoint sets K and L with $K \approx 2$ and $L \approx 2$, and then checking that $K \cup L \approx 4$. (This is not the same as the verification of $2 + 2 = 4$ in Chapter 4, because now we are using the addition operation for cardinal numbers. But Theorem 6J assures us that the answer is unchanged.)

2. For m and n in ω, $m \cdot n = \text{card } m \times n$ and $m^n = \text{card } {}^n m$.

3. For any natural number n,

$$n + \aleph_0 = \aleph_0 \qquad \text{and} \qquad n \cdot \aleph_0 = \aleph_0$$

(unless $n = 0$). Also

$$\aleph_0 + \aleph_0 = \aleph_0 \qquad \text{and} \qquad \aleph_0 \cdot \aleph_0 = \aleph_0.$$

The last equation follows from an earlier example showing that $\omega \times \omega \approx \omega$. To prove that $2 + \aleph_0 = \aleph_0$, it suffices to show that $\{a, b\} \cup \omega \approx \omega$ (where a and b are not in ω). The function

$$f(a) = 0, \qquad f(b) = 1, \qquad f(n) = n^{++}$$

establishes this. The other equations are left for you to check. Observe that cancellation laws fail for infinite cardinals, e.g., $2 + \aleph_0 = 3 + \aleph_0$, but $2 \neq 3$.

4. For any cardinal number κ,

$$\kappa + 0 = \kappa, \qquad \kappa \cdot 0 = 0, \qquad \kappa \cdot 1 = \kappa.$$

5. Recall that ${}^{\varnothing}K = \{\varnothing\}$ for any set K and that ${}^{K}\varnothing = \varnothing$ for nonempty K. In terms of cardinal numbers, these facts become

$$\kappa^0 = 1 \qquad \text{for any } \kappa,$$
$$0^\kappa = 0 \qquad \text{for any nonzero } \kappa.$$

In particular, $0^0 = 1$.

6. For any set A, the cardinality of its power set is $2^{\text{card } A}$. This is because by the definition of exponentiation,

$$2^{\text{card } A} = \text{card}({}^A 2).$$

And we have shown that ${}^A 2 \approx \mathscr{P}A$, and hence

$$2^{\text{card } A} = \text{card}({}^A 2) = \text{card } \mathscr{P}A.$$

In particular, $\text{card } \mathscr{P}\omega = 2^{\aleph_0}$. (The term "power set" is rooted in the fact that $\text{card } \mathscr{P}A$ equals 2 raised to the power $\text{card } A$.)

7. By Cantor's theorem and the preceding example, $\kappa \neq 2^\kappa$ for any κ. In particular $\aleph_0 \neq 2^{\aleph_0}$.

8. For any cardinal κ,

$$\kappa + \kappa = 2 \cdot \kappa.$$

You are invited to explain how this fact is obtained.

The following theorem lists some of the elementary facts of cardinal arithmetic. (Parts of the theorem will later be seen to be rather trivial.)

Theorem 6I For any cardinal numbers κ, λ, and μ:

1. $\kappa + \lambda = \lambda + \kappa$ and $\kappa \cdot \lambda = \lambda \cdot \kappa$.
2. $\kappa + (\lambda + \mu) = (\kappa + \lambda) + \mu$ and $\kappa \cdot (\lambda \cdot \mu) = (\kappa \cdot \lambda) \cdot \mu$.
3. $\kappa \cdot (\lambda + \mu) = \kappa \cdot \lambda + \kappa \cdot \mu$.
4. $\kappa^{\lambda+\mu} = \kappa^\lambda \cdot \kappa^\mu$.
5. $(\kappa \cdot \lambda)^\mu = \kappa^\mu \cdot \lambda^\mu$.
6. $(\kappa^\lambda)^\mu = \kappa^{\lambda \cdot \mu}$.

Proof Take sets K, L, and M with card $K = \kappa$, card $L = \lambda$, and card $M = \mu$; for convenience choose them in such a way that any two are disjoint. Then each of the equations reduces to a corresponding statement about equinumerous sets. For example, $\kappa \cdot \lambda = \text{card } K \times L$ and $\lambda \cdot \kappa = \text{card}(L \times K)$; consequently showing that $\kappa \cdot \lambda = \lambda \cdot \kappa$ reduces to showing that $K \times L \approx L \times K$. Listed in full, the statements to be verified are:

1. $K \cup L \approx L \cup K$ and $K \times L \approx L \times K$.
2. $K \cup (L \cup M) \approx (K \cup L) \cup M$ and $K \times (L \times M) \approx (K \times L) \times M$.
3. $K \times (L \cup M) \approx (K \times L) \cup (K \times M)$.
4. ${}^{(L \cup M)}K \approx {}^LK \times {}^MK$.
5. ${}^M(K \times L) \approx {}^MK \times {}^ML$.
6. ${}^M({}^LK) \approx {}^{(L \times M)}K$.

Most of the verifications are left as exercises.

In the case of item 6 we want a one-to-one function H from ${}^M({}^LK)$ onto ${}^{(L \times M)}K$. For $f \in {}^M({}^LK)$, let $H(f)$ be the function whose value at $\langle l, m\rangle$ equals the value of the function $f(m)$ at l.

To see that H is one-to-one, observe that if $f \neq g$ (both belonging to ${}^M({}^LK)$), then for some m, the functions $f(m)$ and $g(m)$ differ. This in turn implies that for some l, $f(m)(l) \neq g(m)(l)$. Hence

$$H(f)(l, m) = f(m)(l) \neq g(m)(l) = H(g)(l, m)$$

so that $H(f) \neq H(g)$.

Finally, to see that $\operatorname{ran} H$ is all of ${}^{(L \times M)}K$, consider any function $j \in {}^{(L \times M)}K$. Then $j = H(f)$, where (for $m \in M$) $f(m)$ is the function whose value at $l \in L$ is $j(l, m)$. ⊣

The next theorem reassures us that for finite cardinals (i.e., for natural numbers), the present arithmetic operations agree with those of Chapter 4. In Chapter 4, exponentiation was defined only in passing; for completeness we include it here.

Theorem 6J Let m and n be finite cardinals. Then

$$m + n = m +_{\omega} n,$$
$$m \cdot n = m \cdot_{\omega} n,$$
$$m^n = m^n,$$

where on the right side we use the operations of Chapter 4 (defined by recursion) and on the left side we use the operations of cardinal arithmetic.

Proof We use induction on n. First we claim that for any cardinal numbers κ and λ the following identities are correct.

(a1) $\kappa + 0 = \kappa$.
(a2) $\kappa + (\lambda + 1) = (\kappa + \lambda) + 1$.
(m1) $\kappa \cdot 0 = 0$.
(m2) $\kappa \cdot (\lambda + 1) = \kappa \cdot \lambda + \kappa$.
(e1) $\kappa^0 = 1$.
(e2) $\kappa^{\lambda + 1} = \kappa^{\kappa} \cdot \kappa$.

In each case, the equation is either trivial or is an immediate consequence of Theorem 6I (or both).

The second piece of information we need is that for a finite cardinal n,

$$n + 1 = n^{+}$$

(with cardinal addition). This holds because n and $\{n\}$ are disjoint sets of cardinality n and 1, respectively, and hence

$$n + 1 = \operatorname{card}(n \cup \{n\}) = \operatorname{card} n^{+} = n^{+}.$$

It remains only to go through the motions of the induction. Consider any $m \in \omega$ and let

$$T = \{n \in \omega \mid m + n = m +_{\omega} n\}.$$

Then $0 \in A$ since $n + 0 = n = n +_\omega 0$, by (a1) and (A1), where (A1) is given in Theorem 4I. Suppose that $k \in T$. Then

$$\begin{aligned} m + k^+ &= m + (k + 1) & \\ &= (m + k) + 1 & \text{by (a2)} \\ &= (m +_\omega k) + 1 & \text{since} \quad k \in T \\ &= (m +_\omega k)^+ & \\ &= m +_\omega k^+ & \text{by (A2).} \end{aligned}$$

Hence $k^+ \in T$, T is inductive, and $T = \omega$.

For multiplication and exponentiation the argument is identical. ⊣

Corollary 6K If A and B are finite, then $A \cup B$, $A \times B$, and ${}^BA$ are also finite.

Proof Let $m = \operatorname{card} A$ and $n = \operatorname{card} B$. Then we calculate: $\operatorname{card}(A \times B) = \operatorname{card} A \cdot \operatorname{card} B = m \cdot n = m \cdot_\omega n \in \omega$. A similar argument applies to ${}^BA$ and m^n.

For union we must use disjoint sets:

$$A \cup B = A \cup (B - A).$$

$B - A$ is a subset of a finite set and hence (by Corollary 6G) is finite. Let $k = \operatorname{card}(B - A)$. Then $\operatorname{card}(A \cup B) = m + k = m +_\omega k \in \omega$. ⊣

The above corollary can also be proved without using arithmetic (Exercises 8 and 9).

Exercises

10. Prove part 4 of Theorem 6I.

11. Prove part 5 of Theorem 6I.

12. The proof to Theorem 6I involves eight instances of showing two sets to be equinumerous. (The eight are listed in the proof of the theorem as statements numbered 1–6.) In which of these eight cases does equality actually hold?

13. Show that a finite union of finite sets is finite. That is, show that if B is a finite set whose members are themselves finite sets, then $\bigcup B$ is finite.

14. Define a *permutation* of K to be any one-to-one function from K onto K. We can then define the factorial operation on cardinal numbers by the equation

$$\kappa! = \operatorname{card}\{f \mid f \text{ is a permutation of } K\},$$

where K is any set of cardinality κ. Show that $\kappa!$ is well defined, i.e., the value of $\kappa!$ is independent of just which set K is chosen.

ORDERING CARDINAL NUMBERS

We can use the concept of equinumerosity to tell us when two sets A and B are of the same size. But when should we say that B is larger than A?

Definition A set A is *dominated* by a set B (written $A \preccurlyeq B$) iff there is a one-to-one function from A *into* B.

For example, any set dominates itself. If $A \subseteq B$, then A is dominated by B, since the identity function on A maps A one-to-one into B. More generally we have: $A \preccurlyeq B$ iff A is equinumerous to some subset of B. This is just a restatement of the definition, since f is a function from A *into* B iff it is a function from A *onto* a subset of B (see Fig. 38).

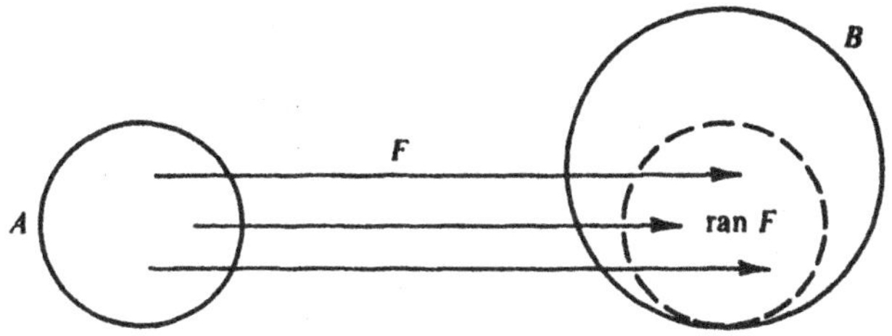

Fig. 38. F shows that $A \preccurlyeq B$.

We define the ordering of cardinal numbers by utilizing the concept of dominance:

$$\operatorname{card} A \leq \operatorname{card} B \quad \text{iff} \quad A \preccurlyeq B.$$

As with the operations of cardinal arithmetic, it is necessary to check that the ordering relation is well defined. For suppose we start with two cardinal numbers, say κ and λ. In order to determine whether or not $\kappa \leq \lambda$, our definition demands that we employ selected representatives K and L for which $\kappa = \operatorname{card} K$ and $\lambda = \operatorname{card} L$. Then

$$\kappa \leq \lambda \quad \text{iff} \quad K \preccurlyeq L.$$

But the truth or falsity of "$\kappa \leq \lambda$" must be independent of *which* selected representatives are chosen. Suppose also that $\kappa = \operatorname{card} K'$ and $\lambda = \operatorname{card} L'$. To avoid embarrassment, we must be certain that

$$K \preccurlyeq L \quad \text{iff} \quad K' \preccurlyeq L'.$$

To prove this, note first that $K \approx K'$ and $L \approx L'$ (because $\operatorname{card} K = \operatorname{card} K'$ and $\operatorname{card} L = \operatorname{card} L'$). If $K \preccurlyeq L$, then we have one-to-one maps (i) from K' onto K, (ii) from K into L, and (iii) from L onto L'. By composing the three functions, we can map K' one-to-one into L', and hence $K' \preccurlyeq L'$.

We further define

$$\kappa < \lambda \quad \text{iff} \quad \kappa \le \lambda \quad \text{and} \quad \kappa \ne \lambda.$$

Thus in terms of sets we have

$$\text{card } K < \text{card } L \quad \text{iff} \quad K \preccurlyeq L \quad \text{and} \quad K \not\approx L.$$

Notice that this condition is stronger than just saying that K is equinumerous to a proper subset of L. After all, ω is equinumerous to a proper subset of itself, but we certainly do not want to have card $\omega <$ card ω. The definition of "$<$" has the expected consequence that

$$\kappa \le \lambda \quad \text{iff} \quad \text{either} \quad \kappa < \lambda \quad \text{or} \quad \kappa = \lambda.$$

Examples 1. If $A \subseteq B$, then card $A \le$ card B. Conversely, whenever $\kappa \le \lambda$, then there exist sets $K \subseteq L$ with card $K = \kappa$ and card $L = \lambda$. To prove this, start with any sets C and L of cardinality κ and λ, respectively. Then $C \preccurlyeq L$, so there is a one-to-one function f from C into L. Let $K = \text{ran } f$; then $C \approx K \subseteq L$.

2. For any cardinal κ, we have $0 \le \kappa$.

3. For any finite cardinal n, we have $n < \aleph_0$. (Why?) For any two finite cardinals m and n, we have

$$m \in n \Rightarrow m \subseteq n \Rightarrow m \le n.$$

Furthermore the converse implications hold. For if $m \le n$, then $m \preccurlyeq n$ and there is a one-to-one function $f: m \to n$. By the pigeonhole principle, it is impossible to have n less than m, so by trichotomy $m \in n$. Thus our new ordering on finite cardinals agrees with the epsilon ordering of Chapter 4.

4. $\kappa < 2^\kappa$ for any cardinal κ. For if A is any set of cardinality κ, then $\mathscr{P}A$ has cardinality 2^κ. Then $A \preccurlyeq \mathscr{P}A$ (map $x \in A$ to $\{x\} \in \mathscr{P}A$), but by Cantor's theorem (Theorem 6B) $A \not\approx \mathscr{P}A$. Hence $\kappa \le 2^\kappa$ but $\kappa \ne 2^\kappa$, i.e., $\kappa < 2^\kappa$. In particular, there is no largest cardinal number.

The first thing to prove about the ordering we have defined for cardinals is that it actually behaves like something we would be willing to call an ordering. After all, just using the symbol "$\le$" does not confer any special properties, but it does indicate the expectation that special properties will be forthcoming. For a start, we ask if the following are valid for all cardinals κ, λ, and μ:

1. $\kappa \le \kappa$.
2. $\kappa \le \lambda \le \mu \Rightarrow \kappa \le \mu$.
3. $\kappa \le \lambda \;\&\; \lambda \le \kappa \Rightarrow \kappa = \lambda$.
4. Either $\kappa \le \lambda$ or $\lambda \le \kappa$.

The first is obvious, since $A \preccurlyeq A$ holds for any set A. The second item follows at once from the fact that

$$A \preccurlyeq B \;\&\; B \preccurlyeq C \quad \Rightarrow \quad A \preccurlyeq C.$$

(We prove this by taking the composition of two functions.) The third item is nontrivial. But the assertion is correct, and is called the *Schröder–Bernstein theorem*. We will also prove the fourth item, but that proof will require the axiom of choice.

First we will prove the Schröder–Bernstein theorem, which will be a basic tool in calculating the cardinalities of sets. Typically when we want to calculate card S for a given set S, we try to squeeze card S between upper and lower bounds. If possible, we try to get these bounds to coincide,

$$\kappa \leq \text{card } S \leq \kappa,$$

whereupon the Schröder–Bernstein theorem asserts that card $S = \kappa$. We will see examples of this technique after proving the theorem.

Schröder–Bernstein Theorem (a) If $A \preccurlyeq B$ and $B \preccurlyeq A$, then $A \approx B$.
(b) For cardinal numbers κ and λ, if $\kappa \leq \lambda$ and $\lambda \leq \kappa$, then $\kappa = \lambda$.

Proof It is done with mirrors (see Fig. 39). We are given one-to-one functions $f: A \to B$ and $g: B \to A$. Define C_n by recursion, using the formulas

$$C_0 = A - \text{ran } g \qquad \text{and} \qquad C_{n^+} = g[\![f[\![C_n]\!]]\!].$$

Thus C_0 is the troublesome part that keeps g from being a one-to-one correspondence between B and A. We bounce it back and forth, obtaining $C_1, C_2, \ldots$. The function showing that $A \approx B$ is the function $h: A \to B$ defined by

$$h(x) = \begin{cases} f(x) & \text{if } \; x \in C_n \quad \text{for some } n, \\ g^{-1}(x) & \text{otherwise.} \end{cases}$$

Note that in the second case ($x \in A$ but $x \notin C_n$ for any n) it follows that $x \notin C_0$ and hence $x \in \text{ran } g$. So $g^{-1}(x)$ makes sense in this case.

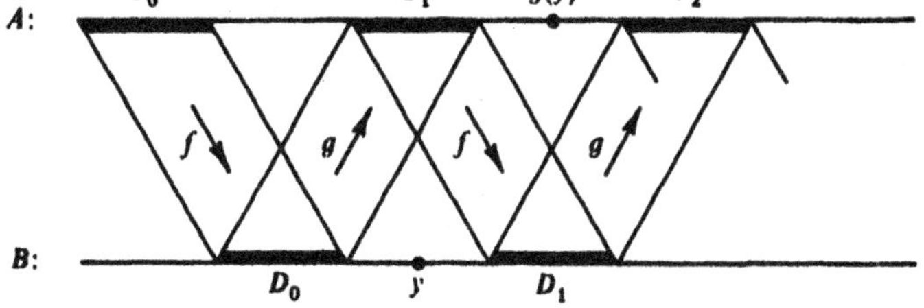

Fig. 39. The Schröder–Bernstein theorem.

Does it work? We must verify that h is one-to-one and has range B. Define $D_n = f[\![C_n]\!]$, so that $C_{n^+} = g[\![D_n]\!]$. To show that h is one-to-one, consider distinct x and x' in A. Since both f and g^{-1} are one-to-one, the only possible problem arises when, say, $x \in C_m$ and $x' \notin \bigcup_{n \in \omega} C_n$. In this case,

$$h(x) = f(x) \in D_m,$$

whereas

$$h(x') = g^{-1}(x') \notin D_m,$$

lest $x' \in C_{m^+}$. So $h(x) \neq h(x')$.

Finally we must check that ran h exhausts B. Certainly each $D_n \subseteq$ ran h, because $D_n = h[\![C_n]\!]$. Consider then a point y in $B - \bigcup_{n \in \omega} D_n$. Where is $g(y)$? Certainly $g(y) \notin C_0$. Also $g(y) \notin C_{n^+}$, because $C_{n^+} = g[\![D_n]\!]$, $y \notin D_n$, and g is one-to-one. So $g(y) \notin C_n$ for any n. Therefore $h(g(y)) = g^{-1}(g(y)) = y$. This shows that $y \in$ ran h, thereby proving part (a).

Part (b) is a restatement of part (a) in terms of cardinal numbers. ⊣

The Schröder–Bernstein theorem is sometimes called the "Cantor–Bernstein theorem." Cantor proved the theorem in his 1897 paper, but his proof utilized a principle that is equivalent to the axiom of choice. Ernst Schröder announced the theorem in an 1896 abstract. His proof, published in 1898, was imperfect, and he published a correction in 1911. The first fully satisfactory proof was given by Felix Bernstein and was published in an 1898 book by Borel.

Examples The usefulness of the Schröder–Bernstein theorem in calculating cardinalities is indicated by the following examples.

1. If $A \subseteq B \subseteq C$ and $A \approx C$, then all three sets are equinumerous. To prove this, let $\kappa =$ card $A =$ card C and let $\lambda =$ card B. Then by hypothesis $\kappa \leq \lambda \leq \kappa$, so by the Schröder–Bernstein theorem $\kappa = \lambda$.

2. The set $\mathbb{R}$ of real numbers is equinumerous to the closed unit interval $[0, 1]$. For we have

$$(0, 1) \subseteq [0, 1] \subseteq \mathbb{R},$$

and (as noted previously) $\mathbb{R} \approx (0, 1)$. Thus by the preceding example, all three sets are equinumerous. (For a more direct construction of a one-to-one correspondence between $\mathbb{R}$ and $[0, 1]$, we suggest trying Exercise 4.)

3. If $\kappa \leq \lambda \leq \mu$, then, as we observed before, $\kappa \leq \mu$. We can now give an improved version:

$$\kappa \leq \lambda < \mu \quad \Rightarrow \quad \kappa < \mu,$$
$$\kappa < \lambda \leq \mu \quad \Rightarrow \quad \kappa < \mu.$$

For by the earlier observation we obtain $\kappa \leq \mu$; if equality held, then (as in the first example) all three cardinal numbers would coincide.

4. $\mathbb{R} \approx {}^{\omega}2$, and hence $\mathbb{R} \approx \mathscr{P}\omega$. Thus the set of real numbers is equinumerous to the power set of ω. To prove this it suffices, by the Schröder–Bernstein theorem, to show that $\mathbb{R} \preccurlyeq {}^{\omega}2$ and ${}^{\omega}2 \preccurlyeq \mathbb{R}$.

To show that $\mathbb{R} \preccurlyeq {}^{\omega}2$, we construct a one-to-one function from the open unit interval $(0, 1)$ into ${}^{\omega}2$. The existence of such a function, together with the fact that $\mathbb{R} \approx (0, 1)$, gives us

$$\mathbb{R} \approx (0, 1) \preccurlyeq {}^{\omega}2.$$

The function is defined by use of binary expansions of real numbers; map the real whose binary expansion is 0.1100010... to the function in ${}^{\omega}2$ whose successive values are 1, 1, 0, 0, 0, 1, 0, ... In general, for a real number z in $(0, 1)$, let $H(z)$ be the function $H(z)\colon \omega \to 2$ whose value at n equals the $(n + 1)$st bit (binary digit) in the binary expansion of z. Clearly H is one-to-one. (But H does not have all of ${}^{\omega}2$ for its range. Note that $0.1000\ldots = 0.0111\ldots = \frac{1}{2}$. For definiteness, always select the nonterminating binary expansion.)

To show that ${}^{\omega}2 \preccurlyeq \mathbb{R}$ we use decimal expansions. The function in ${}^{\omega}2$ whose successive values are 1, 1, 0, 0, 0, 1, 0, ... is mapped to the real number with decimal expansion 0.1100010.... This maps ${}^{\omega}2$ one-to-one into the closed interval $[0, \frac{1}{9}]$.

5. By virtue of the above example,

$$\operatorname{card} \mathbb{R} = 2^{\aleph_0}.$$

Consequently the plane $\mathbb{R} \times \mathbb{R}$ has cardinality

$$2^{\aleph_0} \cdot 2^{\aleph_0} = 2^{\aleph_0 + \aleph_0} = 2^{\aleph_0}.$$

Thus the line $\mathbb{R}$ is equinumerous to the plane $\mathbb{R} \times \mathbb{R}$. This will not come as a surprise if you have heard of "space-filling" curves.

The next theorem shows that the operations of cardinal arithmetic have the expected order-preserving properties.

Theorem 6L Let κ, λ, and μ be cardinal numbers.

(a) $\kappa \leq \lambda \Rightarrow \kappa + \mu \leq \lambda + \mu$.
(b) $\kappa \leq \lambda \Rightarrow \kappa \cdot \mu \leq \lambda \cdot \mu$.
(c) $\kappa \leq \lambda \Rightarrow \kappa^{\mu} \leq \lambda^{\mu}$.
(d) $\kappa \leq \lambda \Rightarrow \mu^{\kappa} \leq \mu^{\lambda}$; if not both κ and μ equal zero.

Proof Let K, L, and M be sets of cardinality κ, λ, and μ, respectively. Then ${}^{M}K$ has cardinality κ^{μ}, etc. We assume that $\kappa \leq \lambda$; hence we may select K and L such that $K \subseteq L$. And we may select M so that $L \cap M = \varnothing$.

Parts (a), (b), and (c) are now immediate, since

$$K \cup M \subseteq L \cup M, \qquad K \times M \subseteq L \times M, \qquad {}^M K \subseteq {}^M L.$$

For part (d), first consider the case in which $\mu = 0$. Then $\kappa \neq 0$ and $\mu^\kappa = 0 \leq \mu^\lambda$. There remains the case in which $\mu \neq 0$, i.e., $M \neq \varnothing$. Take some fixed $a \in M$. We need a one-to-one function G from ${}^K M$ into ${}^L M$. For any $f \in {}^K M$, define $G(f)$ to be the function with domain L for which

$$G(f)(x) = \begin{cases} f(x) & \text{if } x \in K, \\ a & \text{if } x \in L - K. \end{cases}$$

In one line, $G(f) = f \cup ((L - K) \times \{a\})$. Then $G\colon {}^K M \to {}^L M$ and G is clearly one-to-one. ⊣

Example We can calculate the product $\aleph_0 \cdot 2^{\aleph_0}$ by the method of upper and lower bounds:

$$2^{\aleph_0} \leq \aleph_0 \cdot 2^{\aleph_0} \leq 2^{\aleph_0} \cdot 2^{\aleph_0} = 2^{\aleph_0},$$

whence equality holds throughout.

We would like to show that $\aleph_0$ is the *least* infinite cardinal; that is, that $\aleph_0 \leq \kappa$ for any infinite cardinal κ. This amounts (by the definition of $\leq$) to showing that $\omega \preccurlyeq A$ for any infinite A. We might attempt to define a one-to-one function $g\colon \omega \to A$ by recursion:

$$g(0) = \text{some member of } A,$$
$$g(n^+) = \text{some member of } A - g[\![n^+]\!].$$

Here $A - g[\![n^+]\!]$ is nonempty, lest A be finite. A minor difficulty here is that $g(n^+)$ is being defined not from $g(n)$ but from $g[\![n^+]\!] = \{g(0), g(1), \ldots, g(n)\}$. This difficulty is easily circumvented, and will be circumvented in the proof of Theorem 6N. A major difficulty is the phrase "some member." Unless we say *which* member, the above cannot possibly be construed as defining g.

What is needed here is the axiom of choice, which will enable us to convert "some member" into a more acceptable phrase.

Exercises

15. Show that there is no set $\mathscr{A}$ with the property that for every set there is some member of $\mathscr{A}$ that dominates it.

16. Show that for any set S we have $S \preccurlyeq {}^S2$, but $S \not\approx {}^S2$. (This should be done directly, without use of $\mathscr{P}S$ or cardinal numbers. If $F: S \to {}^S2$, then define $g(x) = 1 - F(x)(x)$.)

17. Give counterexamples to show that we cannot strengthen Theorem 6L by replacing "$\leq$" by "$<$" throughout.

AXIOM OF CHOICE

At several points in this book we have already encountered the need for a principle asserting the possibility of selecting members from nonempty sets. We can no longer postpone a systematic discussion of such a principle. There are numerous equivalent formulations of the axiom of choice. The following theorem lists six of them. Others will be found in the exercises.

Theorem 6M The following statements are equivalent.

(1) Axiom of choice, I. For any relation R, there is a function $F \subseteq R$ with dom F = dom R.

(2) Axiom of choice, II; multiplicative axiom. The Cartesian product of nonempty sets is always nonempty. That is, if H is a function with domain I and if $(\forall i \in I)\, H(i) \neq \varnothing$, then there is a function f with domain I such that $(\forall i \in I)\, f(i) \in H(i)$.

(3) Axiom of choice, III. For any set A there is a function F (a "choice function" for A) such that the domain of F is the set of nonempty subsets of A, and such that $F(B) \in B$ for every nonempty $B \subseteq A$.

(4) Axiom of choice, IV. Let $\mathscr{A}$ be a set such that (a) each member of $\mathscr{A}$ is a nonempty set, and (b) any two distinct members of $\mathscr{A}$ are disjoint. Then there exists a set C containing exactly one element from each member of $\mathscr{A}$ (i.e., for each $B \in \mathscr{A}$ the set $C \cap B$ is a singleton $\{x\}$ for some x).

(5) Cardinal comparability. For any sets C and D, either $C \preccurlyeq D$ or $D \preccurlyeq C$. For any two cardinal numbers κ and λ, either $\kappa \leq \lambda$ or $\lambda \leq \kappa$.

(6) Zorn's lemma. Let $\mathscr{A}$ be a set such that for every chain $\mathscr{B} \subseteq \mathscr{A}$, we have $\bigcup \mathscr{B} \in \mathscr{A}$. ($\mathscr{B}$ is called a *chain* iff for any C and D in $\mathscr{B}$, either $C \subseteq D$ or $D \subseteq C$.) Then $\mathscr{A}$ contains an element M (a "maximal" element) such that M is not a subset of any other set in $\mathscr{A}$.

Statements (1)–(4) are synoptic ways of saying that there exist uniform methods for selecting elements from sets. On the other hand, statements (5) and (6) appear to be rather different.

Proof in part First we will prove that (1)–(4) are equivalent.

(1) ⇒ (2) To prove (2), we assume that H is a function with domain I and that $H(i) \neq \varnothing$ for each $i \in I$. In order to utilize (1), define the relation

$$R = \{\langle i, x\rangle \mid i \in I \;\&\; x \in H(i)\}.$$

Then (1) provides us with a function $F \subseteq R$ such that $\operatorname{dom} F = \operatorname{dom} R = I$. Then since $\langle i, F(i)\rangle \in F \subseteq R$, we must have $F(i) \in H(i)$. Thus the conclusion of (2) holds.

(2) ⇒ (4) Let $\mathscr{A}$ be a set meeting conditions (a) and (b) of (4). Let H be the identity function on $\mathscr{A}$; then $(\forall B \in \mathscr{A})\, H(B) \neq \varnothing$. Hence by (2) there is a function f with domain $\mathscr{A}$ such that $(\forall B \in \mathscr{A}) f(B) \in H(B) = B$. Let $C = \operatorname{ran} f$. Then for $B \in \mathscr{A}$ we have $B \cap C = \{f(B)\}$. (Nothing else could belong to $B \cap C$ by condition (b).)

(4) ⇒ (3) Given a set A, define

$$\mathscr{A} = \{\{B\} \times B \mid B \text{ is a nonempty subset of } A\}.$$

Then each member of $\mathscr{A}$ is nonempty, and any two distinct members are disjoint (if $\langle x, y\rangle \in (\{B\} \times B) \cap (\{B'\} \times B')$, then $x = B = B'$). Let C be a set (provided by (4)) whose intersection with each member of $\mathscr{A}$ is a singleton:

$$C \cap (\{B\} \times B) = \{\langle B, x\rangle\},$$

where $x \in B$. It is *a priori* possible that C contains extraneous elements not belonging to any member of $\mathscr{A}$. So discard them by letting $F = C \cap (\bigcup \mathscr{A})$. We claim that F is a choice function for A. Any member of F belongs to some $\{B\} \times B$, and hence is of the form $\langle B, x\rangle$ for $x \in B$. For any one nonempty set $B \subseteq A$, there is a unique x such that $\langle B, x\rangle \in F$, because $F \cap (\{B\} \times B)$ is a singleton. This x is of course $F(B)$ and it is a member of B.

(3) ⇒ (1) Consider any relation R. Then (3) provides us with a choice function G for $\operatorname{ran} R$; thus $G(B) \in B$ for any nonempty subset B of $\operatorname{ran} R$. Then define a function F with $\operatorname{dom} F = \operatorname{dom} R$ by

$$F(x) = G(\{y \mid xRy\}).$$

Then $F(x) \in \{y \mid xRy\}$, i.e., $\langle x, F(x)\rangle \in R$. Hence $F \subseteq R$.

It remains to include parts (5) and (6) of the theorem.

(6) ⇒ (1) The strategy behind this application (and others) of Zorn's lemma is to form a collection $\mathscr{A}$ of pieces of the desired object, and then to show that a maximal piece serves the intended purpose. In the present case, we are given a relation R and we choose to define

$$\mathscr{A} = \{f \subseteq R \mid f \text{ is a function}\}.$$

Before we can appeal to Zorn's lemma, we must check that $\mathscr{A}$ is closed under unions of chains. So consider any chain $\mathscr{B} \subseteq \mathscr{A}$. Since every member of $\mathscr{B}$ is a subset of R, $\bigcup\mathscr{B}$ is a subset of R. To see that $\bigcup\mathscr{B}$ is a function, we use the fact that $\mathscr{B}$ is a chain. If $\langle x, y\rangle$ and $\langle x, z\rangle$ belong to $\bigcup\mathscr{B}$, then

$$\langle x, y\rangle \in G \in \mathscr{B} \qquad \text{and} \qquad \langle x, z\rangle \in H \in \mathscr{B}$$

for some functions G and H in $\mathscr{A}$. Either $G \subseteq H$ or $H \subseteq G$; in either event both $\langle x, y\rangle$ and $\langle x, z\rangle$ belong to a single function, so $y = z$. Hence $\bigcup\mathscr{B}$ is in $\mathscr{A}$. Now we can appeal to (6), which provides us with a maximal function F in $\mathscr{A}$. We claim that $\operatorname{dom} F = \operatorname{dom} R$. For otherwise take any $x \in \operatorname{dom} R - \operatorname{dom} F$. Since $x \in \operatorname{dom} R$, there is some y with xRy. Define

$$F' = F \cup \{\langle x, y\rangle\}.$$

Then $F' \in \mathscr{A}$, contradicting the maximality of F. Hence $\operatorname{dom} F = \operatorname{dom} R$.

(6) ⇒ (5) Let C and D be any sets; we will show that either $C \preccurlyeq D$ or $D \preccurlyeq C$. In order to utilize (6), define

$$\mathscr{A} = \{f \mid f \text{ is a one-to-one function \& } \operatorname{dom} f \subseteq C \text{ \& } \operatorname{ran} f \subseteq D\}.$$

Consider any chain $\mathscr{B} \subseteq \mathscr{A}$. As in the preceding paragraph, $\bigcup\mathscr{B}$ is a function, and a similar argument shows that $\bigcup\mathscr{B}$ is one-to-one. Next consider $\langle x, y\rangle \in \bigcup\mathscr{B}$; then $\langle x, y\rangle \in f \in \mathscr{A}$. Consequently $x \in C$ and $y \in D$. Thus $\operatorname{dom} \bigcup\mathscr{B} \subseteq C$ and $\operatorname{ran} \bigcup\mathscr{B} \subseteq D$. Hence $\bigcup\mathscr{B} \in \mathscr{A}$ and we can apply (6) to obtain a maximal $\hat{f} \in \mathscr{A}$. We claim that either $\operatorname{dom} \hat{f} = C$ (in which case $C \preccurlyeq D$) or $\operatorname{ran} \hat{f} = D$ (in which case $D \preccurlyeq C$ since $\hat{f}^{-1}$ is then a one-to-one function from D into C). Suppose to the contrary that neither condition holds, so that there exist elements $c \in C - \operatorname{dom} \hat{f}$ and $d \in D - \operatorname{ran} \hat{f}$. Then

$$f' = \hat{f} \cup \{\langle c, d\rangle\}$$

is in $\mathscr{A}$, contradicting the maximality of $\hat{f}$. (You will observe that the strategy underlying this application of Zorn's lemma is the same one as in the preceding paragraph.)

At this point we have proved that part of Theorem 6M shown in Fig. 40. The proof will be completed in Chapter 7.

Remarks Zorn's lemma first appeared in a 1922 paper by Kuratowski. Earlier maximality principles that were similar in spirit had been published by Felix Hausdorff. The importance of Zorn's lemma lies in the fact that it is well suited to many applications. in mathematics. For example, to prove in linear algebra that every vector space has a basis requires some form of choice, and Zorn's lemma is a convenient form to use here. (Take $\mathscr{A}$ to be the collection of all linearly independent sets; then a maximal element will be a basis.) Similarly in proving that there exist maximal proper ideals in a ring with identity or in proving that there exist maximal Abelian subgroups in a group, Zorn's lemma provides a suitable tool.

We can give a plausibility argument for Zorn's lemma as follows. $\mathscr{A}$ cannot be empty, because $\varnothing$ is a chain and so $\varnothing = \bigcup\varnothing \in \mathscr{A}$. Probably $\varnothing$ is not maximal, so we can choose a larger set. If that larger set is not maximal, we can choose a still larger one. After infinitely many steps, even if we have not found a maximal set we have at least formed a chain. So we can take its union and continue. The procedure can stop only when we finally reach a maximal set. So if only we are patient enough, we should reach such a set.

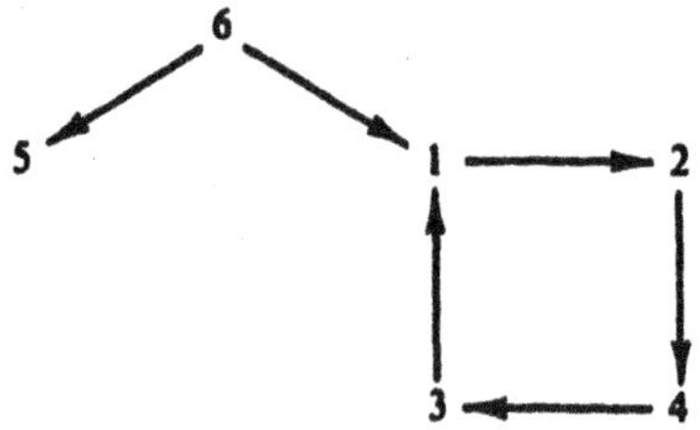

Fig. 40. This much of Theorem 6M we prove now; the rest is postponed.

We now formally add the axiom of choice to our list of axioms. And we do it without a marginal stripe. This is because historically the axiom of choice has had a unique status. Initially some mathematicians objected to the axiom, because it asserts the existence of a set without specifying exactly what is in it. To this extent it is less "constructive" than the other axioms. Gradually the axiom has won acceptance (at least acceptance by most mathematicians willing to accept classical logic). But it retains a slightly tarnished image, from the days when it was not quite respectable. Consequently it has become wide-spread practice, each time the axiom of choice is used, to make explicit mention of the fact. (No such gesture is extended to the other axioms, which are used extensively and mentioned rarely.)

In Chapter 7 we will complete our axiomatization by adding the replacement axioms and the regularity axiom.

Without the axiom of choice, we can still prove that for a *finite* set I, if $H(i) \neq \varnothing$ for all $i \in I$, then there is a function f with domain I such that $f(i) \in H(i)$ for all $i \in I$ (Exercise 19). But for an infinite set I, the axiom of choice is indispensable. (If we are making seven choices, we can explicitly mention each one; for $\aleph_0$ choices this is no longer possible.)

In particular, if we are choosing *one* thing, then we do not need the axiom of choice. (We mention this because people sometimes overreact when first hearing of the axiom of choice.) For example, assume that $A \neq \varnothing$. Then there exists some y_0 in A; we can use y_0 in further arguments (to show, e.g., that some singleton $\{y_0\}$ is a subset of A). This makes no use of the axiom of choice, even when phrased as "we can choose some fixed y_0 in A."

For a second example, assume we are given a relation R. For any particular x in dom R, there exists some y_0 for which xRy_0. And we can conclude that for any x in dom R, there is a singleton $\{y_0\}$ included in $\{t \mid xRt\}$. We have not yet used the axiom of choice. What *does* require the axiom of choice is saying that for any x in dom R there is some y_x for which xRy_x, and then putting all these y_x's together into a set, e.g.,

$$\{y_x \mid x \in \operatorname{dom} R\}.$$

This in effect is making many choices, one for each x in dom R, which may be an infinite set. (A human being cannot perform infinitely many actions. But the axiom of choice asserts that, despite our practical limitations, there exists, in theory, a set produced by making infinitely many choices.)

Another situation in which the axiom of choice can be avoided is where we can specify exactly which object we want to choose. For example, we can show, without using the axiom of choice, that there is a choice function for ω. Namely, define $F\colon (\mathscr{P}\omega - \{\varnothing\}) \to \omega$ by

$$F(A) = \text{the least member of } A$$

for $\varnothing \neq A \subseteq \omega$. The crucial point here is that we can write down an expression *defining* the selected member of A. If ω is replaced by $\mathbb{R}$, for example, there is no longer the possibility of defining a way to select numbers from arbitrary nonempty subsets of $\mathbb{R}$.

Since there is a choice function on ω, it follows that there is a choice function on any set equinumerous to ω. (Why?) And similarly any finite set has a choice function.

The following example is due to Russell. If we have $\aleph_0$ pairs of shoes, then we can select one shoe from each pair *without* using the axiom of choice. We simply select the left shoe from each pair. But if we have $\aleph_0$ pairs of socks, then we must use the axiom of choice if we are to select one sock from each pair. For there is no definable difference between the two socks in a pair.

Example Assume that f is a function from A *onto* B. We claim that $B \preccurlyeq A$. To verify this, recall that by Theorem 3J(b), the proof of which used choice, there is a right inverse $g\colon B \to A$ such that $f \circ g = I_B$. And g is one-to-one since it has a left inverse. The function g shows that $B \preccurlyeq A$.

Conversely assume that $B \preccurlyeq A$ and B is nonempty. Then (since $B \preccurlyeq A$) there is a one-to-one function $g: B \to A$. By Theorem 3J(a) there is a left inverse $f: A \to B$ such that $f \circ g = I_B$. And f maps A *onto* B since it has a right inverse.

Hence we can conclude: If B is nonempty, then $B \preccurlyeq A$ iff there is a function from A onto B. As a special case, we can conclude that a nonempty set B is dominated by ω iff there is a function from ω onto B. (This special case can be proved without the axiom of choice.)

We now utilize the axiom of choice to prove that $\aleph_0$ is the least infinite cardinal number.

Theorem 6N (a) For any infinite set A, we have $\omega \preccurlyeq A$.
(b) $\aleph_0 \leq \kappa$ for any infinite cardinal κ.

Proof Part (b) is merely a restatement of part (a) in terms of cardinals; it suffices to prove (a). So consider an infinite set A. The idea is to select $\aleph_0$ things from A: a first, then a second, then a third, ... Let F be a choice function for A (as provided by the axiom of choice, III). We must somehow employ F repeatedly; in fact $\aleph_0$ times.

As a first try, we could attempt to define a function $g: \omega \to A$ by recursion. We know A is nonempty; fix some element $a \in A$. Maybe we can take

$$\begin{aligned} g(0) &= a, \\ g(n^+) &= \text{the chosen member of } A - \{g(0), \ldots, g(n)\} \\ &= F(A - g[\![n^+]\!]). \end{aligned}$$

No, we cannot. This would require a stronger recursion theorem than the one proved in Chapter 4. We are attempting to define $g(n^+)$ not just from $g(n)$, but from the entire list $g(0), g(1), \ldots, g(n)$.

But our second try will succeed. We will define by recursion a function h from ω into the set of *finite* subsets of A. The guiding idea is to have, for example, $h(7) = \{g(0), \ldots, g(6)\}$. Formally,

$$\begin{aligned} h(0) &= \varnothing, \\ h(n^+) &= h(n) \cup \{F(A - h(n))\}. \end{aligned}$$

Thus we start with $\varnothing$, and successively add chosen new elements from A. $A - h(n)$ is nonempty because A is infinite and $h(n)$ is a finite subset. And $h(n^+)$ is again a finite subset of A.

Now we can go back and recover g. Define

$$g(n) = F(A - h(n))$$

so that $g(n)$ is the thing we added to the set $h(n)$ to make $h(n^+)$. Thus $g: \omega \to A$; we must check that g is one-to-one.

Suppose then that $m \neq n$. One number is less than the other; say $m \in n$. Then $m^+ \in n$ and so

$$g(m) \in h(m^+) \subseteq h(n).$$

But $g(n) \notin h(n)$, since

$$g(n) = F(A - h(n)) \in A - h(n).$$

Hence $g(m) \neq g(n)$. ⊣

We can now list some consequences of the foregoing theorem.

1. Any infinite subset of ω is equinumerous to ω. In this special case we can avoid using the axiom of choice, since we have a choice function for ω anyway.
2. We have by part (b) of the theorem that if $\kappa < \aleph_0$, then κ is finite. Since the converse is clear, we have

$$\kappa < \aleph_0 \quad \text{iff} \quad \kappa \text{ is finite.}$$

3. We get another proof that subsets of finite sets are finite (Corollary 6G). For if

$$\operatorname{card} A \leq \operatorname{card} n < \aleph_0,$$

then by the Schröder–Bernstein theorem card $A < \aleph_0$, and hence A is finite.

Another consequence of the preceding theorem is the following characterization of the infinite sets (proposed by Dedekind in his 1888 book as a definition of "infinite").

Corollary 6P A set is infinite iff it is equinumerous to a proper subset of itself.

Proof Half of this result is contained in Corollary 6D, where we showed that if a set was equinumerous to a proper subset of itself, then it was infinite. Conversely, consider an infinite set A. Then by the above theorem, there is a one-to-one function f from ω into A. Define a function g from A into A by

$$g(f(n)) = f(n^+) \qquad \text{for} \quad n \in \omega,$$
$$g(x) = x \qquad \text{for} \quad x \notin \operatorname{ran} f$$

(see Fig. 41). Then g is a one-to-one function from A onto $A - \{f(0)\}$. ⊣

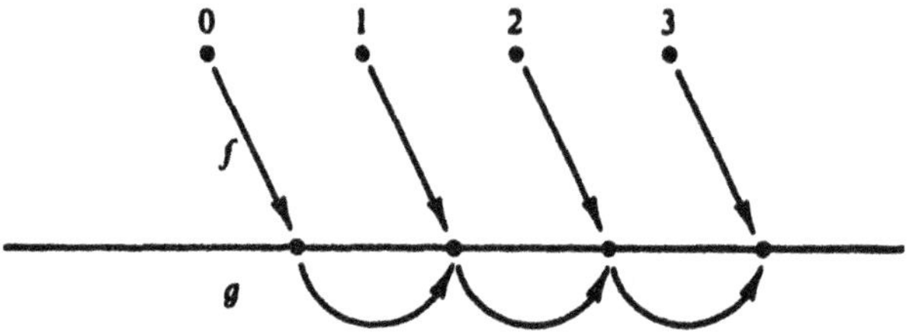

Fig. 41. $f(0)$ is not in ran g.

Exercises

18. Prove that the following statement is equivalent to the axiom of choice: For any set $\mathscr{A}$ whose members are nonempty sets, there is a function f with domain $\mathscr{A}$ such that $f(X) \in X$ for all X in $\mathscr{A}$.

19. Assume that H is a function with finite domain I and that $H(i)$ is nonempty for each $i \in I$. Without using the axiom of choice, show that there is a function f with domain I such that $f(i) \in H(i)$ for each $i \in I$. [*Suggestion*: Use induction on card I.]

20. Assume that A is a nonempty set and R is a relation such that $(\forall x \in A)(\exists y \in A)\, yRx$. Show that there is a function $f\colon \omega \to A$ with $f(n^+)Rf(n)$ for all n in ω.

21. (Teichmüller–Tukey lemma) Assume that $\mathscr{A}$ is a nonempty set such that for every set B,

$$B \in \mathscr{A} \quad \Leftrightarrow \quad \text{every finite subset of } B \text{ is a member of } \mathscr{A}.$$

Show that $\mathscr{A}$ has a maximal element, i.e., an element that is not a subset of any other element of $\mathscr{A}$.

22. Show that the following statement is another equivalent version of the axiom of choice: For any set A there is a function F with dom $F = \bigcup A$ and such that $x \in F(x) \in A$ for all $x \in \bigcup A$.

23. Show that in the proof to Theorem 6N, we have $g[\![n]\!] = h(n)$.

24. How would you define the sum of infinitely many cardinal numbers? Infinite products?

25. Assume that S is a function with domain ω such that $S(n) \subseteq S(n^+)$ for each $n \in \omega$. (Thus S is an increasing sequence of sets.) Assume that B is a subset of the union $\bigcup_{n \in \omega} S(n)$ such that for every infinite subset B' of B there is some n for which $B' \cap S(n)$ is infinite. Show that B is a subset of some $S(n)$.

COUNTABLE SETS

The definition below applies the word "countable" to those sets whose elements can, in a sense, be counted by means of the natural numbers. A "counting" of a set can be taken to be a one-to-one correspondence between the members of the set and the natural numbers (or the natural numbers less than some number n). This requires that the set be no larger than ω.

Definition A set A is *countable* iff $A \preccurlyeq \omega$, i.e., iff card $A \leq \aleph_0$.

Since we have recently found that

$$\kappa < \aleph_0 \quad \Leftrightarrow \quad \kappa \text{ is finite,}$$

we can also formulate the definition as follows: A set A is countable iff either A is finite or A has cardinality $\aleph_0$.

For example, the set ω of natural numbers, the set $\mathbb{Z}$ of integers, and the set $\mathbb{Q}$ of rational numbers are all infinite countable sets. But the set $\mathbb{R}$ of real numbers is uncountable (Theorem 6B).

Any subset of a countable set is obviously countable. The union of two countable sets is countable, as is their Cartesian product. (The union has at most cardinality $\aleph_0 + \aleph_0$, the product at most $\aleph_0 \cdot \aleph_0$. But both of these numbers equal $\aleph_0$.) On the other hand, $\mathscr{P}A$ is uncountable for any infinite set A. (If $2^\kappa \leq \aleph_0$, then $\kappa < \aleph_0$.)

Recall (from the example preceding Theorem 6N) that a nonempty set B is countable iff there is a function from ω onto B. This fact is used in the proof of the next theorem.

Theorem 6Q A countable union of countable sets is countable. That is, if $\mathscr{A}$ is countable and if every member of $\mathscr{A}$ is a countable set, then $\bigcup\mathscr{A}$ is countable.

Proof We may suppose that $\varnothing \notin \mathscr{A}$, for otherwise we could simply remove it without affecting $\bigcup\mathscr{A}$. We may further suppose that $\mathscr{A} \neq \varnothing$, since $\bigcup\varnothing$ is certainly countable. Thus $\mathscr{A}$ is a countable (but nonempty) from $\omega \times \omega$ onto $\bigcup\mathscr{A}$. We already know of functions from ω onto $\omega \times \omega$, and the composition will map ω onto $\bigcup\mathscr{A}$, thereby showing that $\bigcup\mathscr{A}$ is countable.

Since $\mathscr{A}$ is countable but nonempty, there is a function G from ω onto $\mathscr{A}$. Informally, we may write

$$\mathscr{A} = \{G(0), G(1), \ldots\}.$$

(Here G might not be one-to-one, so there may be repetitions in this enumeration.) We are given that each set $G(m)$ is countable and nonempty.

Hence for each m there is a function from ω onto $G(m)$. We must use the axiom of choice to select such a function for each m.

Because the axiom of choice is a recent addition to our repertoire, we will describe its use here in some detail. Let $H: \omega \to {}^{\omega}(\bigcup \mathscr{A})$ be defined by

$$H(m) = \{g \mid g \text{ is a function from } \omega \text{ onto } G(m)\}.$$

We know that $H(m)$ is nonempty for each m. Hence there is a function F with domain ω such that for each m, $F(m)$ is a function from ω onto $G(m)$.

To conclude the proof we have only to let $f(m, n) = F(m)(n)$. Then f is a function from $\omega \times \omega$ onto $\bigcup \mathscr{A}$. ⊣

Example For any set A, define a *sequence* in A to be a function from some natural number into A. Let $\mathrm{Sq}(A)$ be the set of all sequences in A:

$$\begin{aligned}\mathrm{Sq}(A) &= \{f \mid (\exists n \in \omega)\ f \text{ maps } n \text{ into } A\} \\ &= {}^0A \cup {}^1A \cup {}^2A \cup \cdots.\end{aligned}$$

The *length* of a sequence is simply its domain.

In order to verify that $\mathrm{Sq}(A)$ is a legal set, note that if $f: n \to A$, then

$$f \subseteq n \times A \subseteq \omega \times A,$$

so that $f \in \mathscr{P}(\omega \times A)$. Hence $\mathrm{Sq}(A) \subseteq \mathscr{P}(\omega \times A)$.

We now list some observations that establish the existence of transcendental real numbers.

1. $\mathrm{Sq}(\omega)$ has cardinality $\aleph_0$. This can be proved by using primarily the fact that $\omega \times \omega \approx \omega$. Another very direct proof is the following. Consider any $f \in \mathrm{Sq}(\omega)$; say the length of f is n. Then define

$$H(f) = 2^{f(0)+1} \cdot 3^{f(1)+1} \cdot \ldots \cdot p_{n-1}^{f(n-1)+1},$$

where p_i is the $(i + 1)$st prime. (If the length of f is 0, then $H(f) = 1$.) Thus $H: \mathrm{Sq}(\omega) \to \omega$ and by the fundamental theorem of arithmetic (which states that prime factorizations are unique) H is one-to-one. Hence card $\mathrm{Sq}(\omega) \leq \aleph_0$, and the opposite inequality is clear. (In Chapter 4 we did not actually develop the theory of prime numbers. But there are no difficulties in embedding any standard development of the subject into set theory.)

2. $\mathrm{Sq}(A)$ is countable for any countable set A. By the countability of A there is a one-to-one function g from A into ω. This function naturally induces a one-to-one map from $\mathrm{Sq}(A)$ into $\mathrm{Sq}(\omega)$. Hence card $\mathrm{Sq}(A) \leq$ card $\mathrm{Sq}(\omega) = \aleph_0$. (An alternative proof writes $\mathrm{Sq}(A) = \bigcup\{{}^nA \mid n \in \omega\}$, a countable union of countable sets.)

We can think of this set A as an alphabet, and the elements of $\mathrm{Sq}(A)$ as being *words* on the alphabet A. In this terminology, the present example can be stated: On any countable alphabet, there are countably many words.

3. There are $\aleph_0$ algebraic numbers. (Recall that an algebraic number is a real number that is the root of some polynomial with integer coefficients. For this purpose we exclude from the polynomials the function that is identically equal to 0.) As a first step in counting the algebraic numbers, note that the set $\mathbb{Z}$ of integers has cardinality $\aleph_0 + \aleph_0 = \aleph_0$. Next we calculate the cardinality of the set P of polynomials with integer coefficients. We can assign to each polynomial (of degree n) its sequence (of length $n+1$) of coefficients, e.g., $1 + 7x - 5x^2 + 3x^4$ is assigned the sequence of length 5 whose successive values are 1, 7, -5, 0, 3. This defines a one-to-one map from P into $\operatorname{Sq}(\mathbb{Z})$, a countable set. Hence P is countable. Since each polynomial in P has only finitely many roots, the set of algebraic numbers is a countable union of finite sets. Hence it is countable, by Theorem 6Q. Since the set of algebraic numbers is certainly infinite, it has cardinality $\aleph_0$.

4. There are uncountably many transcendental numbers. (Recall that a transcendental number is defined to be a real number that is not algebraic.) Since the set of algebraic numbers is countable, the set of transcendental numbers cannot also be countable lest the set $\mathbb{R}$ be countable. (Soon we will be able to show that the set of transcendental numbers has cardinality $2^{\aleph_0}$.)

Exercises

26. Prove the following generalization of Theorem 6Q: If every member of a set $\mathscr{A}$ has cardinality κ or less, then

$$\operatorname{card} \bigcup \mathscr{A} \leq (\operatorname{card} \mathscr{A}) \cdot \kappa.$$

27. (a) Let A be a collection of circular disks in the plane, no two of which intersect. Show that A is countable.
(b) Let B be a collection of circles in the plane, no two of which intersect. Need B be countable?
(c) Let C be a collection of figure eights in the plane, no two of which intersect. Need C be countable?

28. Find a set $\mathscr{A}$ of open intervals in $\mathbb{R}$ such that every rational number belongs to one of those intervals, but $\bigcup \mathscr{A} \neq \mathbb{R}$. [*Suggestion*: Limit the sum of the lengths of the intervals.]

29. Let A be a set of positive real numbers. Assume that there is a bound b such that the sum of any finite subset of A is less than b. Show that A is countable.

30. Assume that A is a set with at least two elements. Show that $\operatorname{Sq}(A) \preccurlyeq {}^{\omega}A$.

ARITHMETIC OF INFINITE CARDINALS

We can use the axiom of choice to show that adding or multiplying two infinite cardinals is a more trivial matter than it first appeared to be.

Lemma 6R For any infinite cardinal κ, we have $\kappa \cdot \kappa = \kappa$.

Proof Let B be a set of cardinality κ. It would suffice to show that $B \times B \approx B$, and we will almost do this. Define

$$\mathscr{H} = \{f \mid f = \varnothing \text{ or for some infinite } A \subseteq B, f \text{ is a one-to-one correspondence between } A \times A \text{ and } A\}.$$

Our strategy is to use Zorn's lemma to obtain a maximal function f_0 in $\mathscr{H}$. Although f_0 might not quite show that $B \times B \approx B$, it will come close enough. (The reason for including $\varnothing$ as a member of $\mathscr{H}$ is that our statement of Zorn's lemma requires that $\mathscr{H}$ contain the union of any chain, even the empty chain.)

Before applying Zorn's lemma to $\mathscr{H}$, we must check for closure under unions of chains. Let $\mathscr{C}$ be a chain included in $\mathscr{H}$. We may suppose that $\mathscr{C}$ contains some nonempty function, since otherwise $\bigcup\mathscr{C} = \varnothing \in \mathscr{H}$. As we have seen before, $\bigcup\mathscr{C}$ is again a one-to-one function. Define

$$A = \bigcup\{\operatorname{ran} f \mid f \in \mathscr{C}\} = \operatorname{ran} \bigcup\mathscr{C}$$

(compare Exercise 8 of Chapter 3). A is infinite since $\mathscr{C}$ contains some nonempty function. We claim that $\bigcup\mathscr{C}$ is a one-to-one correspondence between $A \times A$ and A. The only part of this claim not yet verified is that $\operatorname{dom} \bigcup\mathscr{C} = A \times A$. First consider any $\langle a_1, a_2\rangle \in A \times A$. Then $a_1 \in \operatorname{ran} f_1$ and $a_2 \in \operatorname{ran} f_2$ for some f_1 and f_2 in $\mathscr{C}$. Either $f_1 \subseteq f_2$ or $f_2 \subseteq f_1$; by symmetry we may suppose that $f_1 \subseteq f_2$. Then

$$\langle a_1, a_2\rangle \in \operatorname{ran} f_2 \times \operatorname{ran} f_2 = \operatorname{dom} f_2 \subseteq \bigcup\{\operatorname{dom} f \mid f \in \mathscr{C}\} = \operatorname{dom} \bigcup\mathscr{C}.$$

Conversely any member of $\operatorname{dom} \bigcup\mathscr{C}$ belongs to $\operatorname{dom} f$ for some $f \in \mathscr{C}$. But $\operatorname{dom} f = \operatorname{ran} f \times \operatorname{ran} f \subseteq A \times A$. Thus $\operatorname{dom} \bigcup\mathscr{C} = A \times A$. So $\bigcup\mathscr{C}$ is a one-to-one correspondence between $A \times A$ and A; hence $\bigcup\mathscr{C} \in \mathscr{H}$.

Zorn's lemma now provides us with a maximal $f_0 \in \mathscr{H}$. First, we must check that $f_0 \neq \varnothing$. Since B is infinite, it has a subset A of cardinality $\aleph_0$. Because $\aleph_0 \cdot \aleph_0 = \aleph_0$, there is a one-to-one correspondence g between $A \times A$ and A. Thus $g \in \mathscr{H}$; since g properly extends $\varnothing$, it is impossible for $\varnothing$ to be maximal is $\mathscr{H}$. Hence $f_0 \neq \varnothing$. So by the definition of $\mathscr{H}$, f_0 is a one-to-one correspondence between $A_0 \times A_0$ and A_0, where A_0 is some infinite subset of B.

Let $\lambda = \operatorname{card} A_0$; then λ is infinite and $\lambda \cdot \lambda = \lambda$. One might hope that by virtue of the maximality, A_0 would be all of B. This may not be true, but we will show that $\lambda = \kappa$ and that $B - A_0$ has smaller cardinality.

In order to show that card $(B - A_0) < \lambda$, suppose that to the contrary $\lambda \le \text{card}(B - A_0)$. Then $B - A_0$ has a subset D of cardinality λ. We will show that this contradicts the maximality of f_0 by finding a proper extension of f_0 that is a one-to-one correspondence between the sets $(A_0 \cup D) \times (A_0 \cup D)$ and $A_0 \cup D$. We have

$$(A_0 \cup D) \times (A_0 \cup D) = (A_0 \times A_0) \cup (A_0 \times D) \cup (D \times A_0) \cup (D \times D)$$

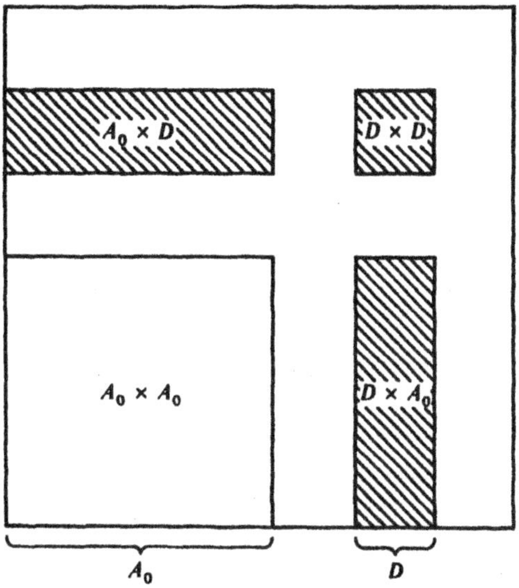

Fig. 42. The set $B \times B$ (in Lemma 6R).

(see Fig. 42). Of the four sets on the right side of this equation, $A_0 \times A_0$ is already paired with A_0 by f_0. The remainder

$$(A_0 \times D) \cup (D \times A_0) \cup (D \times D)$$

(shaded in Fig. 42) clearly has cardinality

$$\begin{aligned} \lambda \cdot \lambda + \lambda \cdot \lambda + \lambda \cdot \lambda &= \lambda + \lambda + \lambda \\ &= 3 \cdot \lambda \\ &\le \lambda \cdot \lambda \\ &= \lambda. \end{aligned}$$

Hence there is a one-to-one correspondence g between the shaded sets and D. Then $f_0 \cup g \in \mathscr{H}$ and properly extends f_0, contradicting the maximality of f_0. Hence our supposition that $\lambda \le \text{card}(B - A_0)$ is false. By cardinal comparability, $\text{card}(B - A_0) < \lambda$.

Finally we have

$$\begin{aligned}\kappa &= \text{card } A_0 + \text{card}(B - A_0) \\ &\leq \lambda + \lambda = 2 \cdot \lambda \leq \lambda \cdot \lambda = \lambda \leq \kappa,\end{aligned}$$

whence $\lambda = \kappa$. Hence $\kappa \cdot \kappa = \kappa$. ⊣

Absorption Law of Cardinal Arithmetic Let κ and λ be cardinal numbers, the larger of which is infinite and the smaller of which is nonzero. Then

$$\kappa + \lambda = \kappa \cdot \lambda = \max(\kappa, \lambda).$$

Proof By the symmetry we may suppose that $\lambda \leq \kappa$. Then

$$\kappa \leq \kappa + \lambda \leq \kappa + \kappa = \kappa \cdot 2 \leq \kappa \cdot \kappa = \kappa$$

and

$$\kappa \leq \kappa \cdot \lambda \leq \kappa \cdot \kappa = \kappa.$$

Hence equality holds throughout. ⊣

Example We do not have a well-defined subtraction operation for infinite cardinal numbers. If you start with $\aleph_0$ things and take away $\aleph_0$ things, then the number of remaining things can be anywhere from 0 to $\aleph_0$, depending on which items were removed. But if you start with κ things (where κ is infinite) and remove λ things where λ is *strictly* less than κ, then exactly κ things remain. To prove this, let μ be the cardinality of the remaining items. Then $\kappa = \lambda + \mu = \max(\lambda, \mu)$, so that we must have $\mu = \kappa$.

Example There are exactly $2^{\aleph_0}$ transcendental numbers. This follows from the preceding example. If from the $2^{\aleph_0}$ real numbers we remove the countable set of algebraic numbers, then $2^{\aleph_0}$ numbers remain.

Example For any infinite cardinal κ, we have $\kappa^\kappa = 2^\kappa$. To prove this, observe that

$$\kappa^\kappa \leq (2^\kappa)^\kappa = 2^{\kappa \cdot \kappa} = 2^\kappa \leq \kappa^\kappa,$$

whence equality holds throughout.

We conclude this section with two counting problems:

1. How many functions from $\mathbb{R}$ into $\mathbb{R}$ are there?
2. How many of these are continuous?

The first question asks for the cardinality of the set ${}^{\mathbb{R}}\mathbb{R}$. The cardinal number of this set is

$$(2^{\aleph_0})^{2^{\aleph_0}} = 2^{\aleph_0 \cdot (2^{\aleph_0})} = 2^{2^{\aleph_0}}.$$

This last expression cannot be further simplified; it provides the answer to the first question. (As with finite numbers, κ^{λ^μ} means $\kappa^{(\lambda^\mu)}$ and not $(\kappa^\lambda)^\mu$.)

Now consider the second question; let $C(\mathbb{R})$ be the set of continuous functions in ${}^{\mathbb{R}}\mathbb{R}$. It is easy to see that

$$2^{\aleph_0} \leq \operatorname{card} C(\mathbb{R}) \leq 2^{2^{\aleph_0}},$$

but we need an exact answer. We claim that $\operatorname{card} C(\mathbb{R}) = 2^{\aleph_0}$. To prove this, we will consider the restriction of the continuous functions to the set $\mathbb{Q}$ of rational numbers (where $\mathbb{Q}$ is regarded as a subset of $\mathbb{R}$). If f and g are two distinct continuous functions, then $f - g$ is not identically zero. (Here $f - g$ is the result of subtracting g from f, not the relative complement.) Hence by continuity, there is an open interval throughout which $f - g$ is nonzero. In this interval lies some rational, so $f \restriction \mathbb{Q} \neq g \restriction \mathbb{Q}$. Hence the map from $C(\mathbb{R})$ into ${}^{\mathbb{Q}}\mathbb{R}$ assigning to each continuous function f its restriction $f \restriction \mathbb{Q}$ is a one-to-one map. Thus $C(\mathbb{R}) \preccurlyeq {}^{\mathbb{Q}}\mathbb{R}$ and so

$$\operatorname{card} C(\mathbb{R}) \leq \operatorname{card} {}^{\mathbb{Q}}\mathbb{R} = (2^{\aleph_0})^{\aleph_0} = 2^{\aleph_0}.$$

Exercises

31. In the proof of Lemma 6R we utilized a certain set $\mathscr{H}$. How do we show from the axioms that such a set exists?

32. Let $\mathscr{F}A$ be the collection of all finite subsets of A. Show that if A is infinite, then $A \approx \mathscr{F}A$.

33. Assume that A is an infinite set. Prove that $A \approx \operatorname{Sq}(A)$.

34. Assume that $2 \leq \kappa \leq \lambda$ and λ is infinite. Show that $\kappa^\lambda = 2^\lambda$.

35. Find a collection $\mathscr{A}$ of $2^{\aleph_0}$ sets of natural numbers such that any two distinct members of $\mathscr{A}$ have finite intersection. [*Suggestion*: Start with the collection of infinite sets of primes.]

36. Show that for an infinite cardinal κ, we have $\kappa! = 2^\kappa$, where $\kappa!$ is defined as in Exercise 14.

CONTINUUM HYPOTHESIS

We have in this chapter given some examples of countable sets and uncountable sets. But every uncountable set examined thus far has had cardinality $2^{\aleph_0}$ or more. This raises the question: Are there any sets with cardinality between $\aleph_0$ and $2^{\aleph_0}$? The "continuum hypothesis" is the assertion that the answer is negative, i.e., that there is no κ with $\aleph_0 < \kappa < 2^{\aleph_0}$. Or equivalently, the continuum hypothesis can be stated: Every uncountable set of real numbers is equinumerous to the set of all real numbers.

Cantor conjectured that the continuum hypothesis was true. And David Hilbert later published a purported proof. But the proof was incorrect, and more recent work has cast doubt on the continuum hypothesis. In 1939 Gödel proved that on the basis of our axioms for set theory (which we here assume to be consistent) the continuum hypothesis could not be *disproved*. Then in 1963 Paul Cohen showed that the continuum hypothesis could not be *proved* from our axioms either.

But since the continuum hypothesis is neither provable nor refutable from our axioms, what can we say about its truth or falsity? We have some informal ideas about what sets are like, but our intuition might not assign a definite answer to the continuum hypothesis. Indeed, one might well question whether there is any meaningful sense in which one can say that the continuum hypothesis is either true *or* false for the "real" sets. Among those set-theorists nowadays who feel that there is such a meaningful sense, the majority seems to feel that the continuum hypothesis is false.

The "generalized continuum hypothesis" is the assertion that for every infinite cardinal κ, there is no cardinal number between κ and 2^κ. Gödel's 1939 work shows that even the generalized continuum hypothesis cannot be disproved from our axioms. And of course Cohen's result shows that it cannot be proved from our axioms (even in the special case $\kappa = \aleph_0$).

There is the possibility of extending the list of axioms beyond those in this book. And the new axioms might conceivably allow us to prove or to refute the continuum hypothesis. But to be acceptable as an *axiom*, a statement must be in clear accord with our informal ideas of the concepts being axiomatized. It would not do, for example, simply to adopt the generalized continuum hypothesis as a new axiom. It remains to be seen whether any acceptable axioms will be found that settle satisfactorily the correctness or incorrectness of the continuum hypothesis.

The work of Gödel and Cohen also shows that the axiom of choice can neither be proved nor refuted from the other axioms (which we continue to assume are consistent). But unlike the continuum hypothesis, the axiom of choice conforms to our informal view of how sets should behave. For this reason, we have adopted it as an axiom.

Results such as those by Gödel and Cohen belong to the *metamathematics* of set theory. That is, they are results that speak of set theory itself, in contrast to theorems within set theory that speak of sets.

CHAPTER 7

ORDERINGS AND ORDINALS

In this chapter we will begin by discussing both linear orderings (mentioned briefly in Chapter 3) and, more generally, partial orderings. But we will soon focus our attention on a special case of the linear orderings, namely the so-called well orderings. The well orderings will lead us to the study of ordinal numbers and to the fulfillment of promises (e.g., the definition of cardinal numbers) that have been made.

PARTIAL ORDERINGS

A partial ordering is a special sort of relation. Before making any definitions, it will be helpful to consider a few examples.

1. Let S be any fixed set, and let $\subset_S$ be the relation of strict inclusion on subsets of S:

$$\subset_S = \{\langle A, B\rangle \mid A \subseteq B \subseteq S \;\&\; A \neq B\}.$$

Of course we write "$A \subset_S B$" in place of the more awkward "$\langle A, B\rangle \in \subset_S$."

2. Let P be the set of positive integers. The strict divisibility relation on P is

$$\{\langle a, b\rangle \in P \times P \mid a \cdot q = b \text{ for some } q \neq 1\}.$$

For example, 5 divides 60 (here $q = 12$), but 2 does not divide 3, nor does 3 divide 2. No number in P strictly divides itself.

3. For the set $\mathbb{R}$ of real numbers, we have the usual ordering relation $<$. For any distinct real numbers x and y, either $x < y$ or $y < x$.

Definition A *partial ordering* is a relation R meeting the following two conditions:

(a) R is a transitive relation: $xRy \;\&\; yRz \Rightarrow xRz$.
(b) R is irreflexive: It is never the case that xRx.

In the foregoing examples, it is easy to see that $\subset_s$, strict divisibility, and $<$ are all partial ordering relations. The preferred symbols for partial ordering relations are $<$ and similar symbols, e.g., $\prec$, $\subset$, and the like. If $<$ is such a relation, then we can define:

$$x \leq y \quad \text{iff} \quad \text{either} \quad x < y \quad \text{or} \quad x = y.$$

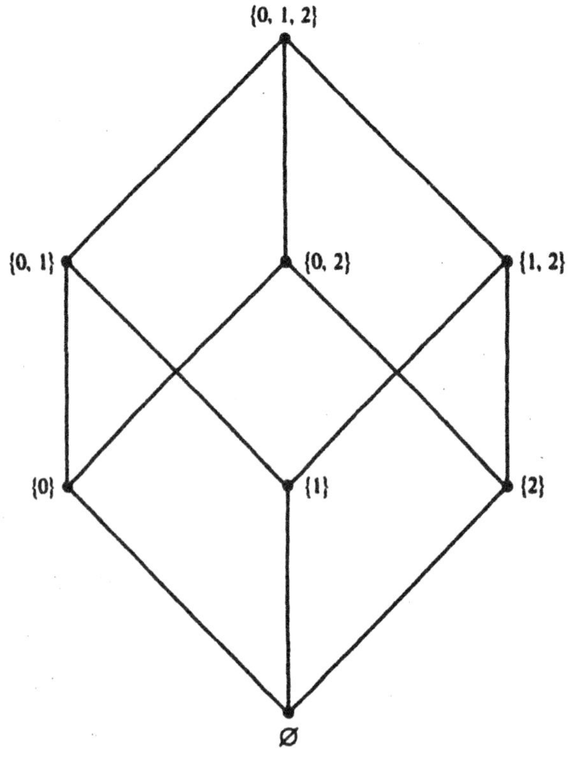

Fig. 43. The inclusion ordering on $\{0, 1, 2\}$.

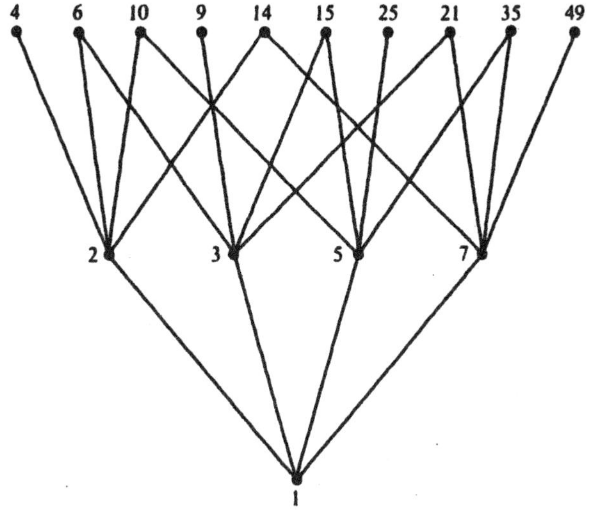

Fig. 44. A finite part of an infinite picture.

It is easy to see that, for example,

$$x \leq y < z \quad \Rightarrow \quad x < z.$$

At the end of Chapter 3 we drew some pictures (Fig. 15) of orderings. For partial orderings the pictures show more variety. Again we represent the members of A by dots, placing the dot for x below the dot for y whenever $x < y$, and drawing lines to connect the dots. For example, Fig. 43 is a picture of the partial ordering $\subset_S$ on the set $S = \{0, 1, 2\}$. There is no need to add a direct nonstop line from $\varnothing$ to $\{0, 1\}$, because there are lines from $\varnothing$ to $\{0\}$ and from $\{0\}$ to $\{0, 1\}$. Figure 44 is a picture of a small piece of the divisibility ordering on the positive integers; the full picture would have to be infinite.

The following theorem lists some easy consequences of our definitions.

Theorem 7A Assume that $<$ is a partial ordering. Then for any x, y, and z:

(a) *At most* one of the three alternatives,

$$x < y, \qquad x = y, \qquad y < x,$$

can hold.

(b) $x \leq y \leq x \Rightarrow x = y$.

Proof In part (a), if we had both $x < y$ and $x = y$, then we would have $x < x$, contradicting irreflexivity. And if both $x < y$ and $y < x$, then by transitivity $x < x$, again contradicting irreflexivity.

In part (b) if $x \neq y$ then we would have $x < y < x$, contradicting part (a). ⊣

We are particularly interested in cases where part (a) of this theorem can be strengthened to read "exactly one," i.e., trichotomy holds.

Definition R is a *linear ordering* on A iff R is a binary relation on A that is a transitive relation and satisfies trichotomy on A, i.e., for any x and y in A exactly one of the three alternatives,

$$xRy, \qquad x = y, \qquad yRx,$$

holds.

Note that we have to specify the set A in this definition. We can also speak of a partial ordering on A, this being just a binary relation on A that is a partial ordering.

Examples The usual ordering relation on $\mathbb{R}$ is linear. Strict divisibility is a partial ordering, but is not a linear ordering on the positive integers. And inclusion $\subset_S$ is nonlinear if S has two or more members.

Notice that if R is a linear ordering on A, then R is also a partial ordering (since trichotomy implies irreflexivity—recall Theorem 3R).

Digression In the study of partial orderings, there is always the question whether to use strict orderings ($<$) or weak orderings ($\leq$) as the basic concept. We have taken the "$<$" course, by demanding that a partial ordering be irreflexive. For the "$\leq$" version, one demands that a partial ordering on A be reflexive on A. Yet a third course is to specify neither extreme, but to allow any pairs $\langle x, x\rangle$ to belong to the ordering or not, as they please. Each alternative has its own minor advantages and its own drawbacks. One feature of demanding reflexivity is that whenever $\leq$ is a partial ordering on A, then $A = \operatorname{fld} \leq$. Consequently we can treat just the ordering $\leq$; the set A is then encoded inside this relation. For strict orderings this feature is lost; for example, $\varnothing$ is a partial ordering on *any* set. Often we will want to refer to a set A *together with* an ordering $<$ on A.

Definition A *structure* is a pair $\langle A, R\rangle$ consisting of a set A and a binary relation R on A (i.e., $R \subseteq A \times A$).

In particular, we can speak of a *partially* (or *linearly*) *ordered structure* if R is a partial (or linear) ordering relation on A. (The terms *poset* and *loset* are sometimes used here.)

There is a certain amount of terminology for orderings that has proved sufficiently useful to have become standard. Let $<$ be a partial ordering; let D be a set. An element m of D is said to be a *minimal* element of D

iff there is no x in D with $x < m$ (or equivalently, iff for all x in D, $x \nless m$). And m is a *least* (or *smallest* or *minimum*) element of D iff $m \leq x$ for all x in D. A least element is also minimal. For a linear ordering on a set that includes D the two concepts coincide, since

$$x \nless m \quad \Rightarrow \quad m \leq x$$

for a linear ordering. But in the nonlinear case, minimality is weaker than leastness.

Example Consider the strict divisibility relation on the set P of positive integers. Then 1 is the least element of P. But let D be $\{a \in P \mid a \neq 1\}$. Then every prime is a minimal element of D, and D has no least element.

A least element of a set (if one exists) is automatically unique. (If m_1 and m_2 are both least, then $m_1 \leq m_2$ and $m_2 \leq m_1$, whence equality holds.) In this event, the least element is the only minimal element. But a set can have many minimal elements, as the above example shows. It is also possible for a set to have no minimal elements at all. (What is an example of such a set?)

Similarly an element m of D is a *maximal* element of D iff there is no x in D with $m < x$ (or equivalently, iff for all x in D, $m \nless x$). And m is a *greatest* (or *largest* or *maximum*) element of D iff $x \leq m$ for all x in D. The remarks we have made concerning minimal and least elements carry over to maximal and greatest elements. That is, a greatest element is also maximal. For a linear ordering (on a set that includes D) the two concepts coincide. A set can have only one greatest element. But if a set has no greatest element, it can have many maximal elements (or it might have none).

If R is a partial ordering, then R^{-1} is also a partial ordering (Exercise 2). And for a set D, the minimal elements of D with respect to R^{-1} are exactly the maximal elements with respect to R. When it is necessary to specify the ordering, we can speak of "R-minimal" elements.

Suppose that $<$ is a partial ordering on A and consider a subset C of A. An *upper bound* of C is an element $b \in A$ such that $x \leq b$ for all $x \in C$. Here b may or may not belong to C; if it belongs to C, then it is clearly the greatest element of C. If b is the least element of the set of all upper bounds for C, then b is the *least upper bound* (or *supremum*) of C. *Lower bound* and *greatest lower bound* (or *infimum*) are defined analogously.

Example Consider a fixed set S and the partial ordering $\subset_S$ on $\mathscr{P}S$. For A and B in $\mathscr{P}S$, the set $\{A, B\}$ has a least upper bound (with respect to $\subset_S$), namely $A \cup B$. Similarly $A \cap B$ is the greatest lower bound of $\{A, B\}$. If $\mathscr{A} \subseteq \mathscr{P}S$, then $\bigcup\mathscr{A}$ is the least upper bound of $\mathscr{A}$. And $\bigcap\mathscr{A}$ (if $\mathscr{A} \neq \varnothing$) is the greatest lower bound.

Example Next consider the set $\mathbb{Q}$ of rational numbers with its usual linear ordering relation $<$. The set of positive rational numbers has no upper bounds at all. The set $\{x \in \mathbb{Q} \mid x^2 < 2\}$ has upper bounds in $\mathbb{Q}$, but no least upper bound in $\mathbb{Q}$.

Example In the set of positive integers ordered by divisibility, greatest lower bounds are called "greatest common divisors" (g.c.d.'s). And least upper bounds are called "least common multiples" (l.c.m.'s). In this ordering, any finite set has an l.c.m., but infinite sets have no upper bounds. Any nonempty set of positive integers has a g.c.d.

Exercises

1. Assume that $<_A$ and $<_B$ are partial orderings on A and B, respectively, and that f is a function from A into B satisfying

$$x <_A y \Rightarrow f(x) <_B f(y)$$

for all x and y in A.
   (a) Can we conclude that f is one-to-one?
   (b) Can we conclude that

$$x <_A y \Leftrightarrow f(x) <_B f(y)?$$

2. Assume that R is a partial ordering. Show that R^{-1} is also a partial ordering.
3. Assume that S is a finite set having n elements. Show that a linear ordering on S contains $\frac{1}{2}n(n-1)$ pairs.

WELL ORDERINGS

Definition A *well ordering*[1] on A is a linear ordering on A with the further property that every nonempty subset of A has a least element.

For example, we proved in Chapter 4 that the usual ordering on ω is a well ordering. But the usual ordering on the set $\mathbb{Z}$ of integers is not a well ordering; $\mathbb{Z}$ has no least element.

Well orderings are significant because they can be used to index constructions that proceed "from the bottom up," where at every stage in the construction (except the last) there is a unique next step. This idea is

[1] This is a grammatically unfortunate phrase, in that it uses "well" as an adjective. Some authors insert a hyphen, which helps a little, but not much.

made precise in the transfinite recursion theorem, which we will encounter presently.

We can very informally describe what well orderings are like; later we will give exact statements and proofs. Assume that $<$ is a well ordering on A. Then the set A itself, if nonempty, has a least element t_0. And then what is left, $A - \{t_0\}$, has (if it is nonempty) a least element t_1. Next consider $A - \{t_0, t_1\}$, and so forth. We obtain

$$t_0 < t_1 < t_2 < \cdots.$$

There may be still more in $A - \{t_0, t_1, \ldots\}$, in which case there is a least element t_ω. And this might continue:

$$t_0 < t_1 < \cdots < t_\omega < t_{\omega+1} < \cdots < t_{\omega\cdot 2} < t_{\omega\cdot 2+1} < \cdots.$$

Eventually we use up all the elements of A, and things grind to a halt.

We can use the axiom of choice to obtain another way of characterizing the well orderings.

Theorem 7B Let $<$ be a linear ordering on A. Then it is a well ordering iff there does *not* exist any function $f: \omega \to A$ with $f(n^+) < f(n)$ for every $n \in \omega$.

A function $f: \omega \to A$ for which $f(n^+) < f(n)$ (for all $n \in \omega$) is sometimes called a *descending chain*, but this terminology should not be confused with other uses we make of the word "chain." The theorem asserts that, for a linear ordering $<$, it is a well ordering iff there are no descending chains.

Proof If there is a descending chain f, then $\operatorname{ran} f$ is a nonempty subset of A. And clearly $\operatorname{ran} f$ has no least element; for each element $f(n)$ there is a smaller element $f(n^+)$. Hence the existence of a descending chain implies that $<$ is not a well ordering.

Conversely, assume that $<$ is not a well ordering on A, so that some nonempty $B \subseteq A$ lacks a least element. Then $(\forall x \in B)(\exists y \in B)\, y < x$. By Exercise 20 of Chapter 6, there is a descending chain $f: \omega \to B$ with $f(n^+) < f(n)$ for each n in ω. ⊣

If $<$ is some sort of ordering on A (at least a partial ordering) and $t \in A$, then the set

$$\operatorname{seg} t = \{x \mid x < t\}$$

is called the *initial segment up to t*. (A less ambiguous notation would be $\operatorname{seg}_< t$, but in practice the simpler notation suffices.) For example, ω was ordered by $\in$, and consequently we had for $n \in \omega$,

$$\operatorname{seg} n = \{x \mid x \in n\} = n.$$

Transfinite Induction Principle Assume that $<$ is a well ordering on A. Assume that B is a subset of A with the special property that for every t in A,

$$\operatorname{seg} t \subseteq B \quad \Rightarrow \quad t \in B.$$

Then B coincides with A.

Before proving this, let us define a subset B of A to be a *$<$-inductive* subset of A iff it has this special property, i.e., for every t in A,

$$\operatorname{seg} t \subseteq B \quad \Rightarrow \quad t \in B.$$

This condition, in words, states that t's membership in B is guaranteed once it is known that all things less than t are already in B. The transfinite induction principle can be restated: If $<$ is a well ordering on A, then any $<$-inductive subset of A must actually coincide with A.

Proof If B is a proper subset of A, then $A - B$ has a least element m. By the leastness, $y \in B$ for any $y < m$. But this is to say that $\operatorname{seg} m \subseteq B$, so by assumption $m \in B$ after all. ⊣

The above proof is so short that we might take a second glance at the situation. First take the least element t_0 of A. Then $\operatorname{seg} t_0 = \varnothing$, and so automatically $\operatorname{seg} t_0 \subseteq B$. The assumption on B then tells us that $t_0 \in B$. Next we proceed to the least element t_1 of $A - \{t_0\}$. Then $\operatorname{seg} t_1 = \{t_0\} \subseteq B$, so we obtain $t_1 \in B$. And so forth and so forth. But to make the "and so forth" part secure, we have the foregoing actual proof of the principle.

The transfinite induction principle is a generalization of the strong induction principle for ω, encountered at the end of Chapter 4. The following theorem is somewhat of a converse to transfinite induction. It asserts that the *only* linear orderings for which the transfinite induction principle is valid are the well orderings. Although we make no later use of this theorem, it is of interest in showing why one might choose to study well orderings.

Theorem 7C Assume that $<$ is a linear ordering on A. Further assume that the only $<$-inductive subset of A is A itself. That is, assume that for any $B \subseteq A$ satisfying the condition

$$(\text{☆}) \qquad (\forall t \in A)(\operatorname{seg} t \subseteq B \quad \Rightarrow \quad t \in B),$$

we have $B = A$. Then $<$ is a well ordering on A.

Proof Let C be any subset of A; we will show that either C has a least element or C is empty. We decide (for reasons that may seem mysterious at the moment) to consider the set B of "strict lower bounds" of C:

$$B = \{t \in A \mid t < x \text{ for every } x \in C\}.$$

Note that $B \cap C = \varnothing$, lest we have $t < t$. We ask ourselves whether condition (☆) holds for B.

Case I Condition (☆) fails. Then there exists some $t \in A$ with $\text{seg } t \subseteq B$ but $t \notin B$. We claim that t is a least element of C. Since $t \notin B$, there is some $x \in C$ with $x \leq t$. But x cannot belong to $\text{seg } t$, which is disjoint from C. Thus $x = t$ and $t \in C$. And t is least in C, since anything smaller than t is in $\text{seg } t$ and hence not in C.

Case II Condition (☆) holds. Then by the hypothesis of the theorem, $A = B$. Consequently in this case C is empty. ⊣

Next we turn to the important business of defining a function on a well-ordered structure by transfinite recursion. Assume that $<$ is a well ordering on A. Conceivably we might possess some rule for defining a function value $F(t)$ at $t \in A$, where the rule requires knowing first all values $F(x)$ for $x < t$. Then we can start with the least element t_0 of A, apply our rule to find $F(t_0)$, go on the next element, and so forth. That phrase "and so forth" has to cover a great deal of ground. But, because $<$ is a well ordering, after defining F on any proper subset B of A, we always have a unique next element (the least element of $A - B$) to which we can apply our rule so as to continue.

Now let us try to make these ideas more concrete. The "rule" of the preceding paragraph might be provided by a function G. Then $F(t)$ for $t \in A$ is to be found by applying G to the values $F(x)$ for $x < t$:

$$F(t) = G(F \restriction \text{seg } t).$$

We will say that F is "G-constructed" if the above equation holds for every $t \in \text{dom } F$. For the right side of this equation to succeed, the domain of G must contain all functions of the form $F \restriction \text{seg } t$. This leads us to define, for a set B, the set ${}^{<A}B$ of all functions from initial segments of $<$ into B:

$${}^{<A}B = \{f \mid \text{for some } t \in A, f \text{ is a function from seg } t \text{ into } B\}.$$

To check that there is indeed such a set, observe that if $f\colon \text{seg } t \to B$, then $f \subseteq A \times B$. Hence ${}^{<A}B$ is obtainable by applying a subset axiom to $\mathscr{P}(A \times B)$.

Transfinite Recursion Theorem, Preliminary Form Assume that $<$ is a well ordering on A, and that $G\colon {}^{<A}B \to B$. Then there is a unique function $F\colon A \to B$ such that for any $t \in A$,

$$F(t) = G(F \restriction \text{seg } t).$$

Example In the case of the well-ordered set ω, we have for each n in ω the equation seg $n = \{x \mid x \in n\} = n$. Hence the above theorem asserts the existence of a unique $F: \omega \to B$ satisfying for every n in ω

$$F(n) = G(F \upharpoonright n).$$

In particular, we have

$$\begin{aligned} F(0) &= G(F \upharpoonright 0) = G(\varnothing), \\ F(1) &= G(F \upharpoonright 1) = G(\{\langle 0, F(0)\rangle\}), \\ F(2) &= G(F \upharpoonright 2) = G(\{\langle 0, F(0)\rangle, \langle 1, F(1)\rangle\}). \end{aligned}$$

This can be compared with the recursion theorem of Chapter 4. There is the difference that in Chapter 4 the value of $F(n)$ was constructed by using only the one immediately preceding value of the function. But now in constructing $F(n)$ we can use all of the previous values, which are given to us by $F \upharpoonright n$. (This alteration is not made solely for reasons of generosity. It is forced on us by the fact that in an arbitrary well ordering, there may not always be an "immediately preceding" element.)

Before proving the transfinite recursion theorem, we want to state it in a stronger form. In our informal comments we spoke of a "rule" for forming $F(t)$ from the restriction $F \upharpoonright \operatorname{seg} t$. We want to be broad-minded and to allow the case in which the rule is *not* given by a function. We have in mind such rules as

$$F(t) = \{F(x) \mid x < t\} = \operatorname{ran}(F \upharpoonright \operatorname{seg} t).$$

There is no function G such that $G(a) = \operatorname{ran} a$ for *every* set a; such a function would have to have everything in its domain, but its domain is merely a set. To be sure, for any fixed set B there is indeed a function G with $G(a) = \operatorname{ran} a$ for every $a \in B$. But our desire is to avoid having to produce in advance the set B.

In terms of proper classes, we can formulate the improved version of transfinite recursion as follows. Let $<$ be a well ordering on A and let G be a "function-class," i.e., a class of ordered pairs that satisfies the definition of a function except for not being a set. Further suppose that the domain of G includes ${}^{<A}\mathsf{V}$, where ${}^{<A}\mathsf{V}$ is the *class*

$$\{f \mid \text{for some } t \text{ in } A, f \text{ is a function with dom } f = \operatorname{seg} t\}.$$

Then there exists a unique function F with domain A that is "G-constructed" in the sense that

$$F(t) = \mathsf{G}(F \upharpoonright \operatorname{seg} t)$$

for all $t \in A$. (Here F *is* a set.)

Because we are working in Zermelo–Fraenkel set theory, we must reword the above formulation of transfinite recursion so as to avoid reference to the class G. Instead we allow ourselves only a formula $\gamma(x, y)$ that defines G:

$$\mathsf{G} = \{\langle x, y\rangle \mid \gamma(x, y)\}.$$

Applying this rewording (the standard rewording for these predicaments), we obtain the statement below.

Transfinite Recursion Theorem Schema For any formula $\gamma(x, y)$ the following is a theorem:

Assume that $<$ is a well ordering on a set A. Assume that for any f there is a unique y such that $\gamma(f, y)$. Then there exists a unique function F with domain A such that

$$\gamma(F \upharpoonright \operatorname{seg} t, F(t))$$

for all t in A.

This is a theorem *schema*; that is, it is an infinite package of theorems, one for each formula $\gamma(x, y)$. (The formula $\gamma(x, y)$ is allowed to mention other fixed sets in addition to x and y.) We will say that F is *γ-constructed* if the condition $\gamma(F \upharpoonright \operatorname{seg} t, F(t))$ holds for every t in dom F.

Example We obtain one instance of transfinite recursion by taking $\gamma(x, y)$ to be: $y = \operatorname{ran} x$. Now it is automatically true that for any f there is a unique y such that $y = \operatorname{ran} f$. So we are left with the theorem:

Assume that $<$ is a well ordering on A. Then there exists a unique function with domain A such that

$$F(t) = \operatorname{ran}(F \upharpoonright \operatorname{seg} t)$$

for all t in A.

This is the instance of transfinite recursion we will need in order to make ordinal numbers.

Example The preliminary form of transfinite recursion is obtained by choosing the formula $\gamma(x, y)$:

$$\gamma(x, y) \iff \begin{array}{ll} \text{either} & \text{(i)}\quad x \in {}^{<A}B \text{ and } y = \mathsf{G}(x) \\ \text{or} & \text{(ii)}\quad x \notin {}^{<A}B \text{ and } y = \varnothing. \end{array}$$

Then for any f there is a unique y such that $\gamma(f, y)$. We can conclude from the transfinite recursion theorem that there is a unique function F with domain A that is γ-constructed: For $t \in A$, either (i) $(F \upharpoonright \operatorname{seg} t) \in {}^{<A}B$

and $F(t) = G(F \upharpoonright \text{seg } t)$ or (ii) $(F \upharpoonright \text{seg } t) \notin {}^{<A}B$ and $F(t) = \varnothing$. Since G: ${}^{<A}B \to B$, we can see (by transfinite induction) that we have alternative (i) for every t in A. Thus F is G-constructed.

One momentary drawback to the above statement of transfinite recursion is that we are unable to prove it with the axioms now at our disposal. Actually this drawback is shared by a number of other statements that are intuitively true under our informal ideas about sets. For example, we are unable to prove that

$$\varnothing \cup \mathscr{P}\varnothing \cup \mathscr{P}\mathscr{P}\varnothing \cup \mathscr{P}\mathscr{P}\mathscr{P}\varnothing \cup \cdots$$

is a set. In fact we cannot prove the existence of any inductive set other than ω. These deficiencies will be eliminated in the next section by the replacement axioms.

Exercises

4. Let $<$ be the usual ordering on the set P of positive integers. For n in P, let $f(n)$ be the number of distinct prime factors of n. Define the binary relation R on P by

$$mRn \iff \text{either } f(m) < f(n) \text{ or } [f(m) = f(n) \;\&\; m < n].$$

Show that R is a well ordering on P. Does $\langle P, R\rangle$ resemble any of the pictures in Fig. 45 (p. 185)?

5. Assume that $<$ is a well ordering on A, and that f: $A \to A$ satisfies the condition

$$x < y \Rightarrow f(x) < f(y)$$

for all x and y in A. Show that $x \leq f(x)$ for all x in A. [*Suggestion*: Consider $f(f(x))$.]

6. Assume that S is a subset of the real numbers that is well ordered (under the usual ordering on reals). Show that S is countable. [*Suggestion*: For each x in S, choose a rational number between x and the next member of S, if any.]

7. Let C be some fixed set. Apply transfinite recursion to ω (with its usual well ordering), using for $\gamma(x, y)$ the formula

$$y = C \cup \bigcup\bigcup \operatorname{ran} x.$$

Let F be the γ-constructed function on ω.

(a) Calculate $F(0)$, $F(1)$, and $F(2)$. Make a good guess as to what $F(n)$ is.

(b) Show that if $a \in F(n)$, then $a \subseteq F(n^+)$.

(c) Let $\bar{C} = \bigcup \operatorname{ran} F$. Show that $\bar{C}$ is a transitive set and that $C \subseteq \bar{C}$. (The set $\bar{C}$ is called the *transitive closure* of C, denoted TC C.)

REPLACEMENT AXIOMS

If H is a function and A is a set, then $H[\![A]\!]$ is a set, simply because it is included in ran H. (Lemma 3D guarantees that ran H is a set.) But now consider a "function-class" H, i.e., a *class* of ordered pairs that satisfies the definition of being a function, except that it may not be a set. Is it still true that $\mathsf{H}[\![A]\!]$ is a set? With the axioms stated thus far, we are unable to prove that it is. (The unprovability of this can actually be *proved*; we will return to this point in Chapter 9.) But certainly $\mathsf{H}[\![A]\!]$ cannot be too big to be a set, for it is no larger than the set A. So if we adopt the principle that what distinguishes the sets from the proper classes is the property of being limited in size, then it is eminently reasonable to adopt an axiom that, in a way, asserts that $\mathsf{H}[\![A]\!]$ is a set. Now the axiom in Zermelo–Fraenkel set theory cannot legally refer to a class H; it must instead involve a formula φ that defines H. (You will recall that there was a similar situation in the case of the subset axioms.) For each formula that might define a function-class, we get an axiom.

Replacement Axioms For any formula $\varphi(x, y)$ not containing the letter B, the following is an axiom:

$$\begin{aligned}\forall A[(\forall x \in A)\, \forall y_1\, \forall y_2(\varphi(x, y_1)\ \&\ \varphi(x, y_2) \ \Rightarrow\ y_1 = y_2)\\ \Rightarrow\ \exists B\, \forall y(y \in B \ \Leftrightarrow\ (\exists x \in A)\, \varphi(x, y))].\end{aligned}$$

Here the formula $\varphi(x, y)$ is permitted to name sets other than x and y; we could have emphasized this fact by writing $\varphi(x, y, t_1, \ldots, t_k)$ and inserting the phrase "$\forall t_1 \cdots \forall t_k$" at the beginning of the axiom. But $\varphi(x, y)$ must not mention the set B whose existence is being asserted by the axiom.

To translate the replacement axioms into words, define the class

$$\mathsf{H} = \{\langle x, y\rangle \mid x \in A\ \&\ \varphi(x, y)\}.$$

Then the hypothesis of the axiom,

$$(\forall x \in A)\, \forall y_1\, \forall y_2(\varphi(x, y_1)\ \&\ \varphi(x, y_2) \ \Rightarrow\ y_1 = y_2),$$

asserts that H is a function-class. And the second line,

$$\exists B\, \forall y(y \in B \ \Leftrightarrow\ (\exists x \in A)\, \varphi(x, y)),$$

asserts that if we let

$$B = \{y \mid (\exists x \in A)\, \varphi(x, y)\} = \mathsf{H}[\![A]\!],$$

then B is a set. (It then follows that H, being included in $A \times B$, is also a set.)

An alternative way of translating replacement into words is as follows. Read $\varphi(x, y)$ as "x nominates y." Then the hypothesis of the axiom says, "Each member of A nominates at most one object." And the conclusion says, "The collection of all nominees is a set."

The name "replacement" reflects the idea of replacing each x in the set A by its nominee (if any) to obtain the set B.

Example If A is a set, then $\{\mathscr{P}a \mid a \in A\}$ is also a set, by Exercise 10 of Chapter 2. But we now have an easy proof of this fact. Take $\varphi(x, y)$ to be $y = \mathscr{P}x$. That is, let each x nominate its own power set. Then replacement tells us that the collection of all power sets of members of A forms a set.

Example Let S be any set. Then replacement tells us that

$$\{\text{card } x \mid x \in S\}$$

is a set. We take the formula $\varphi(x, y)$ to be $y = \text{card } x$.

Proof of the Transfinite Recursion Theorem The proof is similar, in its general outline, to the proof of the recursion theorem on ω. Again we construct the desired function F as the union of many approximating functions. For t in A, say that a function v is *γ-constructed up to t* iff $\text{dom } v = \{x \mid x \preccurlyeq t\}$ and for any x in dom v,

$$\gamma(v \upharpoonright \text{seg } x, v(x)).$$

1. First we claim that if $t_1 \preccurlyeq t_2$, v_1 is γ-constructed up to t_1 and v_2 is γ-constructed up to t_2, then $v_1(x) = v_2(x)$ for all $x \preccurlyeq t_1$. Should this fail, there is a least $x \preccurlyeq t_1$ with $v_1(x) \neq v_2(x)$. By the leastness of x, we have $v_1 \upharpoonright \text{seg } x = v_2 \upharpoonright \text{seg } x$. Also we have

$$\gamma(v_1 \upharpoonright \text{seg } x, v_1(x)) \qquad \text{and} \qquad \gamma(v_2 \upharpoonright \text{seg } x, v_2(x)),$$

whence by our assumption on γ we conclude that $v_1(x) = v_2(x)$ after all. This establishes the claim.

In particular, by taking $t_1 = t_2$ we see that for any $t \in A$ there is at most one function v that is γ-constructed up to t. We next want to form the set $\mathscr{K}$ of all functions v that are, for some t in A, γ-constructed up to t:

$$\mathscr{K} = \{v \mid (\exists t \in A)\ v \text{ is a function } \gamma\text{-constructed up to } t\}.$$

$\mathscr{K}$ is provided by a replacement axiom. Take for $\varphi(t, v)$ the formula

$$v \text{ is a function that is } \gamma\text{-constructed up to } t.$$

We have shown that

$$t \in A \;\&\; \varphi(t, v_1) \;\&\; \varphi(t, v_2) \quad \Rightarrow \quad v_1 = v_2.$$

Hence by replacement there is a set $\mathscr{K}$ such that for any v,

$$\begin{aligned} v \in \mathscr{K} \quad &\Leftrightarrow \quad (\exists t \in A)\, \varphi(t, v) \\ &\Leftrightarrow \quad (\exists t \in A)\, v \text{ is } \gamma\text{-constructed up to } t. \end{aligned}$$

Now let F be $\bigcup \mathscr{K}$, the union of all the v's. Thus

$$(\text{☆}) \qquad \langle x, y\rangle \in F \quad \Leftrightarrow \quad v(x) = y \quad \text{for some } v \text{ in } \mathscr{K}.$$

First observe that F is a function. For suppose that $\langle x, y_1\rangle$ and $\langle x, y_2\rangle$ belong to F. By (☆) there exist v_1, t_1, v_2, and t_2 such that $v_i(x) = y_i$ and v_i is γ-constructed up to t_i for $i = 1, 2$. Either $x \le t_1 \le t_2$ or $x \le t_2 \le t_1$, and in either event we have by our earlier claim $y_1 = v_1(x) = v_2(x) = y_2$.

2. Next we claim that for any $x \in \operatorname{dom} F$, $\gamma(F \restriction \operatorname{seg} x, F(x))$. For if $x \in \operatorname{dom} F$, there exists v in $\mathscr{K}$ with $x \in \operatorname{dom} v$. Then we have

$\gamma(v \restriction \operatorname{seg} x, v(x))$	since $v \in \mathscr{K}$,
$v \restriction \operatorname{seg} x = F \restriction \operatorname{seg} x$	by (☆) and part 1,
$v(x) = F(x)$	by (☆),

from which we conclude that $\gamma(F \restriction \operatorname{seg} x, F(x))$.

3. We now claim that $\operatorname{dom} F = A$. If this fails, then there is a least $t \in A - \operatorname{dom} F$. Then $\operatorname{seg} t \subseteq \operatorname{dom} F$; in fact $\operatorname{seg} t = \operatorname{dom} F$. Take the unique y such that $\gamma(F, y)$ and let

$$v = F \cup \{\langle t, y\rangle\}.$$

We want to show that v is γ-constructed up to t. Clearly v is a function and $\operatorname{dom} v = \{x \mid x \le t\}$. For any $x < t$ we have $v \restriction \operatorname{seg} x = F \restriction \operatorname{seg} x$ and $v(x) = F(x)$, and so by part 2 we conclude that $\gamma(v \restriction \operatorname{seg} x, v(x))$. For the case $x = t$ we have $v \restriction \operatorname{seg} t = F$ and $v(t) = y$, and so by our choice of y we obtain $\gamma(v) \restriction \operatorname{seg} t, v(t))$. Hence v is γ-constructed up to t. But this implies that $t \in \operatorname{dom} F$ after all.

4. Finally we claim that F is unique. For suppose that F_1 and F_2 both satisfy the conclusion of the theorem. We can apply transfinite induction; let B be the set on which F_1 and F_2 agree:

$$B = \{t \in A \mid F_1(t) = F_2(t)\}.$$

It suffices to show that for any $t \in A$,

$$\operatorname{seg} t \subseteq B \quad \Rightarrow \quad t \in B.$$

But this is easy. If $\operatorname{seg} t \subseteq B$, then we have $F_1 \restriction \operatorname{seg} t = F_2 \restriction \operatorname{seg} t$. Also we have

$$\gamma(F_1 \restriction \operatorname{seg} t, F_1(t)) \qquad \text{and} \qquad \gamma(F_2 \restriction \operatorname{seg} t, F_2(t)),$$

whence by our assumption on γ we conclude that $F_1(t) = F_2(t)$. Hence $t \in B$ and we are done. ⊣

Exercises

8. Show that the subset axioms are provable from the other axioms.
9. Show that the pairing axiom is provable from the other axioms.

EPSILON-IMAGES[2]

Our first application of transfinite recursion will be to the construction of the ordinal numbers. Assume that $<$ is a well ordering on A and take for $\gamma(x, y)$ the formula: $y = \operatorname{ran} x$. The transfinite recursion theorem then presents us with a unique function E with domain A such that for any $t \in A$:

$$\begin{aligned} E(t) &= \operatorname{ran}(E \restriction \operatorname{seg} t) \\ &= E[\![\operatorname{seg} t]\!] \\ &= \{E(x) \mid x < t\}. \end{aligned}$$

Let $\alpha = \operatorname{ran} E$; we will call α the *∈-image* of the well-ordered structure $\langle A, <\rangle$. (Later α will also be called an *ordinal number*, but we want to postpone introducing that terminology.)

Example To get some idea of what the ∈-image might look like, take the three-element set $A = \{r, s, t\}$, where $r < s < t$. Then we calculate

$$\begin{aligned} E(r) &= \{E(x) \mid x < r\} = \varnothing, \\ E(s) &= \{E(x) \mid x < s\} = \{E(r)\} = \{\varnothing\}, \\ E(t) &= \{E(x) \mid x < t\} = \{E(r), E(s)\} = \{\varnothing, \{\varnothing\}\}. \end{aligned}$$

Thus $E(r) = 0$, $E(s) = 1$, $E(t) = 2$, and the ∈-image α of $\langle A, <\rangle$ is 3. You are invited to contemplate whether natural numbers will always appear in α like this.

In the above example, α was a set equinumerous with A. Moreover, while $<$ was a well ordering on A, the membership relation

$$\in_\alpha = \{\langle x, y\rangle \in \alpha \times \alpha \mid x \in y\}$$

provided a well ordering on α. We will show that this always happens, whence the name "∈-image."

[2] The membership symbol (∈) is not typographically the letter epsilon but originally it was, and the name "epsilon" persists.

Theorem 7D Let $<$ be a well ordering on A and let E and α be as described as above.

(a) $E(t) \notin E(t)$ for any $t \in A$.
(b) E maps A one-to-one onto α.
(c) For any s and t in A,

$$s < t \quad \text{iff} \quad E(s) \in E(t).$$

(d) α is a transitive set.

Proof We prove that $E(t) \notin E(t)$ by the "least counterexample" method. That is, let S be the set of counterexamples:

$$S = \{t \in A \mid E(t) \in E(t)\}.$$

We hope that S is empty. But if not, then there is a least $\hat{t} \in S$. Since $E(\hat{t}) \in E(\hat{t})$, there is (by the definition of $E(\hat{t})$) some $s < \hat{t}$ with $E(\hat{t}) = E(s)$. But then $E(s) \in E(s)$, contradicting the leastness of $\hat{t}$. Hence $S = \varnothing$, which proves part (a).

For part (b), it is obvious that E maps A onto α; we must prove that E is one-to-one. If s and t are distinct members of A, then one is smaller than the other; assume $s < t$. Then $E(s) \in E(t)$, but, by part (a), $E(t) \notin E(t)$. Hence $E(s) \neq E(t)$.

In part (c) we have

$$s < t \quad \Rightarrow \quad E(s) \in E(t)$$

by definition. Conversely if $E(s) \in E(t)$, then (by the definition of $E(t)$) there is some $x < t$ with $E(s) = E(x)$. Since E is one-to-one, we must have $s = x$ and hence $s < t$.

Finally for part (d) it is easy to see that α is a transitive set. If $u \in E(t) \in \alpha$, then there is some $x < t$ with $u = E(x)$; consequently $u \in \alpha$. ⊣

Define the binary relation on α:

$$\in_\alpha = \{\langle x, y\rangle \in \alpha \times \alpha \mid x \in y\}.$$

Under the assumptions of the preceding theorem, $\in_\alpha$ is a well ordering on α. We will postpone verifying this fact until we can do so in a more general setting (see Corollary 7H). Since α is a transitive set, we can characterize $\in_\alpha$ by the condition

$$x \in_\alpha y \quad \Leftrightarrow \quad x \in y \in \alpha.$$

Exercises

10. For any set S, we can define the relation $\in_S$ by the equation:

$$\in_S = \{\langle x, y\rangle \in S \times S \mid x \in y\}$$

(a) Show that for any natural number n, the $\in$-image of $\langle n, \in_n\rangle$ is n.
(b) Find the $\in$-image of $\langle \omega, \in_\omega\rangle$.

11. (a) Although the set $\mathbb{Z}$ of integers is not well ordered by its normal ordering, show that the ordering

$$0, 1, 2, \ldots, -1, -2, -3, \ldots$$

is a well ordering on $\mathbb{Z}$.
(b) Suppose that we define the usual function E on $\mathbb{Z}$, using the well ordering of part (a). Calculate $E(3)$, $E(-1)$, and $E(-2)$. Describe ran E.

ISOMORPHISMS

Theorem 7D told us that a well-ordered structure $\langle A, <\rangle$ looks a great deal like its $\in$-image $\langle \alpha, \in_\alpha\rangle$. We had a one-to-one correspondence E between A and α with the order-preserving property:

$$s < t \quad \text{iff} \quad E(s) \in E(t)$$

for s and t in A. In formalizing this concept of "looking alike," we need not restrict attention to well orderings. Instead we might as well formulate our definition in broader terms.

Definition Consider structures $\langle A, R\rangle$ and $\langle B, S\rangle$. An *isomorphism* from $\langle A, R\rangle$ onto $\langle B, S\rangle$ is a one-to-one function f from A onto B such that

$$xRy \quad \text{iff} \quad f(x)Sf(y)$$

for x and y in A. If such an isomorphism exists, then we say that $\langle A, R\rangle$ is *isomorphic* to $\langle B, S\rangle$, written: $\langle A, R\rangle \cong \langle B, S\rangle$.

Example Let $<$ be a well ordering on A and let α be the $\in$-image of $\langle A, <\rangle$. Then Theorem 7D states that $\langle A, <\rangle$ is isomorphic to $\langle \alpha, \in_\alpha\rangle$, and that E is an isomorphism.

Example Isomorphic structures really do look alike. We can draw pictures (at least in the simple cases) as in Figs. 43 and 44. Figure 45 consists of pictures of four well-ordered structures. Figure 45a shows a well ordering on $\{r, s, t\}$, and Fig. 45b shows a well ordering on $\{x, y, z\}$. The two structures are isomorphic, and the two pictures are obviously very much alike. Figures 45c and 45d provide pictures of infinite structures. Figure 45c

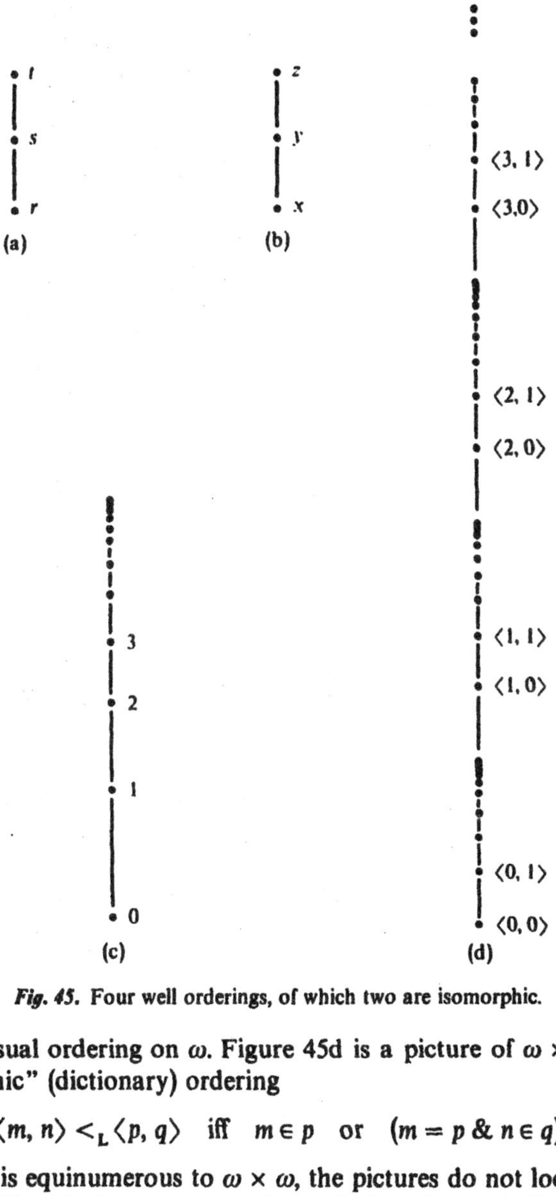

Fig. 45. Four well orderings, of which two are isomorphic.

shows the usual ordering on ω. Figure 45d is a picture of $\omega \times \omega$ with the "lexicographic" (dictionary) ordering

$$\langle m, n\rangle <_L \langle p, q\rangle \quad \text{iff} \quad m \in p \quad \text{or} \quad (m = p \;\&\; n \in q).$$

Although ω is equinumerous to $\omega \times \omega$, the pictures do not look alike. For example, in Fig. 45c the set seg n is always finite, whereas in Fig. 45d the set seg$\langle m, n\rangle$ is infinite for $m \neq 0$. And the two structures are *not* isomorphic.

Actually we have used the word "isomorphic" before, e.g., in Theorem 4H. Both there and here we are using special cases of the general concept of isomorphism of structures. In Theorem 4H we had "structures" $\langle N, S, e\rangle$ consisting of a set N together with a function $S: N \to N$ and an element $e \in N$. Now we have structures $\langle A, R\rangle$, where R is a binary relation on A. We will have no need here for greater generality, but you are nonetheless invited to devise general concepts of structure and isomorphism.

We have the easy theorem stating that isomorphism is an "equivalence concept." That is, it obeys laws of reflexivity, symmetry, and transitivity.

Theorem 7E Let $\langle A, R\rangle$, $\langle B, S\rangle$, and $\langle C, T\rangle$ be any structures. Then

$$\langle A, R\rangle \cong \langle A, R\rangle,$$

$$\langle A, R\rangle \cong \langle B, S\rangle \quad \Rightarrow \quad \langle B, S\rangle \cong \langle A, R\rangle,$$

$$\langle A, R\rangle \cong \langle B, S\rangle \cong \langle C, T\rangle \quad \Rightarrow \quad \langle A, R\rangle \cong \langle C, T\rangle.$$

Proof For the first of these we can use the identity function on A; it is an isomorphism of $\langle A, R\rangle$ onto itself. To prove the second we take the inverse of the given isomorphism; to prove the third we take the composition of the two given isomorphisms. ⊣

If two structures are isomorphic, then for any statement that is true of one, there is a corresponding statement that is true of the other. This is because the structures look exactly alike. But to nail down this general principle in precise terms, we would have to say formally what we meant by "true statement" and "corresponding statement." This would lead us away from set theory into logic. Instead we will just list (in Theorem 7G) some particular instances of the general principle.

Lemma 7F Assume that f is a one-to-one function from A into B, and that $<_B$ is a partial ordering on B. Define the binary relation $<_A$ on A by

$$x <_A y \quad \text{iff} \quad f(x) <_B f(y)$$

for x and y in A.

(a) The relation $<_A$ is a partial ordering on A.
(b) If $<_B$ is a linear ordering on B, then $<_A$ is a linear ordering on A.
(c) If $<_B$ is a well ordering on B, then $<_A$ is a well ordering on A.

Proof The proof is straightforward and is left, for the most part, to the exercises. We will give details here for only the following part: If $<_B$ is a well ordering on B, then any nonempty subset S of A has a least (with respect to $<_A$) element. First note that $f[\![S]\!]$ is a nonempty subset of B. Hence it has a least element, which must be $f(m)$ for some unique $m \in S$. We claim that m is least in S. Let x be any other member of S. Then

$f(x) \in f[\![S]\!]$, and by the leastness of $f(m)$ we have $f(m) \leq_B f(x)$. In fact, $f(m) <_B f(x)$ since f is one-to-one. Hence $m <_A x$, which shows that m is indeed least in S. ⊣

Theorem 7G Assume that structures $\langle A, <_A \rangle$ and $\langle B, <_B \rangle$ are isomorphic. If one is a partially (or linearly or well) ordered structure, so also is the other.

Proof Suppose that one is a partially ordered structure; by symmetry we may suppose that it is $\langle B, <_B \rangle$. We are given an isomorphism f of $\langle A, <_A \rangle$ onto $\langle B, <_B \rangle$. Thus

$$x <_A y \quad \text{iff} \quad f(x) <_B f(y)$$

for x and y in A. Hence we can apply the above lemma. ⊣

Theorem 7D told us that well orderings were isomorphic to their $\in$-images. In light of the above theorem, we can now conclude that the $\in$-images are well ordered by epsilon.

Corollary 7H Assume that $<$ is a well ordering on A and let α be the $\in$-image of $\langle A, < \rangle$. Then α is a transitive set and $\in_\alpha$ is a well ordering on α.

Exercises

12. Complete the proof of Lemma 7F.

13. Assume that two well-ordered structures are isomorphic. Show that there can be only one isomorphism from the first onto the second.

14. Assume that $\langle A, < \rangle$ is a partially ordered structure. Define the function F on A by the equation

$$F(a) = \{x \in A \mid x \leq a\},$$

and let $S = \operatorname{ran} F$. Show that F is an isomorphism from $\langle A, < \rangle$ onto $\langle S, \subset_S \rangle$.

ORDINAL NUMBERS

It will be extremely useful in our discussion of well orderings if we can assign a "number" to each well-ordered structure that measures its "length." Two well-ordered structures should receive the same number iff they look alike, i.e., iff they are isomorphic. The situation is much like that in Chapter 6, where we assigned cardinal numbers to sets so as to measure their size. Two sets received the same cardinal number iff they were equinumerous. The key to the present situation lies in the following theorem.

Theorem 7I Two well-ordered structures are isomorphic iff they have the same $\in$-image. That is, if $<_1$ and $<_2$ are well orderings on A_1 and A_2, respectively, then $\langle A_1, <_1\rangle \cong \langle A_2, <_2\rangle$ iff the $\in$-image of $\langle A_1, <_1\rangle$ is the same as the $\in$-image of $\langle A_2, <_2\rangle$.

Proof In one direction, this follows at once from the fact that well-ordered structures are isomorphic to their $\in$-images. That is, if $\langle A_1, <_1\rangle$ and $\langle A_2, <_2\rangle$ have the same $\in$-image α, then

$$\langle A_1, <_1\rangle \cong \langle \alpha, \in_\alpha\rangle \cong \langle A_2, <_2\rangle.$$

For the other direction, assume that f is an isomorphism from $\langle A_1, <_1\rangle$ onto $\langle A_2, <_2\rangle$. Let $E_1: A_1 \to \alpha_1$ and $E_2: A_2 \to \alpha_2$ be the usual isomorphisms of the well orderings onto their $\in$-images:

$$E_1(s) = \{E_1(x) \mid x <_1 s\} \qquad \text{and} \qquad E_2(t) = \{E_2(y) \mid y <_2 t\}.$$

Then we can use transfinite induction to show that $E_1(s) = E_2(f(s))$ for all $s \in A_1$. For let B be the set on which this equation holds:

$$B = \{s \in A_1 \mid E_1(s) = E_2(f(s))\}.$$

Suppose that seg $s \subseteq B$. Then we calculate

$$\begin{aligned} E_1(s) &= \{E_1(x) \mid x <_1 s\} && \text{by definition} \\ &= \{E_2(f(x)) \mid x <_1 s\} && \text{since seg } s \subseteq B \\ &= \{E_2(y) \mid y <_2 f(s)\} && \text{(see below)} \\ &= E_2(f(s)). \end{aligned}$$

The third line on this calculation requires some thought. Clearly

$$\{E_2(f(x)) \mid x <_1 s\} \subseteq \{E_2(y) \mid y <_2 f(s)\}$$

because we can take $y = f(x)$. Conversely, the other inclusion holds because f maps *onto* A_2 and therefore any y less than $f(s)$ must equal $f(x)$ for some (unique) x less than s.

Thus if seg $s \subseteq B$, then $E_1(s) = E_2(f(s))$ and so $s \in B$. By the transfinite induction principle, $B = A_1$, i.e.,

$$E_1(s) = E_2(f(s))$$

for all $s \in A_1$. Consequently,

$$\begin{aligned} \alpha_1 &= \{E_1(s) \mid s \in A_1\} \\ &= \{E_2(f(s)) \mid s \in A_1\} \\ &= \{E_2(t) \mid t \in A_2\} && \text{since } f \text{ maps } onto\ A_2 \\ &= \alpha_2. \end{aligned}$$

Thus the $\in$-images coincide. ⊣

The solution to our problem of how to assign "numbers" to well orderings is provided by the foregoing theorem. We just use $\in$-images.

Definition Let $<$ be a well ordering on A. The *ordinal number* of $\langle A, <\rangle$ is its $\in$-image. An *ordinal number* is a set that is the ordinal number of some well-ordered structure.

(More generally, we could seek to define an "order type" $\text{ot}\langle A, R\rangle$ whenever R was a *partial* ordering on A. The order type should measure the "shape" of R, in the sense that

$$\text{ot}\langle A, R\rangle = \text{ot}\langle B, S\rangle \quad \text{iff} \quad \langle A, R\rangle \cong \langle B, S\rangle.$$

This can be done; see Exercise 32. But the methods that are required are quite unlike the methods used for ordinal numbers.)

We can generate examples of ordinal numbers by calculating $\in$-images. The example preceding Theorem 7D shows that 3 is an ordinal number. And study of Fig. 45c will show that ω is an ordinal number. More examples will appear presently. First we need some additional information about well orderings.

If $<$ is any ordering on A and C is a subset of A, then the elements of C are, of course, still ordered by $<$. Now it is not strictly true that $<$ is an ordering on C, because (if $C \neq A$) $<$ is not a binary relation *on* C. But $< \cap (C \times C)$ *is* an ordering on C. This notation is excessively cumbersome for the simplicity of the ideas involved. And so we define

$$\langle C, <^\circ\rangle = \langle C, < \cap (C \times C)\rangle.$$

The symbol $^\circ$ reminds us that the relation $<$ can be cut down to fit the set C. In particular, if $C = \text{seg}\, c$, then we have the ordered structure $\langle \text{seg}\, c, <^\circ\rangle$.

Theorem 7J Assume that $<$ is a partial ordering on A, and that $C \subseteq A$. Then $<^\circ$ is a partial ordering on C. This continues to hold with "partial" replaced by "linear" or "well."

Proof This is an immediate consequence of Lemma 7F, wherein we take f to be the identity function on C. ⊣

The following theorem is a sort of trichotomy law for well-ordered structures.

Theorem 7K For any two well-ordered structures, either they are isomorphic or one is isomorphic to a segment of the other. More precisely, let $<_A$ and $<_B$ be well orderings on A and B. Then one of the following alternatives holds:

$$\langle A, <_A\rangle \cong \langle B, <_B\rangle,$$
$$\langle A, <_A\rangle \cong \langle \operatorname{seg} b, <_B^\circ\rangle \qquad \text{for some} \quad b \in B,$$
$$\langle \operatorname{seg} a, <_A^\circ\rangle \cong \langle B, <_B\rangle \qquad \text{for some} \quad a \in A.$$

Proof The idea is to start pairing elements of A with elements of B in the natural way: We pair the least elements of the two sets, then the next smallest, and so forth. Eventually we run out of elements on one side or the other. If the two sets run out simultaneously, then $\langle A, <_A\rangle \cong \langle B, <_B\rangle$. Otherwise one set runs out first; in this case it is isomorphic to a segment of the other.

The pairing is obtained by transfinite recursion. Let e be some extraneous object belonging neither to A nor to B. By transfinite recursion there is a unique function $F: A \to B \cup \{e\}$ such that for each $t \in A$,

$$F(t) = \begin{cases} \text{the least element of } B - F[\operatorname{seg} t] & \text{if any,} \\ e & \text{if } B - F[\operatorname{seg} t] \text{ is empty.} \end{cases}$$

Case I $e \in \operatorname{ran} F$. Let a be the least element of A for which $F(a) = e$. We claim that $F \restriction \operatorname{seg} a$ is an isomorphism from $\langle \operatorname{seg} a, <_A^\circ\rangle$ onto $\langle B, <_B\rangle$. Let $F^\circ = F \restriction \operatorname{seg} a$.

Clearly $F^\circ: \operatorname{seg} a \to B$, and its range is all of B since $B - F[\operatorname{seg} a] = \varnothing$. We have

$$\begin{aligned} x \leq_A y <_A a \quad &\Rightarrow \quad F[\operatorname{seg} x] \subseteq F[\operatorname{seg} y] \\ &\Rightarrow \quad B - F[\operatorname{seg} y] \subseteq B - F[\operatorname{seg} x] \\ &\Rightarrow \quad F(x) \leq_B F(y). \end{aligned}$$

Furthermore if $x <_A y <_A a$ then $F(x) \neq F(y)$, because $F(x) \in F[\operatorname{seg} y]$ but by construction $F(y) \notin F[\operatorname{seg} y]$. Therefore F° is one-to-one, and in fact

$$x <_A y <_A a \quad \Rightarrow \quad F(x) <_B F(y).$$

Consequently, F° preserves order, since

$$\begin{aligned} F(x) <_B F(y) \quad &\Rightarrow \quad F(y) \not\leq_B F(x) \\ &\Rightarrow \quad y \not\leq_A x \\ &\Rightarrow \quad x <_A y. \end{aligned}$$

Thus F° is an isomorphism as claimed.

Case II $\operatorname{ran} F = B$. Then $F: A \to B$, and just as in Case I (but without the a) F is one-to-one and preserves order. Thus F is an isomorphism from $\langle A, <_A\rangle$ onto $\langle B, <_B\rangle$.

Case III Otherwise ran F is a proper subset of B. Let b be the least element of B − ran F. We claim that ran F = seg b. By the leastness of b, we have seg $b \subseteq$ ran F. Conversely, we cannot have $b <_B F(x)$ for any x, because $F(x)$ is *least* in $B - F[\![\text{seg } x]\!]$, a set to which b belongs. Hence ran F = seg b.

As in Cases I and II, F is one-to-one and preserves order. Hence F is an isomorphism from $\langle A, <_A\rangle$ onto $\langle \text{seg } b, <_B^\circ\rangle$. ⊣

We now return to the study of ordinal numbers. The next theorem will show that any ordinal number is its own $\in$-image. To extract the maximum of information from the proof, the following definition will help.

Definition A set A is *well ordered by epsilon* iff the relation

$$\in_A = \{\langle x, y\rangle \in A \times A \mid x \in y\}$$

is a well ordering on A.

We have shown that any ordinal number is a transitive set that is well ordered by epsilon (Corollary 7H). The converse is also true; the ordinal numbers are exactly the transitive sets that are well ordered by epsilon:

Theorem 7L Let α be any transitive set that is well ordered by epsilon. Then α is an ordinal number; in fact α is the $\in$-image of $\langle \alpha, \in_\alpha\rangle$.

Proof Let E be the usual function from α onto its $\in$-image. We can use transfinite induction to show that E is just the identity function on α. Note that for $t \in \alpha$,

$$x \in t \quad \Leftrightarrow \quad x \in_\alpha t$$

because α is a transitive set. As a consequence we have seg $t = t$.

If the equation $E(x) = x$ holds for all x in seg t, then

$$\begin{aligned} E(t) &= \{E(x) \mid x \in_\alpha t\} \\ &= \{x \mid x \in_\alpha t\} \\ &= \text{seg } t \\ &= t. \end{aligned}$$

Hence by transfinite induction, E is the identity function on α, so the $\in$-image of $\langle \alpha, \in_\alpha\rangle$ is just α itself. ⊣

The class of all ordinal numbers is (as we will prove shortly) too large to be a set. But except for not being a set, it has many of the properties of the ordinal numbers themselves. The next theorem shows that the class of all ordinal numbers is a transitive class well ordered by epsilon.

Theorem 7M The following are valid for any ordinal numbers α, β, and γ.

(a) (transitive class) Any member of α is itself an ordinal number.
(b) (transitivity) $\alpha \in \beta \in \gamma \Rightarrow \alpha \in \gamma$.
(c) (irreflexivity) $\alpha \notin \alpha$.
(d) (trichotomy) Exactly one of the alternatives,

$$\alpha \in \beta, \qquad \alpha = \beta, \qquad \beta \in \alpha,$$

holds.

(e) Any nonempty set S of ordinal numbers has a least element μ (i.e., $\mu \subseteq \alpha$ for every $\alpha \in S$).

Proof For part (a), consider any x in α. Now α is the $\in$-image of some well-ordered structure $\langle A, < \rangle$. So $x = E(t)$ where E is the usual isomorphism and t is some member of A. We will prove that x is an ordinal by showing that it is the $\in$-image of $\langle \text{seg } t, <^\circ \rangle$. By Theorem 7J the restriction $<^\circ$ is a well ordering on seg t. The $\in$-image of $\langle \text{seg } t, <^\circ \rangle$ is easily seen to be $E[\![\text{seg } t]\!]$, but this is just $E(t)$, which is x. Hence x is an ordinal.

Part (b) is true because the ordinal γ is a transitive set.

Part (c) follows from Theorem 7D(a). If $\alpha \in \alpha$, then $\alpha = E(t)$ for some t (in the notation used above in proving part (a)). But then $E(t) \in E(t)$, which contradicts Theorem 7D(a).

In trichotomy, the "at most one" part follows from transitivity and irreflexivity (as in Theorem 7A). The "at least one" half is a consequence of Theorem 7K, which tells us that either $\langle \alpha, \in_\alpha \rangle$ and $\langle \beta, \in_\beta \rangle$ are isomorphic, or else one is isomorphic to a segment of the other.

Case I $\langle \alpha, \in_\alpha \rangle \cong \langle \beta, \in_\beta \rangle$. Then both have the same $\in$-image, by Theorem 7I. But ordinals are their own $\in$-images by Theorem 7L, so $\alpha = \beta$.

Case II By symmetry we may suppose that $\langle \alpha, \in_\alpha \rangle$ is isomorphic to $\langle \text{seg } \delta, \in_\beta^\circ \rangle$, where $\delta \in \beta$. But then δ is an ordinal, seg $\delta = \delta$, and $\in_\beta^\circ = \in_\delta$. Thus we have

$$\langle \alpha, \in_\alpha \rangle \cong \langle \delta, \in_\delta \rangle.$$

Now by Case I, $\alpha = \delta$. Hence $\alpha \in \beta$.

For part (e), first consider some arbitrary $\beta \in S$. If $\beta \cap S = \varnothing$, then we claim that β is least in S. This is because

$$\alpha \in S \quad \Rightarrow \quad \alpha \notin \beta \quad \Rightarrow \quad \beta \subseteq \alpha$$

by trichotomy. There remains the possibility that $\beta \cap S \neq \varnothing$. Then we have here a nonempty subset of β. Hence $\beta \cap S$ has a least (with respect to $\in_\beta$)

member μ. We claim that μ is least in S. To verify this, consider any $\alpha \in S$:

$$\alpha \notin \beta \;\Rightarrow\; \beta \in\!\!\!\!_{=} \alpha \;\Rightarrow\; \beta \subseteq \alpha \;\Rightarrow\; \mu \in \alpha,$$

$$\alpha \in \beta \;\Rightarrow\; \alpha \in \beta \cap S \;\Rightarrow\; \mu \in\!\!\!\!_{=} \alpha.$$

Either way, $\mu \in\!\!\!\!_{=} \alpha$. ⊣

Corollary 7N (a) Any transitive set of ordinal numbers is itself an ordinal number.

(b) 0 is an ordinal number.

(c) If α is any ordinal number, then α^+ is also an ordinal number.

(d) If A is any set of ordinal numbers, then $\bigcup A$ is also an ordinal number.

Proof (a) We can conclude from the preceding theorem that any set of ordinals is well ordered by epsilon. If, in addition, the set is a transitive set, then it is an ordinal number by Theorem 7L.

(b) $\varnothing$ is a transitive set of ordinals. (It is the $\in$-image of the well-ordered structure $\langle \varnothing, \varnothing \rangle$.)

(c) Recall that $\alpha^+ = \alpha \cup \{\alpha\}$. The members of α^+ are all ordinals (by Theorem 7M(a)). And α^+ is a transitive set by Theorem 4E (compare Exercise 2 of Chapter 4).

(d) We can also apply part (a) to prove part (d). Any member of $\bigcup A$ is a member of some ordinal, and so is itself an ordinal. To show that $\bigcup A$ is a transitive set, we calculate

$$\begin{aligned} \delta \in \bigcup A \;&\Rightarrow\; \delta \in \alpha \in A && \text{for some } \alpha \\ &\Rightarrow\; \delta \subseteq \alpha \in A && \text{since } \alpha \text{ is transitive} \\ &\Rightarrow\; \delta \subseteq \bigcup A. \end{aligned}$$

Thus by part (a), $\bigcup A$ is an ordinal. ⊣

In fact more is true here. If A is any set of ordinals, then $\bigcup A$ is an ordinal that is the least upper bound of A. To verify this, first note that the epsilon ordering on the class of all ordinals is also defined by $\subset$. That is, for ordinals α and β we have:

$$\begin{aligned} \alpha \in \beta \;&\Rightarrow\; \alpha \subset \beta && \text{since } \beta \text{ is a transitive set and } \alpha \neq \beta \\ &\Rightarrow\; \beta \notin\!\!\!\!_{=} \alpha && \text{lest } \beta \in \beta \\ &\Rightarrow\; \alpha \in \beta && \text{by trichotomy.} \end{aligned}$$

Thus $\alpha \in \beta$ iff $\alpha \subset \beta$. Since $\bigcup A$ is the least upper bound for A in the inclusion ordering, it is the least upper bound for the epsilon ordering.

In the same vein, we can assert that α^+ is the least ordinal larger than α. Clearly $\alpha \in \alpha^+$, so α^+ *is* larger than α. Suppose $\alpha \in \beta$ for some ordinal β. Then $\alpha \subseteq \beta$ as well, and hence $\alpha^+ \subseteq \beta$. As noted above, this yields the fact that $\alpha^+ \underline{\in} \beta$. This shows that α^+ is the least ordinal larger than α.

Observe also that any ordinal is just the set of all smaller ordinals. That is:

$$\begin{aligned}\alpha &= \{x \mid x \in \alpha\}\\ &= \{x \mid x \text{ is an ordinal } \& \ x \in \alpha\}.\end{aligned}$$

Since our ordering relation is $\in$, this can be read as, "α equals the set of ordinals less than α."

At long last we can get a better picture of what the ordinals are like. The least ordinal is 0. Next come 0^+, 0^{++}, ..., i.e., the natural numbers. The least ordinal greater than these is the least upper bound of the set of natural numbers. This is just $\bigcup\omega$, which equals ω. (Or more directly: ω is an ordinal and the ordinals less than ω are exactly its members.) Then come ω^+, ω^{++}, ... All of these are countable ordinals (i.e., ordinals that are countable sets). The least uncountable ordinal Ω is the set of all smaller ordinals, i.e., is the set of all countable ordinals. But here we are on thin ice, because we have not shown that there *are* any uncountable ordinals. That defect will be corrected in the next section.

Burali-Forti Theorem There is no set to which every ordinal number belongs.

Proof Theorem 7M told us that the class of all ordinals was a transitive class well ordered by epsilon. If it were a set, then by Theorem 7L it would be an ordinal itself. But then it would be a member of itself, and no ordinal has that property. ⊣

This theorem is also called the "Burali-Forti paradox." Like the later paradox of Russell, it showed the inconsistency of Cantor's casual use of the abstraction principle, which permitted speaking of "the set of all ordinals." The theorem was published in 1897 by Burali-Forti, and was the first of the set-theoretic paradoxes to be published.

Exercises

15. (a) Assume that $<$ is a well ordering on A and that $t \in A$. Show that $\langle A, <\rangle$ is never isomorphic to $\langle \text{seg } t, <^\circ\rangle$.
(b) Show that in Theorem 7K, at most one of the three alternatives holds.

16. Assume that α and β are ordinal numbers with $\alpha \in \beta$. Show that $\alpha^+ \in \beta^+$. Conclude that whenever $\alpha \neq \beta$, then $\alpha^+ \neq \beta^+$.

17. Assume that $\langle A, <\rangle$ is a well-ordered structure with ordinal number α and that $B \subseteq A$. Let β be the ordinal number of the well-ordered structure $\langle B, <^{\circ}\rangle$ and show that $\beta \in \alpha$.

18. Assume that S is a set of ordinal numbers. Show that either (i) S has no greatest element, $\bigcup S \notin S$, and $\bigcup S$ is not the successor of another ordinal; or (ii) $\bigcup S \in S$ and $\bigcup S$ is the greatest element of S.

19. Assume that A is a finite set and that $<$ and $\prec$ are linear orderings on A. Show that $\langle A, <\rangle$ and $\langle A, \prec\rangle$ are isomorphic.

20. Show that if R and R^{-1} are both well orderings on the same set S, then S is finite.

21. Prove the following version of Zorn's lemma. Assume that $<$ is a partial ordering on A. Assume that whenever C is a subset of A for which $<^{\circ}$ is a linear ordering on C, then C has an upper bound in A. Then there exists a maximal element of A.

DEBTS PAID

In this section we will pay off the debts incurred in Chapter 6. We will define cardinal numbers as promised, and we will complete the proof of Theorem 6M.

The first theorem assures an adequate supply of ordinal numbers. Recall that B is dominated by A ($B \preccurlyeq A$) iff B is equinumerous to a subset of A.

Hartogs' Theorem For any set A, there is an ordinal not dominated by A.

Proof There is a systematic way to construct the least such ordinal α. Any ordinal that is dominated by A is smaller than (i.e., is a member of) α. And conversely, any ordinal smaller than (i.e., a member of) α should be dominated by A, if α is to be least. Hence we decide to try defining

$$\alpha = \{\beta \mid \beta \text{ is an ordinal } \& \ \beta \preccurlyeq A\}.$$

The essential part of the proof is showing that α is a set. After all, the theorem is equivalent to the assertion that α is not the class of all ordinals.

We begin by defining

$$W = \{\langle B, <\rangle \mid B \subseteq A \ \& \ < \text{ is a well ordering on } B\}.$$

We claim that any member of α is obtainable as the $\in$-image of something in W. (Thus α is, in a sense, no larger than W.) Consider any ordinal β in α. Then some function f maps β one-to-one onto a subset B of A. There is a well ordering $<$ on B such that f is an isomorphism from $\langle \beta, \in_\beta\rangle$ onto $\langle B, <\rangle$, namely

$$< \, = \{\langle f(x), f(y)\rangle \mid x \in y \in \beta\}.$$

Thus $\langle B, < \rangle \in W$ and the $\in$-image of $\langle B, < \rangle$ is the same as the $\in$-image of $\langle \beta, \in_\beta \rangle$, which is just β. Thus β is indeed the $\in$-image of a member of W.

W is a set, because if $\langle B, < \rangle \in W$, then

$$\langle B, < \rangle \in \mathscr{P}A \times \mathscr{P}(A \times A).$$

We can then use a replacement axiom to construct the set $\mathscr{E}$ of $\in$-images of members of W. The preceding paragraph shows that $\alpha \subseteq \mathscr{E}$. (In fact it is not hard to see that $\alpha = \mathscr{E}$.)

Since α is a set, we can conclude from the Burali-Forti theorem that some ordinals do not belong to α. This proves the theorem, but we will continue and show that α itself is the least such ordinal. Clearly α is a set of ordinals, and since

$$\gamma \in \beta \in \alpha \quad \Rightarrow \quad \gamma \subseteq \beta \preccurlyeq A \quad \Rightarrow \quad \gamma \in \alpha,$$

α is transitive. Hence α is an ordinal. We could not have $\alpha \preccurlyeq A$, lest α belong to itself. And α is the least such ordinal, since $\beta \in \alpha \Rightarrow \beta \preccurlyeq A$ by construction. ⊣

Hartogs' theorem does not require the axiom of choice. But the following theorem does.

Well-Ordering Theorem For any set A, there is a well ordering on A.

This theorem is often stated more informally: Any set can be well ordered. As we will observe presently, the well-ordering theorem is equivalent to the axiom of choice. We can prove the well-ordering theorem from Zorn's lemma (Exercise 22). But in order to complete the proof of Theorem 6M, we will instead use the axiom of choice, III.

Proof Let G be a choice function for A, and let α be an ordinal not dominated by A (furnished by Hartogs' theorem). Our strategy is to order A by first choosing a least element, then a next-to-least, and so forth. For the "and so forth" part, we use α as a base; α is large enough that we will exhaust A before coming to the end of α.

Let e be an extraneous object not belonging to A. We can use transfinite recursion to obtain a function $F: \alpha \to A \cup \{e\}$ such that for any $\gamma \in \alpha$,

$$F(\gamma) = \begin{cases} G(A - F[\![\gamma]\!]) & \text{if } A - F[\![\gamma]\!] \neq \varnothing, \\ e & \text{if } A - F[\![\gamma]\!] = \varnothing. \end{cases}$$

In other words, $F(\gamma)$ is the chosen member of $A - F[\![\gamma]\!]$, until A is exhausted.

Suppose $\gamma \in \beta \in \alpha$. If neither $F(\gamma)$ nor $F(\beta)$ equals e, then $F(\gamma) \neq F(\beta)$, since $F(\gamma) \in F[\![\beta]\!]$ but $F(\beta) \notin F[\![\beta]\!]$. That is, F is one-to-one until e appears. It follows that $e \in \operatorname{ran} F$, lest F be a one-to-one function from α into A, contradicting the fact that A does not dominate α.

Let δ be the least element of α for which $F(\delta) = e$. Then $F[\![\delta]\!] = A$. And by the preceding paragraph, $F \upharpoonright \delta$ is a one-to-one map from δ onto A. This produces a well ordering on A, namely

$$< \, = \{\langle F(\beta), F(\gamma)\rangle \mid \beta \in \gamma \in \delta\}.$$

Then $F \upharpoonright \delta$ is an isomorphism from $\langle \delta, \in_\delta\rangle$ onto $\langle A, <\rangle$. ⊣

Example The well-ordering theorem claims that there exists a well ordering W on the set $\mathbb{R}$ of real numbers. Now the usual ordering on $\mathbb{R}$ is definitely *not* a well ordering, so W is some unusual ordering. What is it? We have a theorem asserting the existence of W, but its proof (which used choice) gives no clues as to just what W is like. It is entirely possible that there is no formula of the language of set theory that will explicitly define a well ordering on $\mathbb{R}$.

The following variant of the well-ordering theorem follows easily from it.

Numeration Theorem Any set is equinumerous to some ordinal number.

Proof Consider any set A; by the preceding theorem there is a well ordering $<$ on A. Then A is equinumerous to the $\in$-image α of $\langle A, <\rangle$. ⊣

The name "numeration theorem" reflects the possibility of "counting" the members of A by using the ordinal numbers in α, where a "counting" is a one-to-one correspondence.

The numeration theorem produces a satisfactory definition of cardinal number.

Definition For any set A, define the *cardinal number* of A (card A) to be the least ordinal equinumerous to A.

The following theorem makes good our promise of Chapter 6.

Theorem 7P (a) For any sets A and B,

$$\text{card } A = \text{card } B \quad \text{iff} \quad A \approx B.$$

(b) For a finite set A, card A is the unique natural number equinumerous to A.

Proof First, note that by our definition, $A \approx \text{card } A$ for any set A. Hence if card A = card B, then we have

$$A \approx \text{card } A = \text{card } B \approx B.$$

Conversely if $A \approx B$, then A and B are equinumerous to exactly the same ordinals. Hence the least such ordinal is the same in both cases.

Part (b) follows from the fact that the natural numbers are also ordinal numbers. If A is finite, then (by definition) $A \approx n$ for a unique natural number n. A is not equinumerous to any smaller ordinal, since the smaller ordinals are just the natural numbers in n. Hence card $A = n$. ⊣

Proof of Theorem 6M, concluded Figure 46 shows how Fig. 40 is to be extended. Here "WO" is the well-ordering theorem, which we proved from (3). It remains to establish (5) ⇒ WO and WO ⇒ (6).

Proof of (5) ⇒ WO Given any set A, we obtain from Hartogs' theorem an ordinal α not dominated by A. By (5), we have $A \preccurlyeq \alpha$; i.e., there is a

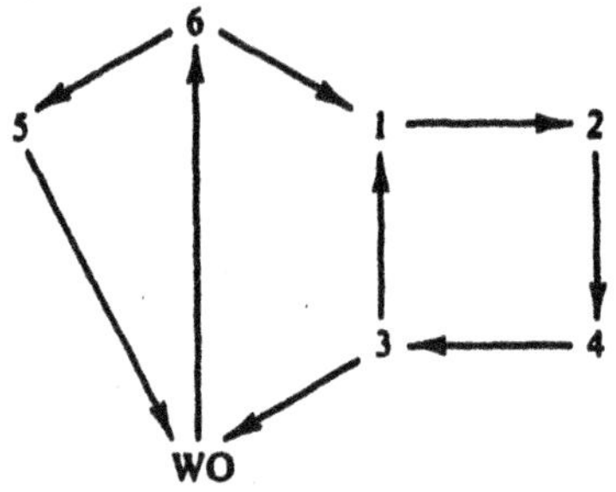

Fig. 46. The proof of Theorem 6M.

one-to-one function f from A into α. This induces a well ordering of A; define

$$s < t \quad \Leftrightarrow \quad f(s) \in f(t)$$

for s and t in A. Then $<$ is a well ordering of A (by Lemma 7F).

Proof of WO ⇒ (6) Consider any $\mathscr{A}$ that is closed under unions of chains. By WO there exists a well ordering $<$ of $\mathscr{A}$. We want to make a very large chain $\mathscr{C} \subseteq \mathscr{A}$ such that $\bigcup \mathscr{C}$ is a maximal element. The idea is to go through the members of $\mathscr{A}$ in order, adding to $\mathscr{C}$ any member of $\mathscr{A}$ that includes each set previously placed in $\mathscr{C}$. This construction uses transfinite recursion. That is, the transfinite recursion theorem gives us a function $F: \mathscr{A} \to 2$ such that for any $A \in \mathscr{A}$,

$$F(A) = \begin{cases} 1 & \text{if } A \text{ includes every set } B < A \text{ for which } F(B) = 1, \\ 0 & \text{otherwise.} \end{cases}$$

Then let $\mathscr{C} = \{A \in \mathscr{A} \mid F(A) = 1\}$. Thus F is the characteristic function of $\mathscr{C}$, and for $A \in \mathscr{A}$,

$$A \in \mathscr{C} \quad \Leftrightarrow \quad A \text{ includes every } B < A \text{ for which } B \in \mathscr{C}.$$

We claim that $\bigcup \mathscr{C}$ is a maximal element of $\mathscr{A}$.

First of all, $\mathscr{C}$ is a chain. For if A and B are in $\mathscr{C}$, then

$$A \leq B \Rightarrow A \subseteq B \quad \text{and} \quad B \leq A \Rightarrow B \subseteq A.$$

Since $\mathscr{C}$ is a chain, $\bigcup\mathscr{C} \in \mathscr{A}$. To prove maximality, suppose that $\bigcup\mathscr{C} \subseteq D \in \mathscr{A}$. Then $D \in \mathscr{C}$, since D includes every member of $\mathscr{C}$. Hence $D \subseteq \bigcup\mathscr{C}$, and so $D = \bigcup\mathscr{C}$. Thus $\bigcup\mathscr{C}$ is not a proper subset of any member of $\mathscr{A}$. ⊣

There are several easy consequences of our definition of cardinal numbers. Say that an ordinal number is an *initial* ordinal iff it is not equinumerous to any smaller ordinal number. Then any initial ordinal is its own cardinal number. Conversely any cardinal number must, by the definition, be an initial ordinal. Thus the cardinal numbers and the initial ordinals are exactly the same things.

Next consider any cardinal number, say $\kappa = \text{card } S$. Then $\kappa \approx S$ (and κ is the least ordinal with this property). Also card $\kappa = \kappa$, since κ is an initial ordinal. That is, κ is itself a set of cardinality κ.

The cardinal numbers, like any other ordinal numbers, are well ordered by the epsilon relation as described in Theorem 7M. It is now easy to verify that this ordering agrees with the ordering of cardinal numbers defined in Chapter 6. For any cardinals κ and λ,

$$\begin{aligned} \kappa \in \lambda &\Rightarrow \kappa \subset \lambda \\ &\Rightarrow \kappa \preccurlyeq \lambda \ \& \ \kappa \neq \lambda \\ &\Rightarrow \kappa \leq \lambda \ \& \ \kappa \neq \lambda \\ &\Rightarrow \kappa < \lambda. \end{aligned}$$

Conversely,

$$\begin{aligned} \kappa \notin \lambda &\Rightarrow \lambda \in \kappa \\ &\Rightarrow \lambda \leq \kappa \qquad \text{by the above} \\ &\Rightarrow \kappa \nless \lambda. \end{aligned}$$

In particular, any nonempty set of cardinal numbers has a smallest element. And the word "smallest" is used unambiguously here, since the epsilon ordering and the cardinality ordering agree. The smallest cardinal after $\aleph_0$ is of course denoted as $\aleph_1$. We can also characterize $\aleph_1$ as the least uncountable ordinal. Next in order of magnitude comes $\aleph_2$. And this continues. In Chapter 8, we define $\aleph_\alpha$ for every ordinal α.

Exercises

22. Prove the well-ordering theorem from the version of Zorn's lemma given in Exercise 21.

23. Assume that A is a set and define (as in Hartogs' theorem) α to be the set of ordinals dominated by A. Show that (i) α is a cardinal number, (ii) card $A < \alpha$, and (iii) α is the least cardinal greater than card A.

24. Show that for any ordinal number α, there is a cardinal number κ that is (as an ordinal) larger than α.

25. (Transfinite induction schema) Let $\varphi(x)$ be a formula and show that the following holds:

Assume that for any ordinal α, we have $(\forall x \in \alpha)\varphi(x) \Rightarrow \varphi(\alpha)$. Then $\varphi(\alpha)$ for any ordinal α.

RANK

In this final section of the chapter we want to return to the informal view of sets mentioned in Chapter 1—and especially to Fig. 3. The description that was somewhat vague at the start of this book can now be made quite precise.

We want to define for every ordinal number α the set V_α. V_0 is to be empty, and, in general, V_α is to contain those sets whose members are all in some V_β for β less than α. Thus we want

$$\begin{aligned} a \in V_\alpha \quad &\Leftrightarrow \quad a \subseteq V_\beta \qquad \text{for some } \beta \in \alpha \\ &\Leftrightarrow \quad a \in \mathscr{P}V_\beta \qquad \text{for some } \beta \in \alpha, \end{aligned}$$

or equivalently,

$$V_\alpha = \bigcup\{\mathscr{P}V_\beta \mid \beta \in \alpha\}.$$

Theorem 7U will show that in particular cases, this equation can be simplified. If $\alpha = \delta^+$, the equation will become

$$V_{\delta^+} = \mathscr{P}V_\delta,$$

whereas if α is not the successor of another ordinal, then the equation will become

$$\begin{aligned} V_\alpha &= \bigcup\{V_\beta \mid \beta \in \alpha\} \\ &= \bigcup_{\beta \in \alpha} V_\beta. \end{aligned}$$

Thus $V_3 = \mathscr{P}\mathscr{P}\mathscr{P}\varnothing$ and $V_\omega = V_0 \cup V_1 \cup V_2 \cup \cdots$.

But we are getting ahead of ourselves. We have yet to show how to define V_α. The equation

$$V_\alpha = \bigcup\{\mathscr{P}V_\beta \mid \beta \in \alpha\}$$

defines V_α in terms of V_β for β less than α. So we need to define V_α by transfinite recursion. The class of all ordinals is well ordered, so transfinite recursion over this class should give us a function-class $\mathbf{F}$ such that

$$\mathbf{F}(\alpha) = \bigcup\{\mathscr{P}\mathbf{F}(\beta) \mid \beta \in \alpha\}$$

for each ordinal α. But there is an obvious hitch here. We cannot work directly with classes that are not sets. Instead we will sneak up gradually on $\mathbf{F}$.

Lemma 7Q For any ordinal number δ there is a function F_δ with domain δ such that

$$F_\delta(\alpha) = \bigcup\{\mathscr{P}F_\delta(\beta) \mid \beta \in \alpha\}$$

for every $\alpha \in \delta$.

Proof We apply transfinite recursion. On δ we have the well ordering $\in_\delta$. Take for $\gamma(x, y)$ the formula

$$y = \bigcup\{\mathscr{P}z \mid z \in \operatorname{ran} x\}.$$

We can show that $\{\mathscr{P}z \mid z \in \operatorname{ran} x\}$ is a set by showing that it is included in $\mathscr{P}\mathscr{P}\bigcup(\operatorname{ran} x)$. But an easier proof uses a replacement axiom: Let $\varphi(z, w)$ be the formula $w = \mathscr{P}z$. Then because $\operatorname{ran} x$ is a set, $\{\mathscr{P}z \mid z \in \operatorname{ran} x\}$ is also a set.

Obviously for any f there is a unique y such that $\gamma(f, y)$, namely $y = \bigcup\{\mathscr{P}z \mid z \in \operatorname{ran} f\}$. Transfinite recursion then gives us a function F_δ such that

$$(\star) \qquad F_\delta(\alpha) = \bigcup\{\mathscr{P}z \mid z \in \operatorname{ran}(F_\delta \restriction \operatorname{seg} \alpha)\}$$

for every $\alpha \in \delta$. Now then,

$$\begin{aligned} \operatorname{seg} \alpha &= \{t \mid t \in_\delta \alpha\} \\ &= \{t \mid t \in \alpha\} \\ &= \alpha. \end{aligned}$$

Thus

$$\begin{aligned} z \in \operatorname{ran}(F_\delta \restriction \operatorname{seg} \alpha) &\Leftrightarrow z \in \operatorname{ran}(F_\delta \restriction \alpha) \\ &\Leftrightarrow z = F_\delta(\beta) \qquad \text{for some } \beta \in \alpha. \end{aligned}$$

So $(\star)$ becomes

$$\begin{aligned} F_\delta(\alpha) &= \bigcup\{\mathscr{P}z \mid z = F_\delta(\beta) \text{ for some } \beta \in \alpha\} \\ &= \bigcup\{\mathscr{P}F_\delta(\beta) \mid \beta \in \alpha\} \end{aligned}$$

as desired. ⊣

Now consider any two ordinals δ and ε. The smaller of the two is $\delta \cap \varepsilon$. (Why?) We have two functions F_δ and F_ε from the preceding lemma. These functions agree on $\delta \cap \varepsilon$:

Lemma 7R Let δ and ε be ordinal numbers; let F_δ and F_ε be functions from Lemma 7Q. Then

$$F_\delta(\alpha) = F_\varepsilon(\alpha)$$

for all $\alpha \in \delta \cap \varepsilon$.

Proof By the symmetry, we suppose that $\delta \subseteq \varepsilon$. Hence $\delta \subseteq \varepsilon$ and $\delta \cap \varepsilon = \delta$. We will establish the equation $F_\delta(\alpha) = F_\varepsilon(\alpha)$ by using transfinite induction in $\langle \delta, \in_\delta \rangle$. Define

$$B = \{\alpha \in \delta \mid F_\delta(\alpha) = F_\varepsilon(\alpha)\}.$$

In order to show that $B = \delta$, it suffices to show that B is "$\in_\delta$-inductive," i.e., that

$$\operatorname{seg} \alpha \subseteq B \quad \Rightarrow \quad \alpha \in B$$

for each $\alpha \in \delta$.

We calculate:

$$\begin{aligned} \operatorname{seg} \alpha \subseteq B \quad &\Rightarrow \quad F_\delta(\beta) = F_\varepsilon(\beta) \qquad \text{for} \quad \beta \in \alpha \\ &\Rightarrow \quad \bigcup\{\mathscr{P} F_\delta(\beta) \mid \beta \in \alpha\} = \bigcup\{\mathscr{P} F_\varepsilon(\beta) \mid \beta \in \alpha\} \\ &\Rightarrow \quad F_\delta(\alpha) = F_\varepsilon(\alpha) \\ &\Rightarrow \quad \alpha \in B \end{aligned}$$

for $\alpha \in \delta$. And so we are done. ⊣

In particular (by taking $\delta = \varepsilon$) we see that the function F_δ from Lemma 7Q is unique. We can now unambiguously define V_α.

Definition Let α be an ordinal number. Define V_α to be the set $F_\delta(\alpha)$, where δ is any ordinal greater than α (e.g., $\delta = \alpha^+$).

Theorem 7S For any ordinal number α,

$$V_\alpha = \bigcup\{\mathscr{P} V_\beta \mid \beta \in \alpha\}.$$

Proof Let $\delta = \alpha^+$. Then $V_\alpha = F_\delta(\alpha)$ and $V_\beta = F_\delta(\beta)$ for $\beta \in \alpha$. Hence the desired equation reduces to Lemma 7Q. ⊣

Lemma 7T For any ordinal number α, V_α is a transitive set.

Proof We would like to prove this by transfinite induction over the class of all ordinals. This can be done by utilizing Exercise 25. But we can also avoid that exercise by proving for each ordinal δ that V_α is a transitive set whenever $\alpha \in \delta$. This requires only transfinite induction over $\langle \delta, \in_\delta \rangle$.

Let B be the set on which the conclusion holds:

$$B = \{\alpha \in \delta \mid V_\alpha \text{ is a transitive set}\}.$$

We want to show that $B = \delta$; this will follow from the transfinite induction theorem if we show that B is "$\in_\delta$-inductive," i.e., that

$$\alpha \subseteq B \quad \Rightarrow \quad \alpha \in B$$

for all $\alpha \in \delta$. (Recall that seg $\alpha = \alpha$.)

Suppose then that $\alpha \subseteq B$. Then for any $\beta \in \alpha$, V_β is a transitive set, as is $\mathscr{P}V_\beta$ (by Exercise 3 of Chapter 4). It follows that V_α is a transitive set, because

$$\begin{aligned} x \in V_\alpha \quad &\Rightarrow \quad x \in \mathscr{P}V_\beta \qquad \text{for some } \beta \in \alpha \\ &\Rightarrow \quad x \subseteq \mathscr{P}V_\beta \qquad \text{for some } \beta \in \alpha \\ &\Rightarrow \quad x \subseteq V_\alpha. \end{aligned}$$

Thus $\alpha \in B$, and we are done. ⊣

There are three sorts of ordinal number. First, there is 0, called *zero*. Second, there are the ordinals of the form α^+ for a smaller ordinal α. These are called the *successor ordinals*. Third, there are all the others, called *limit ordinals*. For example, the least limit ordinal is ω.

If λ is a limit ordinal and $\beta \in \lambda$, then $\beta^+ \in \lambda$. This is because β^+, being the least ordinal larger than β, must satisfy $\beta^+ \in\!\!\!\!\_\ \lambda$. And $\beta^+ \neq \lambda$ since λ is a limit ordinal.

The following theorem describes V_α for each of the three sorts of ordinal number.

Theorem 7U (a) For ordinals $\beta \in \alpha$ we have $V_\beta \subseteq V_\alpha$.
(b) $V_0 = \varnothing$.
(c) $V_{\alpha^+} = \mathscr{P}V_\alpha$ for any ordinal number α.
(d) $V_\lambda = \bigcup_{\beta \in \lambda} V_\beta$ for any limit ordinal λ.

Proof For part (a), first observe that $V_\beta \in \mathscr{P}V_\beta \subseteq V_\alpha$, so that $V_\beta \in V_\alpha$. Since V_α is a transitive set, we also have $V_\beta \subseteq V_\alpha$.

Part (b) is clear.

For part (c), we first obtain from part (a)

$$\beta \in \alpha \quad \Rightarrow \quad V_\beta \subseteq V_\alpha \quad \Rightarrow \quad \mathscr{P}V_\beta \subseteq \mathscr{P}V_\alpha.$$

Now $V_{\alpha^+} = \bigcup\{\mathscr{P}V_\beta \mid \beta \in\!\!\!\!\_\ \alpha\}$ and this union is equal to the largest set, $\mathscr{P}V_\alpha$.

Finally for part (d) we have one inclusion:

$$\begin{aligned} x \in V_\lambda \quad &\Rightarrow \quad x \in \mathscr{P}V_\beta \qquad \text{for some } \beta \in \lambda \\ &\Rightarrow \quad x \in V_{\beta^+} \qquad \text{by part (c)} \\ &\Rightarrow \quad x \in \bigcup_{\beta \in \lambda} V_\beta \qquad \text{since } \beta^+ \in \lambda. \end{aligned}$$

The other inclusion is similar:

$$\begin{aligned} x \in \bigcup_{\beta \in \lambda} V_\beta \quad &\Rightarrow \quad x \in V_\beta \subseteq V_{\beta^+} = \mathscr{P} V_\beta \qquad \text{for some } \beta \in \lambda \\ &\Rightarrow \quad x \in V_\lambda. \end{aligned}$$

Thus equality holds. ⊣

We will say that a set A is *grounded* iff $A \subseteq V_\alpha$ for some ordinal number α. In this event, we define the *rank* of A (rank A) to be the least such ordinal α. Thus for a grounded set A:

$$A \subseteq V_{\text{rank } A} \qquad \text{and} \qquad A \in V_{(\text{rank } A)^+}.$$

The number rank A indicates how many times the power set operation must be used to obtain all the members of A. For example, every ordinal α is grounded and rank $\alpha = \alpha$ (Exercise 26).

There is one methodological objection that can be raised concerning the definition of rank. We want rank A to be the least member of $\{\alpha \mid A \subseteq V_\alpha\}$. But (for a grounded set A) this is not a set of ordinals, but a proper class of ordinals. Nonetheless, it has a least member. Being nonempty, it has some member β. Either β is its least member, or

$$\{\alpha \in \beta \mid A \subseteq V_\alpha\}$$

is a nonempty *set* of ordinals. In the latter case, we can apply Theorem 7M(e) to obtain a least element.

Theorem 7V (a) If $a \in A$ and A is grounded, then a is grounded and

$$\text{rank } a \in \text{rank } A.$$

(b) If every member of A is grounded, then A is also grounded and

$$\text{rank } A = \bigcup \{(\text{rank } a)^+ \mid a \in A\}.$$

Proof For the first part, assume that A is grounded and that $a \in A$. Then

$$\begin{aligned} A \subseteq V_{\text{rank } A} \quad &\Rightarrow \quad a \in V_{\text{rank } A} \\ &\Rightarrow \quad a \in \mathscr{P} V_\beta \qquad \text{for some } \beta \in \text{rank } A \\ &\Rightarrow \quad a \subseteq V_\beta \qquad \text{for some } \beta \in \text{rank } A \\ &\Rightarrow \quad \text{rank } a \in \text{rank } A. \end{aligned}$$

For part (b), assume that every member of A is grounded. Define

$$\alpha = \bigcup \{(\text{rank } a)^+ \mid a \in A\}.$$

To show that α is a set, we use a replacement axiom first to conclude that $\{(\text{rank } a)^+ \mid a \in A\}$ is a set. Then by the union axiom, α is a set.

By Corollary 7N(d), α is an ordinal. First we claim that $A \subseteq V_\alpha$. (This will show that A is grounded and that rank $A \in \alpha$.) We calculate

$$\begin{aligned} a \in A &\Rightarrow a \subseteq V_{\text{rank } a} \\ &\Rightarrow a \in V_{(\text{rank } a)^+} \\ &\Rightarrow a \in V_\alpha \qquad \text{by Theorem 7U(a),} \end{aligned}$$

since $(\text{rank } a)^+ \underline{\in} \alpha$ for each $a \in A$. Thus $A \subseteq V_\alpha$.

To show that $\alpha \underline{\in} \text{rank } A$, we use part (a).

$$\begin{aligned} a \in A &\Rightarrow \text{rank } a \in \text{rank } A \\ &\Rightarrow (\text{rank } a)^+ \underline{\in} \text{rank } A. \end{aligned}$$

Hence rank A is an upper bound for the ordinals $(\text{rank } a)^+$, and so is at least as large as their least upper bound α. ⊣

Part (b) of the foregoing theorem can be translated into words as: The rank of a set is the least ordinal that is strictly greater than the ranks of all members of the set.

Theorem 7W The following two statements are equivalent.

(a) Every set is grounded.

(b) (Regularity) Every nonempty set A has a member m with $m \cap A = \varnothing$.

Proof First assume that every set is grounded, and let A be a nonempty set. The idea is to take $m \in A$ having the least possible rank μ. Thus from

$$\{\text{rank } a \mid a \in A\},$$

we take the least member μ, and then we select $m \in A$ having rank $m = \mu$.

If $x \in m$, then by Theorem 7V(a), rank $x \in \mu$. We cannot have $x \in A$, due to the leastness of μ. Hence $m \cap A = \varnothing$ as desired.

For the converse, assume that (b) holds. Suppose that, contrary to our expectations, some set c is not grounded. Then some set (e.g., $\{c\}$) has a nongrounded member. And hence some transitive set B has a nongrounded member; we can take B to be the transitive closure of $\{c\}$ (see Exercise 7).

Let $A = \{x \in B \mid x \text{ is not grounded}\}$. Since A is nonempty, by (b) there is some $m \in A$ with $m \cap A = \varnothing$. We claim that every member of m is grounded. If $x \in m$, then $x \in B$ since B is transitive. But $x \notin A$ because $m \cap A = \varnothing$, so x must be grounded.

Since every member of m is grounded, we can conclude from Theorem 7V(b) that m is also grounded. This contradicts the fact that $m \in A$. ⊣

Near the bottom of p. 8, we stated as "a fundamental principle" that every set is grounded. So on an informal level at least, we must accept both parts of the above theorem as true.

But neither part is provable from the axioms adopted up to now. To correct this defect, we will now adopt the regularity axiom. For this axiom we could use either (a) or (b) from the above theorem. We select (b) because it has a more elementary formulation.

Regularity Axiom Every nonempty set A has a member m with $m \cap A = \varnothing$.

The regularity axiom is also known as the foundation axiom or the *Fundierungsaxiom*. The idea first appeared in a 1917 paper by Mirimanoff; the axiom was listed explicitly in von Neumann's 1925 paper.

The concept of rank, as well as the concept of regularity, appeared in Mirimanoff's 1917 paper. The idea of rank is a descendant of Russell's concept of *type*.

The following theorem lists some basic consequences of regularity.

Theorem 7X (a) No set is a member of itself.

(b) There do not exist any sets a and b with $a \in b$ and $b \in a$.

(c) There is no function f with domain ω such that

$$\cdots \in f(2) \in f(1) \in f(0).$$

Proof For part (c), suppose to the contrary that $f(n^+) \in f(n)$ for each $n \in \omega$. Let $A = \operatorname{ran} f$. Any member m of A must equal $f(n)$ for some n. Then $f(n^+) \in m \cap A$, so that $m \cap A$ is always nonempty. This contradicts regularity.

Parts (a) and (b) can be proved either by similar arguments or as consequences of part (c). We leave the details as an exercise. ⊣

Since every set is grounded, the sets are arranged in an orderly hierarchy according their rank. This is the situation that Fig. 3 attempts to illustrate. Thus the universe of all sets is, in a sense, determined by two factors:

1. the extent of the class of all ordinals, and
2. the variety of subsets that are assigned to a set by the power set operation.

Exercises

26. Show that every ordinal number α is grounded, and that rank $\alpha = \alpha$.

27. Show that the set $\mathbb{R}$ of real numbers, as constructed in Chapter 5, has rank ω^{+++++}

28. Show that

$$V_\alpha = \{X \mid \text{rank } X \in \alpha\}.$$

29. Prove parts (a) and (b) of Theorem 7X.

30. Show that for any sets

$$\begin{aligned}\text{rank}\{a, b\} &= \max(\text{rank } a, \text{rank } b)^+,\\ \text{rank } \mathscr{P}a &= (\text{rank } a)^+,\\ \text{rank } \bigcup a &\in \text{rank } a.\end{aligned}$$

31. Define kard A to be the collection of all sets B such that (i) A is equinumerous to B, and (ii) nothing of rank less than rank B is equinumerous to B.

(a) Show that kard A is a set.
(b) Show that kard A is nonempty.
(c) Show that for any sets A and B,

$$\text{kard } A = \text{kard } B \quad \text{iff} \quad A \approx B.$$

32. Let $\langle A, R\rangle$ be a structure. Define the *isomorphism type* it$\langle A, R\rangle$ of this structure to be the set of all structures $\langle B, S\rangle$ such that (i) $\langle A, R\rangle$ is isomorphic to $\langle B, S\rangle$, and (ii) nothing of rank less than rank$\langle B, S\rangle$ is isomorphic to $\langle B, S\rangle$.

(a) Show that it$\langle A, R\rangle$ is a set (and not a proper class).
(b) Show that it$\langle A, R\rangle$ is nonempty.
(c) Show that

$$\text{it}\langle A, R\rangle = \text{it}\langle B, S\rangle \quad \text{iff} \quad \langle A, R\rangle \cong \langle B, S\rangle.$$

33. Assume that D is a transitive set. Let B be a set with the property that for any a in D,

$$a \subseteq B \quad \Rightarrow \quad a \in B.$$

Show that $D \subseteq B$.

34. Assume that

$$\{x, \{x, y\}\} = \{u, \{u, v\}\}$$

and show that $x = u$ and $y = v$.

35. Show that if $a^+ = b^+$, then $a = b$.

36. Show that the rank of any set S is the same as the rank of its transitive closure TC S (as defined in Exercise 7).

37. Show that a set α is an ordinal number iff it is a transitive set with the property that for any distinct x and y in α, either $x \in y$ or $y \in x$.

38. Show that whenever λ is a limit ordinal, then $\lambda = \bigcup \lambda$.

39. Prove that a set is an ordinal number iff it is a transitive set of transitive sets.

CHAPTER 8

ORDINALS AND ORDER TYPES

This chapter ends with a discussion of the arithmetic of ordinal numbers. That topic is preceded by material that is useful for ordinal arithmetic, and that also has independent interest.

The logical dependencies among the sections of this chapter are as follows. The section on alephs depends on the preceding section on transfinite recursion. The section on ordinal operations can be independent of the others, but it refers to the section on alephs for examples. The section on isomorphism types is independent of the earlier sections, and in turn forms the basis for the following section on the arithmetic of order types.

The section on ordinal arithmetic refers to the preceding section for addition and multiplication, and to transfinite recursion for exponentiation. But it is indicated how all of the operations can be based on transfinite recursion, thereby avoiding order types. In either event, the section on ordinal operations is utilized here.

TRANSFINITE RECURSION AGAIN

In Chapter 7 we defined a set V_α for each ordinal α. That definition proceeded in a somewhat roundabout manner, defining first a function

F_δ for each δ, and then proving that $F_\delta(\alpha)$ was actually independent of δ (as long as δ was greater than α). We could then define V_α to be $F_\delta(\alpha)$ for any large δ.

It is interesting to note that the goal of the construction was not a function but a *definition.* We could not hope to have a function F with $\mathsf{F}(\alpha) = V_\alpha$ for every α, because the domain of F would have to be the proper class of all ordinals.

We now want a general transfinite recursion theorem that will yield directly the definition of V_α, and will be applicable to other cases as well. In terms of proper classes, we can state this theorem as follows:

Let G be a function-class whose domain is the class V of all sets. Then we claim that there is a function-class F whose domain is the class of all ordinals and such that

$$\mathsf{F}(\alpha) = \mathsf{G}(\mathsf{F} \restriction \alpha)$$

for every α.

Now we must apply the standard rewording. In place of G we have a formula $\gamma(x, y)$ that defines G. The assumption that G is a function-class with domain V is reworded to state that for any x there is a unique y such that $\gamma(x, y)$ holds. In place of F we also have a formula $\varphi(x, y)$ defining F. For every ordinal α there must be a unique y such that $\varphi(\alpha, y)$ holds. The equation

$$\mathsf{F}(\alpha) = \mathsf{G}(\mathsf{F} \restriction \alpha)$$

is reworded to state that whenever f is a function with domain α such that $\varphi(\beta, f(\beta))$ holds for all $\beta \in \alpha$ (i.e., whenever $f = \mathsf{F} \restriction \alpha$) then the unique y such that $\varphi(\alpha, y)$ equals the unique y such that $\gamma(f, y)$.

Transfinite Recursion Schema on the Ordinals Let $\gamma(x, y)$ be any formula. Then we can construct another formula $\varphi(u, v)$ such that the following is a theorem.

Assume that for every f there is a unique set y such that $\gamma(f, y)$. Then for every ordinal number α there is a unique y such that $\varphi(\alpha, y)$. Furthermore whenever f is a function whose domain is an ordinal number α and such that $\varphi(\beta, f(\beta))$ for all β less than α, we then have

$$\varphi(\alpha, y) \quad \text{iff} \quad \gamma(f, y)$$

for every y.

We will give the proof presently. The significance of the theorem is that it justifies making a certain definition. We select some available symbol, say "t," and define the term

$$t_\alpha = \text{the unique } y \text{ such that } \varphi(\alpha, y)$$

for every ordinal α. The operation of going from α to t_α is γ-constructed in the following sense. Whenever f is the restriction of the operation to some ordinal α (i.e., $\operatorname{dom} f = \alpha$ and $f(\beta) = t_\beta$ for each β less than α), then we can conclude from the above theorem that $\gamma(f, t_\alpha)$. We might abbreviate this by writing

$$\gamma(t \restriction \alpha, t_\alpha).$$

Example We will apply the foregoing theorem to construct V_α. For γ use the formula we used before (in Lemma 7Q), so that $\gamma(f, y)$ is

$$y = \bigcup\{\mathscr{P}z \mid z \in \operatorname{ran} f\}.$$

Now plug this into the theorem and the subsequent discussion, but select the symbol "V" instead of "t." We get a set V_α for each α, but we must use the fact that the operation is γ-constructed to be sure that we have the right V_α. So consider any fixed α and let f be the restriction of the operation to α:

$$f(\beta) = V_\beta$$

for β less than α. Then we can conclude that $\gamma(f, V_\alpha)$, which when expanded becomes

$$\begin{aligned} V_\alpha &= \bigcup\{\mathscr{P}z \mid z \in \operatorname{ran} f\} \\ &= \bigcup\{\mathscr{P}V_\beta \mid \beta \in \alpha\}. \end{aligned}$$

Comparing this equation with Theorem 7S, we see (by transfinite induction) that we have the same V_α as in Chapter 7.

Proof of the theorem The formula $\varphi(\alpha, y)$ is the following:

There exists an ordinal δ greater than α and there exists a γ-constructed function F_δ with domain δ such that $F_\delta(\alpha) = y$.

As in Chapter 7, F_δ is said to be γ-constructed iff $\gamma(F_\delta \restriction \beta, F_\delta(\beta))$ for every $\beta \in \delta$.

First we will show that these F_δ functions always exist. We are given that for every f there is a unique y such that $\gamma(f, y)$. Hence for any ordinal δ we can apply transfinite recursion (from Chapter 7) to the well-ordered structure $\langle \delta, \in_\delta \rangle$. By doing so, we obtain the unique γ-constructed function F_δ with domain δ. Now suppose we are given some ordinal α and we want to find some y such that $\varphi(\alpha, y)$. We choose δ to be any larger ordinal (such as α^+) and take $y = F_\delta(\alpha)$. Then clearly $\varphi(\alpha, y)$ holds.

Next we will prove a uniqueness result. Namely, we claim that if F_δ and F_ε are γ-constructed functions with domains δ and ε, respectively, then $F_\delta(\alpha) = F_\varepsilon(\alpha)$ for all α in $\delta \cap \varepsilon$. (This corresponds to Lemma 7R in the special case of V_α.) The claim is proved by transfinite induction. For the inductive step, suppose that $F_\delta(\beta) = F_\varepsilon(\beta)$ for all $\beta \in \alpha$. Then $F_\delta \restriction \alpha = F_\varepsilon \restriction \alpha$. Since both $\gamma(F_\delta \restriction \alpha,\ F_\delta(\alpha))$ and $\gamma(F_\varepsilon \restriction \alpha,\ F_\varepsilon(\alpha))$, we can conclude that $F_\delta(\alpha) = F_\varepsilon(\alpha)$.

It now follows that for any ordinal α there is a *unique* y such that $\varphi(\alpha, y)$, and we know what this unique y is. Finally assume that f is a function whose domain is α and that $\varphi(\beta, f(\beta))$ holds for all $\beta \in \alpha$. We must show that

$$\varphi(\alpha, y) \quad \Leftrightarrow \quad \gamma(f, y)$$

for each y. But what is f? It must be the function F_α, by the above. The only y for which $\gamma(F_\alpha, y)$ is $F_\delta(\alpha)$, where δ is some larger ordinal. And it is also the only y for which $\varphi(\alpha, y)$. ⊣

In the next section we will see further examples of the use of transfinite recursion on the ordinals.

ALEPHS

Suppose that we have defined some class A of ordinals. Now A might or might not be a set. If A is bounded, i.e., if there is an ordinal β such that $\alpha \in \beta$ for each α in A, then $A \subseteq \beta^+$ and consequently A is a set. On the other hand, if A is unbounded, then $\bigcup A$ is the class of all ordinals. (This holds because any β, failing to bound A, is less than some α in A. Thus $\beta \in \alpha \in A$ and $\beta \in \bigcup A$.) By the Burali-Forti theorem, A cannot be a set. Thus we can conclude that A is a set iff it is bounded.

Now focus attention on the case where we have defined an unbounded (and hence proper) class of ordinals. As an example we will take the class of infinite cardinal numbers, but later we can apply our work to other classes. The class of infinite cardinal numbers is unbounded. (One way to prove this fact is to observe that for any ordinal β we obtain from Hartogs' theorem the least ordinal α not dominated by β. Then α is an initial ordinal and is larger than β. An alternative proof uses $2^{\text{card}\,\beta}$.)

The class of infinite cardinal numbers, like any class of ordinals, inherits the well ordering by epsilon on the class of all ordinals, given by Theorem 7M. We want to enumerate its members in ascending order. There is a least infinite cardinal, which we have always called $\aleph_0$. Then there is a next one (the least infinite cardinal greater than $\aleph_0$), which we call $\aleph_1$. And so forth.

We will now expand that "and so forth." Suppose we have worked our way up to α, i.e., we have defined $\aleph_\beta$ for all β less than α and we are ready to try defining $\aleph_\alpha$. (In the preceding paragraph $\alpha = 2$.) Naturally we define

$$\aleph_\alpha = \text{the least infinite cardinal different from } \aleph_\beta \text{ for every } \beta \text{ less than } \alpha.$$

Such a cardinal must exist, because $\{\aleph_\beta \mid \beta \in \alpha\}$ is merely a set, whereas the class of infinite cardinals is unbounded.

Now that we know how to construct $\aleph_\alpha$ from the smaller alephs, we can apply transfinite recursion on the ordinals. Choose for $\gamma(f, y)$ the formula

y is the least infinite cardinal not in ran f.

Then for any set f, there is a unique y such that $\gamma(f, y)$, again because ran f is merely a set whereas the class of infinite cardinals is unbounded. Then transfinite recursion lets us pick a symbol (we pick "$\aleph$") and define $\aleph_\alpha$ for each α in a way that is γ-constructed. That is, if f is the function with domain α defined by

$$f(\beta) = \aleph_\beta \qquad \text{for} \quad \beta \in \alpha,$$

then $\gamma(f, \aleph_\alpha)$. And $\gamma(f, \aleph_\alpha)$, when written out, becomes

$\aleph_\alpha$ is the least infinite cardinal not in ran f,

which by the definition of f becomes

$\aleph_\alpha$ is the least infinite cardinal different from $\aleph_\beta$ for every β less than α.

The following theorem verifies that this construction enumerates the infinite cardinals in ascending order.

Theorem 8A (a) If $\alpha \in \beta$, then $\aleph_\alpha < \aleph_\beta$.
(b) Every infinite cardinal is of the form $\aleph_\alpha$ for some α.

Proof (a) This is a consequence of the fact that $\aleph_\alpha$ is the least cardinal meeting certain conditions that become more stringent as α increases. Both $\aleph_\alpha$ and $\aleph_\beta$ meet the condition of being different from $\aleph_\gamma$ for all γ less than α. Since $\aleph_\alpha$ was the least such candidate (and $\aleph_\beta \neq \aleph_\alpha$), we have $\aleph_\alpha < \aleph_\beta$.

(b) We use transfinite induction on the class of infinite cardinal numbers (a subclass of the ordinals). Suppose, as the inductive hypothesis, that κ is an infinite cardinal for which all smaller infinite cardinals are in the range of the aleph operation. Consider the corresponding set $\{\beta \mid \aleph_\beta < \kappa\}$ of ordinals. This is a set (and not a proper class), being no larger than κ. And it is a transitive set by part (a). So it is itself an ordinal; call it α. By construction, $\aleph_\alpha$ is the least infinite cardinal different from $\aleph_\beta$ whenever $\aleph_\beta < \kappa$. By the inductive hypothesis, this is the least infinite cardinal different from all those less than κ, which is just κ itself. ⊣

The construction up to here is applicable to any unbounded class **A** of ordinals that we might have defined. We take for $\gamma(f, y)$ the formula

y is the least member of **A** not in ran f.

Transfinite recursion then lets us define t_α for every α in such a way that

t_α is the least member of A different from t_β for every β less than α.

The analogue of Theorem 8A holds (and with the same proof), showing that we have constructed an enumeration of A in ascending order.

Now that we have a legal definition of $\aleph_\alpha$, we can try to see what that definition can give us. There are three possibilities for α: zero, a successor ordinal, or a limit ordinal. We already know about $\aleph_0$, so consider the case where α is a successor ordinal, say $\alpha = \beta^+$. By the construction and Theorem 8A, $\aleph_{\beta^+}$ is the least infinite cardinal greater than $\aleph_\gamma$ for every γ less than β^+. We can simplify this since the largest candidate for $\aleph_\gamma$ is $\aleph_\beta$. Thus

$$\aleph_{\beta^+} = \text{the least cardinal greater than } \aleph_\beta.$$

By Exercise 23 of Chapter 7, the least cardinal greater than $\aleph_\beta$, as a set, is the set of all ordinals dominated by the set $\aleph_\beta$.

Now consider the case of limit ordinal λ. As before, $\aleph_\lambda$ is the least infinite cardinal greater than $\aleph_\beta$ for every β less than λ. We claim that in fact

$$\aleph_\lambda = \bigcup_{\beta \in \lambda} \aleph_\beta.$$

From Chapter 7 we know that $\bigcup_{\beta \in \lambda} \aleph_\beta$ is the least *ordinal* greater than $\aleph_\beta$ for every β in λ. The following lemma shows that it is actually a cardinal number.

Lemma 8B The union of any set of cardinal numbers is itself a cardinal number.

Proof Let A be any set of cardinals. Then $\bigcup A$ is an ordinal by Corollary 7N(d). We must show that it is an initial ordinal. So assume that $\alpha \in \bigcup A$; we must show that $\alpha \not\approx \bigcup A$. We know that $\alpha \in \kappa \in A$ for some cardinal number κ. Thus $\alpha \subseteq \kappa \subseteq \bigcup A$, so that if $\alpha \approx \bigcup A$, then $\alpha \approx \kappa$. But it is impossible to have $\alpha \approx \kappa$, since κ is an initial ordinal. Hence $\alpha \not\approx \bigcup A$. ⊣

As another application of transfinite recursion on the ordinals, we can define the beth numbers $\beth_\alpha$. (The letter $\beth$, called *beth*, is the second letter of the Hebrew alphabet.) The following three equations describe the term $\beth_\alpha$ that we want to define:

$$\beth_0 = \aleph_0,$$
$$\beth_{\alpha^+} = 2^{\beth_\alpha},$$
$$\beth_\lambda = \bigcup_{\alpha \in \lambda} \beth_\alpha,$$

where λ is a limit ordinal. The transfinite recursion machinery lets us convert these three equations into a legal definition.

To operate the machinery, let $\gamma(f, y)$ be the formula:

Either (i) f is a function with domain 0 and $y = \aleph_0$,
or (ii) f is a function whose domain is a successor ordinal α^+ and $y = 2^{f(\alpha)}$,
or (iii) f is a function whose domain is a limit ordinal λ, and $y = \bigcup(\text{ran } f)$,
or (iv) none of the above and $y = \varnothing$.

Then transfinite recursion gives us a formula φ with the usual properties. We select the symbol $\beth$ and define

$$\beth_\alpha = \text{the unique } y \text{ such that } \varphi(\alpha, y).$$

Then the machinery tells us that "$\gamma(\beth \restriction \alpha, \beth_\alpha)$." Writing out this condition with $\alpha = 0$ produces the equation

$$\beth_0 = \aleph_0.$$

Similarly, by using a successor ordinal and then a limit ordinal we get the other two equations

$$\beth_{\alpha^+} = 2^{\beth_\alpha},$$
$$\beth_\lambda = \bigcup_{\alpha \in \lambda} \beth_\alpha.$$

The continuum hypothesis (mentioned in Chapter 6) can now be stated by the equation $\aleph_1 = \beth_1$, and the generalized continuum hypothesis is the assertion that $\aleph_\alpha = \beth_\alpha$ for every α. (We are merely stating these hypotheses as objects for consideration; we are not claiming that they are true.)

Exercises

1. Show how to define a term t_α (for each ordinal α) so that $t_0 = 5$, $t_{\alpha^+} = (t_\alpha)^+$, and $t_\lambda = \bigcup_{\alpha \in \lambda} t_\alpha$ for a limit ordinal λ.

2. In the preceding exercise, show that if $\alpha \in \omega$, then $t_\alpha = 5 + \alpha$. Show that if $\omega \subseteq \alpha$, then $t_\alpha = \alpha$.

ORDINAL OPERATIONS

There are several operations on the class of ordinal numbers that are of interest to us. A few of these operations have already been defined: the successor operation assigning to each ordinal β its successor β^+, and the aleph and beth operations assigning to β the numbers $\aleph_\beta$ and $\beth_\beta$. More examples will be encountered when we study ordinal arithmetic.

These operations do not correspond to functions (which are *sets* of ordered pairs), because the class of all ordinals fails to be a set. Instead for each ordinal β we define an ordinal t_β as the unique ordinal meeting certain specified conditions. We will say that the operation is *monotone* iff the condition

$$\alpha \in \beta \quad \Rightarrow \quad t_\alpha \in t_\beta$$

always holds. We will say that the operation is *continuous* iff the equation

$$t_\lambda = \bigcup_{\beta \in \lambda} t_\beta$$

holds for every limit ordinal λ. Finally we will say that the operation is *normal* iff it is both monotone and continuous.

Example If $\alpha \in \beta$, then $\alpha^+ \in \beta^+$ (by Exercise 16 of Chapter 7). Thus the successor operation on the ordinals is monotone. But it lacks continuity, because $\omega^+ \neq \bigcup_{n \in \omega} n^+ = \omega$.

Example The aleph operation is normal by Theorem 8A (for monotonicity) and Lemma 8B with the accompanying discussion (for continuity). More generally, whenever the t_α's enumerate an unbounded class A of ordinals in ascending order, then monotonicity holds (by the analogue of Theorem 8A). Pursuing the analogy, we can assert that continuity holds if A is *closed* in the sense that the union of any nonempty set of members of A is itself a member of A. Conversely whenever we have defined some normal ordinal operation t_α, then its range-class

$$\{t_\alpha \mid \alpha \text{ is an ordinal number}\}$$

is a closed unbounded class of ordinal numbers (Exercise 6).

Recall (from the discussion following Corollary 7N) that whenever S is a set of ordinal numbers, then $\bigcup S$ is an ordinal that is the least upper bound of S. It is therefore natural to define the *supremum* of S (sup S) to be simply $\bigcup S$, with the understanding that we will use this notation only when S is a set of ordinal numbers. For example,

$$\omega = \sup\{0, 2, 4, \ldots\}.$$

The condition for an operation taking β to t_β to be continuous can be stated

$$t_\lambda = \sup\{t_\beta \mid \beta \in \lambda\}$$

for limit ordinals λ.

The following result shows that for a continuous operation, the condition for monotonicity can be replaced by a more "local" version.

Theorem Schema 8C Assume that we have defined a continuous operation assigning an ordinal t_β to each ordinal number β. Then the operation is monotone provided that $t_\gamma \in t_{\gamma^+}$ for every ordinal γ.

Proof We consider a fixed ordinal α and prove by transfinite induction on β that

$$\alpha \in \beta \quad \Rightarrow \quad t_\alpha \in t_\beta .$$

Case I β is zero. Then the above condition is vacuously true, since $\alpha \notin \beta$.

Case II β is a successor ordinal γ^+. Then

$$\begin{aligned} \alpha \in \beta \quad &\Rightarrow \quad \alpha \in\!\!\!\!\_ \ \gamma \\ &\Rightarrow \quad t_\alpha \in\!\!\!\!\_ \ t_\gamma \qquad \text{by the inductive hypothesis} \\ &\Rightarrow \quad t_\alpha \in t_{\gamma^+} \qquad \text{since} \quad t_\gamma \in t_{\gamma^+} \\ &\Rightarrow \quad t_\alpha \in t_\beta . \end{aligned}$$

Case III β is a limit ordinal. Then

$$\begin{aligned} \alpha \in \beta \quad &\Rightarrow \quad \alpha \in \alpha^+ \in \beta \\ &\Rightarrow \quad t_\alpha \in t_{\alpha^+} \in\!\!\!\!\_ \ t_\beta \end{aligned}$$

by continuity. Hence $t_\alpha \in t_\beta$. ⊣

Example The beth operation (assigning $\beth_\beta$ to β) is normal. The continuity is obvious from its definition. And since $\beth_\gamma$ is always less than $2^{\beth_\gamma}$, we have (by the above theorem) monotonicity.

The next results will have useful consequences for the arithmetic of ordinal numbers.

Theorem Schema 8D Assume that we have defined a normal operation assigning an ordinal t_α to each ordinal α. Then for any given ordinal β that is at least as large as t_0, there exists a *greatest* ordinal γ such that $t_\gamma \in\!\!\!\!\_ \ \beta$.

We know that any nonempty set of ordinals has a least member, but the above theorem asserts that the set $\{\gamma \mid t_\gamma \in\!\!\!\!\_ \ \beta\}$ has a *greatest* member.

Proof Consider the set $\{\alpha \mid t_\alpha \in\!\!\!\!\_ \ \beta\}$. This is a set of ordinals and it is (by monotonicity) a transitive set. So it is itself an ordinal. It is not 0, because $t_0 \in\!\!\!\!\_ \ \beta$. Could it be a limit ordinal λ? If so, then

$$t_\lambda = \sup\{t_\alpha \mid \alpha \in \lambda\} \in\!\!\!\!\_ \ \beta,$$

whence $\lambda \in \lambda$. This is impossible, so our set must in fact be a successor ordinal γ^+. Thus γ is the largest member of the set, and so is the largest ordinal for which $t_\gamma \underline{\in} \beta$. ⊣

Theorem Schema 8E Assume that we have defined a normal operation assigning t_α to each ordinal number α. Let S be a nonempty set of ordinal numbers. Then

$$t_{\sup S} = \sup\{t_\alpha \mid \alpha \in S\}.$$

Proof We get the "$\geq$" half by monotonicity:

$$\begin{aligned} \alpha \in S \quad &\Rightarrow \quad \alpha \underline{\in} \sup S \\ &\Rightarrow \quad t_\alpha \underline{\in} t_{\sup S}, \end{aligned}$$

whence $\sup\{t_\alpha \mid \alpha \in S\} \underline{\in} t_{\sup S}$.

For the other inequality, there are two cases. If S has a largest member δ, then $\sup S = \delta$ and so $t_{\sup S} = t_\delta \underline{\in} \sup\{t_\alpha \mid \alpha \in S\}$.

If S has no largest member, then $\sup S$ must be a limit ordinal (since $S \neq \varnothing$). So by continuity,

$$t_{\sup S} = \sup\{t_\beta \mid \beta \in \sup S\}.$$

If $\beta \in \sup S$, then $\beta \in \gamma \in S$ for some γ. Consequently $t_\beta \in t_\gamma$, and so

$$t_\beta \in \sup\{t_\alpha \mid \alpha \in S\}.$$

Since β was an arbitrary member of $\sup S$,

$$\sup\{t_\beta \mid \beta \in \sup S\} \underline{\in} \sup\{t_\alpha \mid \alpha \in S\}$$

as desired. ⊣

For a monotone operation, we always have $\beta \underline{\in} t_\beta$, by Exercise 5. Is it possible to have $\beta = t_\beta$? Yes; the identity operation (assigning β to β) is normal. But even for normal operations with rapid growth (such as the beth operation), there are "fixed points" β with $\beta = t_\beta$. The following theorem, although not essential to our later work, shows that in fact such fixed points form a proper class.

Veblen Fixed-Point Theorem Schema (1907) Assume that we have defined a normal operation assigning t_α to each ordinal number α. Then the operation has arbitrarily large fixed points, i.e., for every ordinal number β we can find an ordinal number γ with $t_\gamma = \gamma$ and $\beta \underline{\in} \gamma$.

Proof From the monotonicity we have $\beta \underline{\in} t_\beta$. If $\beta = t_\beta$, we are done—just take $\gamma = \beta$. So we may assume that $\beta \in t_\beta$. Then by monotonicity

$$\beta \in t_\beta \in t_{t_\beta} \in \cdots.$$

We claim that the supremum of these ordinals is the desired fixed point.

More formally, we define by recursion a function f from ω into the ordinals such that

$$f(0) = \beta \qquad \text{and} \qquad f(n^+) = t_{f(n)}.$$

Thus $f(1) = t_\beta$, $f(2) = t_{t_\beta}$, and so forth. We will write t_β^n for $f(n)$. As mentioned above, by monotonicity $t_\beta^n \in t_\beta^{n^+}$, so that

$$\beta \in t_\beta \in t_\beta^2 \in \cdots.$$

Define $\lambda = \sup \operatorname{ran} f = \sup\{t_\beta^n \mid n \in \omega\}$. We claim that $t_\lambda = \lambda$.

Clearly λ has no largest element, and hence is a limit ordinal. Let $S = \operatorname{ran} f = \{t_\beta^n \mid n \in \omega\}$, so that $\lambda = \sup S$. By Theorem 8E,

$$\begin{aligned} t_\lambda &= \sup\{t_\alpha \mid \alpha \in S\} \\ &= \sup\{t_\beta^{n^+} \mid n \in \omega\} \\ &= \lambda. \end{aligned}$$

Thus we have a fixed point. ⊣

This proof actually gives us the *least* fixed point that is at least as large as β (by Exercise 7). For example, consider the aleph operation with $\beta = 0$. Then

$$0 \in \aleph_0 \in \aleph_{\aleph_0} \in \cdots$$

and the supremum of these numbers is the least fixed point.

Since the class of fixed points of the operation t is an unbounded class of ordinals, we can define the *derived* operation t' enumerating the fixed points in ascending order. The definition of t' is produced by the transfinite recursion machinery; the crucial equation is

$$t'_\alpha = \text{the least fixed point of } t \text{ different from } t'_\beta \text{ for every } \beta \in \alpha.$$

It turns out (Exercise 8) that t' is again a normal operation.

Exercises

3. Assume that we have defined a monotone operation assigning t_α to each ordinal α. Show that

$$\beta \in \gamma \iff t_\beta \in t_\gamma$$

and

$$t_\beta = t_\gamma \implies \beta = \gamma$$

for any ordinals β and γ.

4. Assume that we have defined a normal operation assigning t_α to each α. Show that whenever λ is a limit ordinal, then t_λ is also a limit ordinal.

5. Assume that we have defined a monotone operation assigning t_α to each α. Show that $\beta \subseteq t_\beta$ for every ordinal number β.

6. Assume that we have defined a normal operation assigning t_α to each α. Show that the range-class

$$\{t_\alpha \mid \alpha \text{ is an ordinal number}\}$$

is a closed unbounded class of ordinal numbers.

7. Show that the proof of Veblen's fixed-point theorem produces the *least* fixed point that is at least as large as β.

8. Assume that we have defined a normal operation assigning t_α to each α. Let t' be the derived operation, enumerating the fixed points of t in order. Show that t' is again a normal operation.

ISOMORPHISM TYPES

We have ordinal numbers to measure the length of well orderings. If you know the ordinal number of a well-ordered structure, then you know everything there is to know about that structure, "to within isomorphism." This is an informal way of saying that two well-ordered structures receive the same ordinal iff they are isomorphic (Theorem 7I).

Our next undertaking will be to extend these ideas to handle structures that are *not* well ordered. Although our intent is to apply the extended ideas to linearly ordered structures, the initial definitions will be quite general.

Consider then a structure $\langle A, R\rangle$. The "shape" of this structure is to be measured to within isomorphism by $\text{it}\langle A, R\rangle$, the isomorphism type of $\langle A, R\rangle$. This is to be defined in such a way that two structures receive the same isomorphism type iff they are isomorphic:

$$\text{it}\langle A, R\rangle = \text{it}\langle B, S\rangle \quad \text{iff} \quad \langle A, R\rangle \cong \langle B, S\rangle.$$

As a first guess, we could argue as follows: Isomorphism is an equivalence concept, so why not take $\text{it}\langle A, R\rangle$ to be the equivalence class of $\langle A, R\rangle$? Then as in Lemma 3N, two structures should have the same equivalence class iff they are isomorphic.

In one way, this first guess is a very bad idea. The "equivalence classes" here will fail to be sets. If A is nonempty, then no set contains every structure that is isomorphic to $\langle A, R\rangle$; this is Exercise 9.

With but one modification, this first guess will serve our needs perfectly. Do not take *all* structures isomorphic to $\langle A, R\rangle$, just those of least possible rank α. Then these structures will be in V_{α^+}, and so $\text{it}\langle A, R\rangle$ will be a set. We now will write this down officially.

Definition Let R be a binary relation on A. The *isomorphism type* $\text{it}\langle A, R\rangle$ of the structure $\langle A, R\rangle$ is the set of all structures $\langle B, S\rangle$ such that

(i) $\langle A, R\rangle \cong \langle B, S\rangle$, and
(ii) no structure of rank less than $\text{rank}\langle B, S\rangle$ is isomorphic to $\langle B, S\rangle$.

It will help to bring in (but only for very temporary use) some terminology. Call a structure *pioneering* iff there is no structure of smaller rank isomorphic to it. If two pioneering structures are isomorphic, then clearly they have the same rank.

We claim that any structure $\langle A, R\rangle$ is isomorphic to some (not necessarily unique) pioneering structure. The class

$$\{\alpha \mid \alpha \text{ is the rank of a structure isomorphic to } \langle A, R\rangle\}$$

is nonempty ($\text{rank}\langle A, R\rangle$ is in it). So it has a least element α_0, which we can call the *pioneer ordinal* for $\langle A, R\rangle$. Some structure of rank α_0 is isomorphic to $\langle A, R\rangle$, and it is pioneering by the leastness of α_0.

The definition of $\text{it}\langle A, R\rangle$ can now be phrased: It is the set of all pioneering structures that are isomorphic to $\langle A, R\rangle$. This set has been seen to be nonempty.

Note that $\text{it}\langle A, R\rangle$ is indeed a set, being a subset of V_{α^+}, where α is the pioneer ordinal for $\langle A, R\rangle$.

Theorem 8F Structures $\langle A, R\rangle$ and $\langle B, S\rangle$ have the same isomorphism type iff they are isomorphic:

$$\text{it}\langle A, R\rangle = \text{it}\langle B, S\rangle \quad \text{iff} \quad \langle A, R\rangle \cong \langle B, S\rangle.$$

Proof First assume that $\text{it}\langle A, R\rangle = \text{it}\langle B, S\rangle$. Let $\langle C, T\rangle$ be any structure in this common isomorphism type. Then

$$\langle A, R\rangle \cong \langle C, T\rangle \cong \langle B, S\rangle.$$

For the converse, assume that $\langle A, R\rangle \cong \langle B, S\rangle$. Then the same structures are isomorphic to each of these two; in particular, the same pioneering structures are isomorphic to each. Hence $\text{it}\langle A, R\rangle = \text{it}\langle B, S\rangle$. ⊣

Digression Concerning Cardinal Numbers The crucial property of cardinal numbers,

$$\text{card } A = \text{card } B \quad \text{iff} \quad A \approx B,$$

looks a lot like Theorem 8F, simplified by omission of the binary relations. This indicates the possibility of an alternative definition of cardinal numbers.

Specifically, define kard A to be the set of all sets B equinumerous to A and having the least possible rank. Then (as in Theorem 8F),

$$\text{kard } A = \text{kard } B \quad \text{iff} \quad A \approx B.$$

In comparing card A with kard A, we see that the definition of card A uses the axiom of choice (but not regularity). The definition of kard A relies on regularity but does not require the axiom of choice. (This is a point in favor of kard.) For a finite nonempty set A, kard A fails to be a natural number. (This is a point in favor of card.)

Exercises

9. Assume that $\langle A, R\rangle$ is a structure with $A \neq \varnothing$. Show that no set contains every structure isomorphic to $\langle A, R\rangle$.

10. (a) Show that kard $n = \{n\}$ for $n = 0$, 1, and 2.
    (b) Calculate kard 3.

ARITHMETIC OF ORDER TYPES

We now focus attention solely on the order types.

Definition An *order type* is the isomorphism type of some linearly ordered structure.

We will use Greek letters $\rho, \sigma, \ldots$ for order types. Any member of an order type ρ is said to be a linearly ordered structure of *type* ρ.

First we want to define addition of order types. The basic idea is that $\rho + \sigma$ should be the order type "first ρ and then σ." For the real definition of $\rho + \sigma$, first select $\langle A, R\rangle$ of type ρ and $\langle B, S\rangle$ of type σ with $A \cap B = \varnothing$. (This is possible, by Exercise 11.) Then define the relation $R \oplus S$ by

$$R \oplus S = R \cup S \cup (A \times B).$$

(We reject the cumbersome $R\,{}_A\!\oplus_B S$ notation.) Finally we define $\rho + \sigma$ by

$$\rho + \sigma = \text{it}\langle A \cup B, R \oplus S\rangle.$$

The idea is that we want to order $A \cup B$ in such a way that any member of A is less than any member of B. But within A and B, we order according to R and S.

The next lemma verifies that $\rho + \sigma$ is a well-defined order type.

Lemma 8G Assume that $\langle A, R\rangle$ and $\langle B, S\rangle$ are linearly ordered structures with A and B disjoint.

(a) $R \oplus S$ is a linear ordering on $A \cup B$.
(b) If $\langle A, R\rangle \cong \langle A', R'\rangle$, $\langle B, S\rangle \cong \langle B', S'\rangle$, and $A' \cap B' = \varnothing$, then

$$\langle A \cup B, R \oplus S\rangle \cong \langle A' \cup B', R' \oplus S'\rangle.$$

Proof (a) First of all, $R \oplus S$ is irreflexive because a pair of the form $\langle x, x\rangle$ cannot belong to R or S, nor can it belong to $A \times B$ since A and B are disjoint.

To check that $R \oplus S$ is transitive, suppose that $\langle x, y\rangle$ and $\langle y, z\rangle$ belong to $R \oplus S$. If both of these pairs come from R (or from S), then of course xRz (or xSz). Otherwise one of the pairs comes from $A \times B$. We may suppose that $\langle x, y\rangle \in A \times B$; the other case is similar. Then $x \in A$ and the pair $\langle y, z\rangle$ comes from S (because $y \in B$). Hence $z \in B$ and $\langle x, z\rangle \in A \times B$.

Finally, it is easy to see that $R \oplus S$ is connected on $A \cup B$. Consider any x and y in $A \cup B$. There are three cases: both in A, both in B, or mixed. But all three are trivial.

Part (b) is sufficiently straightforward so that nothing more need be said. ⊣

The role of this lemma is to assure us that when we define

$$\rho + \sigma = \text{it}\langle A \cup B, R \oplus S\rangle,$$

the end result will be independent of just which structures of type ρ and σ are utilized.

Example For each ordinal α, we have the order type it$\langle \alpha, \in_\alpha\rangle$. Distinct ordinals yield distinct order types, since

$$\begin{aligned} \text{it}\langle \alpha, \in_\alpha\rangle = \text{it}\langle \beta, \in_\beta\rangle \quad &\Rightarrow \quad \langle \alpha, \in_\alpha\rangle \cong \langle \beta, \in_\beta\rangle \\ &\Rightarrow \quad \alpha = \beta \end{aligned}$$

for ordinals, by Theorems 7I and 7L. The order type it$\langle \alpha, \in_\alpha\rangle$ will be denoted as $\bar{\alpha}$. In particular, we have the order types $\bar{1}$, $\bar{3}$, and $\bar{\omega}$ (which are it$\langle 1, \in_1\rangle$ and so forth). And $\bar{1} + \bar{3} = \bar{4}$. (Please explain why.) Also $\bar{1} + \bar{\omega} = \bar{\omega}$, whereas $\bar{\omega} + \bar{1} = \overline{\omega^+}$.

Example It is traditional to use η and λ to denote the order type of the rationals and reals, respectively (in their usual ordering):

$$\eta = \text{it}\langle \mathbb{Q}, <_Q\rangle \qquad \text{and} \qquad \lambda = \text{it}\langle \mathbb{R}, <_R\rangle.$$

Then $\bar{1} + \eta$ is an order type with a least element but no greatest element. (Or more pedantically, the ordered structures of type $\bar{1} + \eta$ have least elements but not greatest elements.) But $\eta + \bar{1}$ has a greatest element but no least element; hence $\bar{1} + \eta \neq \eta + \bar{1}$. Also $\lambda + \lambda \neq \lambda$, because in type $\lambda + \lambda$ there is a bounded nonempty set without a least upper bound. On the other hand $\eta + \eta = \eta$. This is not obvious, but see Exercise 19.

Any order type ρ can be run backwards to yield an order type ρ^*. More specifically, we can select a linearly ordered structure $\langle A, R\rangle$ of type ρ. Then we define

$$\rho^* = \text{it}\langle A, R^{-1}\rangle.$$

It is routine to check that ρ^* is a well-defined order type (compare Exercise 43 of Chapter 3). For any *finite* ordinal α, we have $\bar{\alpha}^* = \bar{\alpha}$ by Exercise 19 of Chapter 7. But $\bar{\omega}^* \neq \bar{\omega}$; in fact $\bar{\omega}^*$ is the order type of the negative integers, which are not well ordered. It is easy to see that $\eta^* = \eta$ and $\lambda^* = \lambda$.

Example The sum $\bar{\omega}^* + \bar{\omega}$ is the order type of the set $\mathbb{Z}$ of integers. But $\bar{\omega} + \bar{\omega}^*$ is different; it has both a least element and a greatest element, and infinitely many points between the two.

Now consider the multiplication of order types. The product $\rho \cdot \sigma$ can be described informally as "ρ, σ times." More formally, we select structures $\langle A, R\rangle$ and $\langle B, S\rangle$ of types ρ and σ, respectively. Then define $R * S$ to be "Hebrew lexicographic order" on $A \times B$:

$$\langle a_1, b_1\rangle(R * S)\langle a_2, b_2\rangle \quad \text{iff} \quad \text{either } b_1 S b_2 \quad \text{or } (b_1 = b_2 \text{ and } a_1 R a_2).$$

This orders the pairs in $A \times B$ according to their second coordinates, and then by their first coordinates when the second coordinates coincide. We can now define the product

$$\rho \cdot \sigma = \text{it}\langle A \times B, R * S\rangle.$$

The next lemma verifies that $\rho \cdot \sigma$ is a well-defined order type. This lemma does for multiplication what Lemma 8G does for addition.

Lemma 8H Assume that $\langle A, R\rangle$ and $\langle B, S\rangle$ are linearly ordered structures.

(a) $R * S$ is a linear ordering on $A \times B$.
(b) If $\langle A, R\rangle \cong \langle A', R'\rangle$ and $\langle B, S\rangle \cong \langle B', S'\rangle$, then

$$\langle A \times B, R * S\rangle \cong \langle A' \times B', R' * S'\rangle.$$

Proof (a) It is easy to see from the definition of $R * S$ that it is irreflexive (because both R and S are) and is connected on $A \times B$. It remains to verify that it is a transitive relation. So assume that

$$\langle a_1, b_1\rangle(R * S)\langle a_2, b_2\rangle \qquad \text{and} \qquad \langle a_2, b_2\rangle(R * S)\langle a_3, b_3\rangle.$$

This assumption breaks down into four cases, as illustrated in Fig. 47. But in each case, we have $\langle a_1, b_1\rangle(R * S)\langle a_3, b_3\rangle$.

Part (b) is again sufficiently straightforward that nothing more need be said. ⊣

$\langle a_1, b_1\rangle(R * S)\langle a_2, b_2\rangle$ \ $\langle a_2, b_2\rangle(R * S)\langle a_3, b_3\rangle$	$b_2 S b_3$	$b_2 = b_3$ & $a_2 R a_3$
$b_1 S b_2$	$b_1 S b_3$	$b_1 S b_3$
$b_1 = b_2$ & $a_1 R a_2$	$b_1 S b_3$	$b_1 = b_3$ & $a_1 R a_3$

Fig. 47. Transitivity of $R * S$.

Examples The product $\bar{\omega} \cdot \bar{2}$ is "$\bar{\omega}$, $\bar{2}$ times." The set $\omega \times 2$ under Hebrew lexicographic ordering is:

$$\langle 0, 0\rangle \quad \langle 1, 0\rangle, \quad \langle 2, 0\rangle, \ldots, \langle 0, 1\rangle, \quad \langle 1, 1\rangle, \quad \langle 2, 1\rangle, \ldots.$$

On the other hand $\bar{2} \cdot \bar{\omega}$ is "$\bar{2}$, $\bar{\omega}$ times." The set $2 \times \omega$ under Hebrew lexicographic order is:

$$\langle 0, 0\rangle \quad \langle 1, 0\rangle, \quad \langle 0, 1\rangle, \quad \langle 1, 1\rangle, \quad \langle 0, 2\rangle, \langle 1, 2\rangle, \ldots.$$

This ordering, unlike the other, is isomorphic to the natural numbers, i.e., $\bar{2} \cdot \bar{\omega} = \bar{\omega}$. In particular, $\bar{2} \cdot \bar{\omega} \neq \bar{\omega} \cdot \bar{2}$.

The next theorem gives some of the general laws that addition and multiplication obey. We have already seen that neither operation obeys the commutative law in general; for example,

$$\bar{1} + \bar{\omega} \neq \bar{\omega} + \bar{1} \qquad \text{and} \qquad \bar{2} \cdot \bar{\omega} \neq \bar{\omega} \cdot \bar{2}.$$

Furthermore the right distributive law fails; for example,

$$(\bar{\omega} + \bar{1}) \cdot \bar{2} \neq \bar{\omega} \cdot \bar{2} + \bar{1} \cdot \bar{2}$$

(Exercise 15).

Theorem 8I The following identities hold for any order types.

(a) Associative laws

$$(\rho + \sigma) + \tau = \rho + (\sigma + \tau),$$
$$(\rho \cdot \sigma) \cdot \tau = \rho \cdot (\sigma \cdot \tau).$$

(b) Left distributive law

$$\rho \cdot (\sigma + \tau) = (\rho \cdot \sigma) + (\rho \cdot \tau).$$

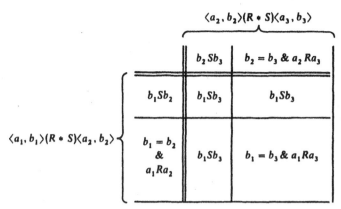

(c) Identity elements

$$\rho + \bar{0} = \bar{0} + \rho = \rho,$$
$$\rho \cdot \bar{1} = \bar{1} \cdot \rho = \rho,$$
$$\rho \cdot \bar{0} = \bar{0} \cdot \rho = \bar{0}.$$

Proof Both $(\rho + \sigma) + \tau$ and $\rho + (\sigma + \tau)$ produce the "first ρ then σ and then τ" ordering. Let $\langle A, R\rangle$, $\langle B, S\rangle$, and $\langle C, T\rangle$ be of type ρ, σ, and τ, respectively, with A, B, and C disjoint. Then both $(R \oplus S) \oplus T$ and $R \oplus (S \oplus T)$ are easily shown to be

$$R \cup S \cup T \cup A \times B \cup A \times C \cup B \times C.$$

Both $(\rho \cdot \sigma) \cdot \tau$ and $\rho \cdot (\sigma \cdot \tau)$ produce Hebrew lexicographic order on $A \times B \times C$. $(R * S) * T$ is an ordering on $(A \times B) \times C$, and $R * (S * T)$ is an ordering on $A \times (B \times C)$, but we can establish an isomorphism. In detail, the condition for $\langle a_1, b_1, c_1\rangle$ to be less than $\langle a_2, b_2, c_2\rangle$ under $(R * S) * T$ is, when expanded,

$$c_1 T c_2 \quad \text{or} \quad (c_1 = c_2 \;\&\; b_1 S b_2) \quad \text{or} \quad (c_1 = c_2 \;\&\; b_1 = b_2 \;\&\; a_1 R a_2).$$

The condition for $\langle a_1, \langle b_1, c_1\rangle\rangle$ to be less than $\langle a_2, \langle b_2, c_2\rangle\rangle$ under $R * (S * T)$ is, when expanded, exactly the same.

In the left distributive law, both $R * (S \oplus T)$ and $(R * S) \oplus (R * T)$ are orderings on the set $(A \times B) \cup (A \times C)$. And in both cases the condition for $\langle a_1, x\rangle$ to be less than $\langle a_2, y\rangle$ turns out to be

$$xSy \quad \text{or} \quad xTy \quad \text{or} \quad (x \in B \;\&\; y \in C) \quad \text{or} \quad (x = y \;\&\; a_1 R a_2).$$

Part (c) is straightforward. ⊣

Exercises

11. Show that for any order types ρ and σ there exist structures $\langle A, R\rangle$ and $\langle B, S\rangle$ of types ρ and σ, respectively, such that $A \cap B = \varnothing$.

12. Prove that for any linearly ordered structures,

$$\text{it}\langle A, R\rangle + \text{it}\langle B, S\rangle = \text{it}\langle(\{0\} \times A) \cup (\{1\} \times B), <_L\rangle,$$

where $<_L$ is lexicographic ordering.

13. Supply proofs for part (b) of Lemma 8G and part (b) of Lemma 8H.

14. Assume that $\rho \cdot \sigma = \bar{0}$. Show that either $\rho = \bar{0}$ or $\sigma = \bar{0}$.

15. Show that $(\bar{\omega} + \bar{1}) \cdot \bar{2}$ is not the same as $(\bar{\omega} \cdot \bar{2}) + (\bar{1} \cdot \bar{2})$.

16. Supply a proof for part (c) of Theorem 8I.

17. A partial ordering R is said to be *dense* iff whenever xRz, then xRy and yRz for some y. For example, the usual ordering $<_Q$ on the set $\mathbb{Q}$ of rational numbers is dense. Assume that $\langle A, R\rangle$ is a linearly ordered structure with A countable and R dense. Show that $\langle A, R\rangle$ is isomorphic to $\langle B, <_Q^\circ\rangle$ for some subset B of $\mathbb{Q}$. [*Suggestion*: Suppose that $A = \{a_0, a_1, \ldots\}$. Define $f(a_i)$ by recursion on i.]

18. Assume that $\langle A, R\rangle$ and $\langle B, S\rangle$ are both linearly ordered structures with dense orderings. Assume that A and B are countable and nonempty. Assume that neither ordering has a first or last element. Show that $\langle A, R\rangle \cong \langle B, S\rangle$. [Suppose that $A = \{a_0, a_1, \ldots\}$ and $B = \{b_0, b_1, \ldots\}$. At stage $2n$ be sure that a_n is paired with some suitable b_j and at stage $2n+1$ be sure that b_n is paired with some suitable a_i.]

19. Use the preceding exercise to show that $\eta + \eta = \eta \cdot \eta = \eta$.

ORDINAL ARITHMETIC

There are two available methods for defining addition and multiplication on the ordinals. One method uses transfinite recursion on the ordinals, extending the definitions by recursion used for the finite ordinals in Chapter 4. The second method uses our more recent work with order types, and defines $\alpha + \beta$ to be the ordinal γ such that $\bar{\alpha} + \bar{\beta} = \bar{\gamma}$ (under addition of order types).

Since both methods have their advantages, we will show that they produce exactly the same operations. We will then be able to draw on either method as the occasion demands.

We will say that an order type ρ is *well ordered* iff the structures of type ρ are well ordered.

Theorem 8J The sum and the product of well-ordered types is again well ordered.

Proof Assume that $\langle A, R\rangle$ and $\langle B, S\rangle$ are well-ordered structures with A and B disjoint. What is to be proved is that $\langle A \cup B, R \oplus S\rangle$ and $\langle A \times B, R * S\rangle$ are well-ordered structures. If C is a nonempty subset of $A \cup B$, then either $C \cap A \neq \varnothing$ (in which case its R-least element is $(R \oplus S)$-least in C) or else $C \subseteq B$ (in which case its S-least element is $(R \oplus S)$-least).Similarly if D is a nonempty subset of $A \times B$, we first take the S-least b_0 in ran D. Let a_0 be the R-least member of $\{a \mid \langle a, b_0\rangle \in D\}$. Then $\langle a_0, b_0\rangle$ is the $(R * S)$-least element of D. ⊣

The above theorem lets us transfer addition and multiplication from well-ordered types directly to the ordinals. For ordinals α and β, we have the well-ordered types $\bar{\alpha} + \bar{\beta}$ and $\bar{\alpha} \cdot \bar{\beta}$. The ordinals of these types are defined to be $\alpha + \beta$ and $\alpha \cdot \beta$.

Definition Let α and β be ordinal numbers. Define the sum $\alpha + \beta$ to be the unique ordinal γ such that $\bar{\alpha} + \bar{\beta} = \bar{\gamma}$. Define the product $\alpha \cdot \beta$ to be the unique ordinal δ such that $\bar{\alpha} \cdot \bar{\beta} = \bar{\delta}$.

To be more specific, observe that $\bar{\alpha} + \bar{\beta}$ is the order type of

$$(\{0\} \times \alpha) \cup (\{1\} \times \beta)$$

in lexicographic order $<_L$ (compare Exercise 12). Consequently $\alpha + \beta$ is the ordinal number of the structure

$$\langle (\{0\} \times \alpha) \cup (\{1\} \times \beta), <_L \rangle.$$

This is the ordinal number that measures the "first α and then β" ordering. Similarly $\bar{\alpha} \cdot \bar{\beta}$ is the order type of $\alpha \times \beta$ with Hebrew lexicographic order $<_H$. Consequently $\alpha \cdot \beta$ is the ordinal number of the structure $\langle \alpha \times \beta, <_H \rangle$. This is the ordinal number that measures the "α, β times" ordering.

We use the same symbols $+$ and $\cdot$ for both order type arithmetic and ordinal arithmetic, but we will try always to be clear about which is intended.

Example To calculate the sum $1 + \omega$, we shift to the order types $\bar{1} + \bar{\omega}$. From a previous example $\bar{1} + \bar{\omega} = \bar{\omega}$, so going back to ordinals we have $1 + \omega = \omega$. Similarly from the known equations $\bar{\omega} + \bar{1} = \overline{\omega^+}$ and $\bar{2} \cdot \bar{\omega} = \bar{\omega}$, we obtain the corresponding equations for ordinals: $\omega + 1 = \omega^+$ and $2 \cdot \omega = \omega$.

Example We claim that $\alpha + 1 = \alpha^+$ for any ordinal α. The sum $\bar{\alpha} + \bar{1}$ is the order type of $\alpha \cup \{s\}$ (where $s \notin \alpha$) under the ordering that makes s the largest element. This is the same as the order type of $\alpha \cup \{\alpha\}$ under the epsilon ordering, which is just $\overline{\alpha^+}$. So we have $\bar{\alpha} + \bar{1} = \overline{\alpha^+}$, whereupon $\alpha + 1 = \alpha^+$.

Note that the definition of ordinal addition and multiplication yields the equations

$$\bar{\alpha} + \bar{\beta} = \overline{\alpha + \beta} \qquad \text{and} \qquad \bar{\alpha} \cdot \bar{\beta} = \overline{\alpha \cdot \beta}.$$

To verify an equation, e.g., $\omega \cdot 2 = \omega + \omega$, we can use the following strategy. If suffices to show that the order types $\overline{\omega \cdot 2}$ and $\overline{\omega + \omega}$ are the same, since the assignment of order types to ordinals is one-to-one. By the above equations, this reduces to verifying that $\bar{\omega} \cdot \bar{2} = \bar{\omega} + \bar{\omega}$. And this can be done by selecting representative structures for each side of the equation and showing them to be isomorphic. In recent notation, the fact that

$$\langle \omega \times 2, <_H \rangle \cong \langle (\{0\} \times \omega) \cup (\{1\} \times \omega), <_L \rangle$$

does the job. (Alternatively, we can appeal to the next theorem to obtain $\omega \cdot 2 = \omega + \omega$.)

The laws previously established for all order types must in particular be valid for ordinals:

Theorem 8K For any ordinal numbers:

$$(\alpha + \beta) + \gamma = \alpha + (\beta + \gamma),$$
$$(\alpha \cdot \beta) \cdot \gamma = \alpha \cdot (\beta \cdot \gamma),$$
$$\alpha \cdot (\beta + \gamma) = (\alpha \cdot \beta) + (\alpha \cdot \gamma),$$
$$\alpha + 0 = 0 + \alpha = \alpha,$$
$$\alpha \cdot 1 = 1 \cdot \alpha = \alpha,$$
$$\alpha \cdot 0 = 0 \cdot \alpha = 0.$$

Proof We know from Theorem 8I that addition of order types is associative, so

$$(\bar{\alpha} + \bar{\beta}) + \bar{\gamma} = \bar{\alpha} + (\bar{\beta} + \bar{\gamma}).$$

Hence the ordinals $(\alpha + \beta) + \gamma$ and $\alpha + (\beta + \gamma)$ have the same order type, and thus are equal. The same argument is applicable to the other parts of the theorem. ⊣

Our work with order types also provides us with *counterexamples* to the commutative laws and the right distributive laws:

$$1 + \omega \neq \omega + 1,$$
$$2 \cdot \omega \neq \omega \cdot 2,$$
$$(\omega + 1) \cdot 2 \neq (\omega \cdot 2) + (1 \cdot 2).$$

The next theorem gives the equations that characterize addition and multiplication by recursion. Recall that for a set S of ordinals, its supremum $\sup S$ is just $\bigcup S$.

Theorem 8L For any ordinal numbers α and β and any limit ordinal λ the following equations hold.

(A1) $$\alpha + 0 = \alpha,$$

(A2) $$\alpha + \beta^+ = (\alpha + \beta)^+,$$

(A3) $$\alpha + \lambda = \sup\{\alpha + \beta \mid \beta \in \lambda\},$$

(M1) $$\alpha \cdot 0 = 0,$$

(M2) $$\alpha \cdot \beta^+ = \alpha \cdot \beta + \alpha,$$

(M3) $$\alpha \cdot \lambda = \sup\{\alpha \cdot \beta \mid \beta \in \lambda\}.$$

Proof (A1) and (M1) are contained in the previous theorem. To prove (A2), recall that $\beta^+ = \beta + 1$. Hence

$$\begin{aligned}\alpha + \beta^+ &= \alpha + (\beta + 1)\\ &= (\alpha + \beta) + 1\\ &= (\alpha + \beta)^+.\end{aligned}$$

Similarly for (M2) we have

$$\begin{aligned}\alpha \cdot \beta^+ &= \alpha \cdot (\beta + 1)\\ &= (\alpha \cdot \beta) + (\alpha \cdot 1)\\ &= (\alpha \cdot \beta) + \alpha.\end{aligned}$$

It remains to prove (A3) and (M3). For that proof we need a lemma on chains of well orderings. Say that a structure $\langle B, <_B\rangle$ is an *end extension* of a structure $\langle A, <_A\rangle$ iff $A \subseteq B$, $<_A \subseteq <_B$, and every element of $B - A$ is larger than anything in A:

$$a \in A \;\&\; b \in B - A \quad\Rightarrow\quad a <_B b.$$

Lemma 8M Assume that $\mathscr{C}$ is a set of well-ordered structures. Further assume that if $\langle A, <_A\rangle$ and $\langle B, <_B\rangle$ are structures in $\mathscr{C}$, then one is an end extension of the other. Let $\langle W, <_W\rangle$ be the union of all these structures in the sense that

$$\begin{aligned}W &= \bigcup\{A \mid \langle A, <_A\rangle \in \mathscr{C} \text{ for some } <_A\},\\ <_W &= \bigcup\{<_A \mid \langle A, <_A\rangle \in \mathscr{C} \text{ for some } A\}.\end{aligned}$$

Then $\langle W, <_W\rangle$ is a well-ordered structure whose ordinal number is the supremum of the ordinals of the members of $\mathscr{C}$.

Proof of the Lemma For each structure $\langle A, <_A\rangle$ in $\mathscr{C}$, we have the usual isomorphism E_A onto its ordinal number, ordered by epsilon. If $\langle B, <_B\rangle$ is an end extension of $\langle A, <_B\rangle$, then E_B is an extension of E_A, i.e., $E_A = E_B \restriction A$. (This is easy to see, but formally we verify $E_A(a) = E_B(a)$ by induction on a in A.) Thus the set of all possible E_A's

$$\{E_A \mid \langle A, <_A\rangle \in \mathscr{C} \text{ for some } <_A\}$$

is a chain of one-to-one functions. So its union (call it E) is a one-to-one function. The domain of E is the union of all possible A's; that is, dom $E = W$. The range of E is the union (supremum) of the ordinals of the structures in $\mathscr{C}$; call this ordinal θ. Thus E maps W one-to-one onto θ.

Furthermore it preserves order:

$$\begin{aligned} x <_W y \quad &\Leftrightarrow \quad x <_A y && \text{for some } A \text{ in } \mathscr{C} \\ &\Leftrightarrow \quad E_A(x) \in E_A(y) && \text{for some } A \text{ in } \mathscr{C} \\ &\Leftrightarrow \quad E(x) \in E(y). \end{aligned}$$

Thus E is an isomorphism of $\langle W, <_W\rangle$ onto $\langle \theta, \in_\theta\rangle$. Hence $\langle W, <_W\rangle$ is well ordered, and its ordinal is θ. ⊣

Now return to the proof of Theorem 8L. Since λ is a limit ordinal, we have $\lambda = \bigcup\lambda$. The sum $\alpha + \lambda$ is the ordinal number of the following set (under lexicographic order):

$$\begin{aligned} (\{0\} \times \alpha) \cup (\{1\} \times \lambda) &= (\{0\} \times \alpha) \cup (\{1\} \times \bigcup\{\beta \mid \beta \in \lambda\}) \\ &= (\{0\} \times \alpha) \cup \bigcup\{\{1\} \times \beta \mid \beta \in \lambda\} \\ &= \bigcup\{(\{0\} \times \alpha) \cup (\{1\} \times \beta) \mid \beta \in \lambda\} \\ &= \bigcup\{A_\beta \mid \beta \in \lambda\}, \end{aligned}$$

where $A_\beta = (\{0\} \times \alpha) \cup (\{1\} \times \beta)$. The ordinal of A_β (under lexicographic order) is $\alpha + \beta$. The ordinal of $\bigcup\{A_\beta \mid \beta \in \lambda\}$ is provided by the lemma. Once we verify that the lemma is applicable, it will tell us that the ordinal of $\bigcup\{A_\beta \mid \beta \in \lambda\}$ is $\sup\{\alpha + \beta \mid \beta \in \lambda\}$, which is what we need for (A3).

For any β and γ in λ, either $\beta \subseteq \gamma$ or $\gamma \subseteq \beta$. If it is the former, then A_γ (under lexicographic order) is clearly an end extension of A_β. Thus the lemma *is* applicable.

The proof of (M3) is similar. The product $\alpha \cdot \lambda$ is the ordinal number of $\alpha \times \lambda$ under $<_H$, Hebrew lexicographic order:

$$\begin{aligned} \alpha \times \lambda &= \alpha \times \bigcup\{\beta \mid \beta \in \lambda\} \\ &= \bigcup\{\alpha \times \beta \mid \beta \in \lambda\} \end{aligned}$$

and the ordinal of $\alpha \times \beta$ is $\alpha \cdot \beta$. Again the lemma is applicable, because whenever $\beta \subseteq \gamma$, then $\alpha \times \gamma$ is an end extension of $\alpha \times \beta$ (under $<_H$). The lemma tells us that the ordinal of $\bigcup\{\alpha \times \beta \mid \beta \in \lambda\}$ is $\sup\{\alpha \cdot \beta \mid \beta \in \lambda\}$, which completes the proof of (M3). ⊣

For finite ordinals (the natural numbers) we have now defined addition and multiplication three times: first in Chapter 4 (by recursion), then in Chapter 6 (as finite cardinals), and now again (by use of order types). All three agree on the natural numbers; Theorem 8L shows that the recent definition is in agreement with Chapter 4.

Theorem 8L also indicates how addition and multiplication could have been (equivalently) defined by transfinite recursion on the ordinals. The six equations in the theorem describe how, for fixed α, to form $\alpha + \beta$ from earlier values $\alpha + \gamma$ for γ less than β.

We will give the details of this approach of exponentiation, but they are easily modified to cover addition and multiplication. Consider then, a fixed nonzero ordinal α. We propose to define α^β in such a way as to satisfy the following three equations:

(E1) $$\alpha^0 = 1,$$

(E2) $$\alpha^{\beta^+} = \alpha^\beta \cdot \alpha,$$

(E3) $$\alpha^\lambda = \sup\{\alpha^\beta \mid \beta \in \lambda\}$$

for a limit ordinal λ.

Now we activate the transfinite recursion machinery. The procedure is similar to that used for the beth operation. Let $\gamma(f, y)$ be the formula:

Either (i) f is a function with domain 0 and $y = 1$,
or (ii) f is a function whose domain is a successor ordinal β^+ and $y = f(\beta) \cdot \alpha$,
or (iii) f is a function whose domain is a limit ordinal λ and $y = \bigcup \operatorname{ran} f$,
or (iv) none of the above and $y = \varnothing$.

Then transfinite recursion replies with a formula φ that will define the exponentiation operation. We select the symbol "${}_\alpha E$" and define

$${}_\alpha E_\beta = \text{the unique } y \text{ such that } \varphi(\beta, y)$$

for every ordinal β. Then we are assured that "$\gamma({}_\alpha E \restriction \beta, {}_\alpha E_\beta)$." For $\beta = 0$, this fact becomes

$${}_\alpha E_0 = 1.$$

For a successor ordinal β^+ in place of β it becomes

$${}_\alpha E_{\beta^+} = {}_\alpha E_\beta \cdot \alpha$$

and for a limit ordinal λ in place of β it becomes

$${}_\alpha E_\lambda = \sup\{{}_\alpha E_\gamma \mid \gamma \in \lambda\}.$$

These three displayed equations are (E1)–(E3), except for notational differences. We henceforth dispense with any special symbol for exponentiation, and instead use the traditional placement of letters:

$$\alpha^\beta = {}_\alpha E_\beta.$$

There is a special problem in defining 0^β. If we were to follow blindly (E1)–(E3), we would have $0^\omega = 1$. This is undesirable, which is our reason for having specified in the foregoing that α is a fixed *nonzero* ordinal. We can simply define 0^β directly: $0^0 = 1$ and $0^\beta = 0$ for $\beta \neq 0$.

Example We have $2^{\omega} = \sup\{2^n \mid n \in \omega\} = \omega$. Thus ordinal exponentiation is very different from cardinal exponentiation, since $2^{\kappa} \neq \kappa$ for cardinals. Ordinal addition and multiplication are also very different from cardinal addition and multiplication. Please do not confuse them. Theorem 6J tells us that the operations agree on *finite* numbers; that theorem does *not* extend to infinite numbers.

We can now apply our remarks on ordinal operations to derive information concerning ordinal arithmetic. Consider a *fixed* ordinal number α. Then the operation of *α-addition* is the operation assigning to each ordinal β the sum $\alpha + \beta$. (In our earlier notation, $t_{\beta} = \alpha + \beta$, where α is fixed.) Similarly, the operation of *α-multiplication* assigns to β the product $\alpha \cdot \beta$, and *α-exponentiation* assigns to β the power α^{β}.

Theorem 8N (a) For any ordinal number α, the operation of α-addition is normal.

(b) If $1 \in \alpha$, then the operation of α-multiplication is normal.

(c) If $2 \in \alpha$, then the operation of α-exponentiation is normal.

Proof Continuity is immediate from (A3) and (M3) of Theorem 8L and from (E3). For monotonicity, we use Theorem 8C, which tells us that it suffices to show that:

$$\begin{aligned} &\alpha + \beta \in \alpha + \beta^{+}, \\ &\alpha \cdot \beta \in \alpha \cdot \beta^{+} \qquad \text{if} \quad 1 \in \alpha, \\ &\alpha^{\beta} \in \alpha^{\beta^{+}} \qquad\quad \text{if} \quad 2 \in \alpha. \end{aligned}$$

The first of these is immediate from (A2) of Theorem 8L:

$$\alpha + \beta \in (\alpha + \beta)^{+} = \alpha + \beta^{+},$$

whence α-addition is normal.

For multiplication we have

$$1 \in \alpha \quad \Rightarrow \quad 0 \in \alpha$$

and so by the monotonicity of $(\alpha \cdot \beta)$-addition,

$$\alpha \cdot \beta \in \alpha \cdot \beta + \alpha.$$

This together with (M2) gives $\alpha \cdot \beta \in \alpha \cdot \beta^{+}$.

Exponentiation is similar;

$$2 \in \alpha \quad \Rightarrow \quad 1 \in \alpha,$$

whence by the monotonicity of α^{β}-multiplication and (E3)

$$\alpha^{\beta} \in \alpha^{\beta} \cdot \alpha = \alpha^{\beta^{+}}.$$

This completes the proof. ⊣

Example Assume that λ is a limit ordinal. By Exercise 4, t_λ (for a normal operation) is also a limit ordinal. So by the preceding theorem, $\alpha + \lambda$ is a limit ordinal for any ordinal α. If $1 \in \alpha$, then both $\alpha \cdot \lambda$ and $\lambda \cdot \alpha$ are limit ordinals. (In the second case we use the fact that $\lambda \cdot \beta^+ = \lambda \cdot \beta + \lambda$.) Similarly if $2 \in \alpha$, then both α^λ and λ^α are limit ordinals.

Corollary 8P (a) The following order-preserving laws hold.

$$\beta \in \gamma \quad \Leftrightarrow \quad \alpha + \beta \in \alpha + \gamma.$$

If $1 \in \alpha$, then

$$\beta \in \gamma \quad \Leftrightarrow \quad \alpha \cdot \beta \in \alpha \cdot \gamma.$$

If $2 \in \alpha$, then

$$\beta \in \gamma \quad \Leftrightarrow \quad \alpha^\beta \in \alpha^\gamma.$$

(b) The following left cancellation laws hold:

$$\alpha + \beta = \alpha + \gamma \quad \Rightarrow \quad \beta = \gamma.$$

If $1 \in \alpha$, then

$$\alpha \cdot \beta = \alpha \cdot \gamma \quad \Rightarrow \quad \beta = \gamma.$$

If $2 \in \alpha$, then

$$\alpha^\beta = \alpha^\gamma \quad \Rightarrow \quad \beta = \gamma.$$

Proof These are consequences of monotonicity. Any monotone operation assigning t_β to β has the properties

$$\beta \in \gamma \quad \Leftrightarrow \quad t_\beta \in t_\gamma,$$
$$t_\beta = t_\gamma \quad \Rightarrow \quad \beta = \gamma,$$

by Exercise 3. ⊣

Part (b) of this theorem gives only left cancellation laws. Right cancellation laws fail in general. For example,

$$2 + \omega = 3 + \omega = \omega,$$
$$2 \cdot \omega = 3 \cdot \omega = \omega,$$
$$2^\omega = 3^\omega = \omega,$$

but we cannot cancel the ω's to get $2 = 3$. There is a weakened version of part (a) that holds:

Theorem 8Q The following weak order-preserving laws hold for any ordinal numbers:

(a) $\beta \in \gamma \Rightarrow \beta + \alpha \subseteq \gamma + \alpha$.
(b) $\beta \in \gamma \Rightarrow \beta \cdot \alpha \subseteq \gamma \cdot \alpha$.
(c) $\beta \in \gamma \Rightarrow \beta^\alpha \subseteq \gamma^\alpha$.

Proof Each part can be proved by transfinite induction on α. But parts (a) and (b) also can be proved using concepts from order types. Assume that $\beta \in \gamma$; then also $\beta \subseteq \gamma$. Thus

$$(\{0\} \times \beta) \cup (\{1\} \times \alpha) \subseteq (\{0\} \times \gamma) \cup (\{1\} \times \alpha)$$

and $\beta \times \alpha \subseteq \gamma \times \alpha$. Furthermore in each case the relevant ordering (lexicographic and Hebrew lexicographic, respectively) on the subset is the restriction of the ordering on the larger set. Hence when we take the ordinal numbers of the sets, we get $\beta + \alpha \in \gamma + \alpha$ and $\beta \cdot \alpha \in \gamma \cdot \alpha$ (compare Exercise 17 of Chapter 7).

We will prove (c) by transfinite induction on α. So suppose that $\beta \in \gamma$ and that $\beta^\delta \in \gamma^\delta$ whenever $\delta \in \alpha$; we must show that $\beta^\alpha \in \gamma^\alpha$. If $\alpha = 0$, then $\beta^\alpha = \gamma^\alpha = 1$. Next take the case where α is a successor ordinal δ^+. Then

$$\begin{aligned} \beta^\alpha &= \beta^\delta \cdot \beta \\ &\in \gamma^\delta \cdot \beta && \text{since} \quad \beta^\delta \in \gamma^\delta \\ &\in \gamma^\delta \cdot \gamma && \text{by Corollary 8P} \\ &= \gamma^\alpha. \end{aligned}$$

Finally take the case where α is a limit ordinal. We may suppose that neither β nor γ is 0, since those cases are clear. Since $\beta^\delta \in \gamma^\delta$ for δ in α, we have

$$\beta^\alpha = \sup\{\beta^\delta \mid \delta \in \alpha\} \in \sup\{\gamma^\delta \mid \delta \in \alpha\} = \gamma^\alpha,$$

whence $\beta^\alpha \in \gamma^\alpha$. ⊣

It is not possible to replace "$\in$" by "$\in$" in the above theorem, since

$$2 + \omega \notin 3 + \omega,$$
$$2 \cdot \omega \notin 3 \cdot \omega,$$
$$2^\omega \notin 3^\omega,$$

despite the fact that $2 \in 3$.

Subtraction Theorem If $\alpha \in \beta$ (for ordinal numbers α and β), then there exists a unique ordinal number δ (their "difference") such that $\alpha + \delta = \beta$.

Proof The uniqueness of δ is immediate from the left cancellation law (Corollary 8P). We will indicate two proofs of existence.

Since α-addition is a normal operation and β is at least as large as $\alpha + 0$, we know from Theorem 8D that there is a largest δ for which $\alpha + \delta \in \beta$. But equality must hold, for if $\alpha + \delta \in \beta$, then

$$\alpha + \delta^+ = (\alpha + \delta)^+ \in \beta$$

contradicting the maximality of δ.

The other proof starts from the fact that $\beta - \alpha$ (i.e., $\{x \in \beta \mid x \notin \alpha\}$) is well ordered by epsilon. Let δ be its ordinal. Then $\alpha + \delta$ is the ordinal of $(\{0\} \times \alpha) \cup (\{1\} \times (\beta - \alpha))$ under lexicographic order, which is isomorphic to β under the epsilon ordering. ⊣

For example, if $\alpha = 3$ and $\beta = \omega$, then the "difference" is ω, since $3 + \omega = \omega$. If $\alpha \in \beta$, we cannot in general find a δ for which $\delta + \alpha = \beta$; for example, there is no δ for which $\delta + 3 = \omega$. (Why?)

Division Theorem Let α and δ be ordinal numbers with δ (the divisor) nonzero. Then there is a unique pair of ordinal numbers β and γ (the quotient and the remainder) such that

$$\alpha = \delta \cdot \beta + \gamma \qquad \text{and} \qquad \gamma \in \delta.$$

Proof First we will prove existence. Since δ-multiplication is a normal operation, there is a largest β for which $\delta \cdot \beta \in\!\!\!\!= \alpha$. Then by the subtraction theorem there is some γ for which $\delta \cdot \beta + \gamma = \alpha$. We must show that $\gamma \in \delta$. If to the contrary $\delta \in\!\!\!\!= \gamma$, then

$$\delta \cdot \beta^+ = \delta \cdot \beta + \delta \in\!\!\!\!= \delta \cdot \beta + \gamma = \alpha,$$

contradicting the maximality of β.

Having established existence, we now turn to uniqueness. Assume that

$$\alpha = \delta \cdot \beta_1 + \gamma_1 = \delta \cdot \beta_2 + \gamma_2,$$

where γ_1 and γ_2 are less than δ. To show that $\beta_1 = \beta_2$, it suffices to eliminate the alternatives. Suppose, contrary to our hopes, that $\beta_1 \in \beta_2$. Then $\beta_1^+ \in\!\!\!\!= \beta_2$ and so $\delta \cdot (\beta_1^+) \in\!\!\!\!= \delta \cdot \beta_2$. Since $\gamma_1 \in \delta$, we have

$$\alpha = \delta \cdot \beta_1 + \gamma_1 \in \delta \cdot \beta_1 + \delta = \delta \cdot (\beta_1^+) \in\!\!\!\!= \delta \cdot \beta_2 \in\!\!\!\!= \delta \cdot \beta_2 + \gamma_2 = \alpha.$$

But this is impossible, so $\beta_1 \notin \beta_2$. By symmetry, $\beta_2 \notin \beta_1$. Hence $\beta_1 = \beta_2$; call it simply β. We now have

$$\alpha = \delta \cdot \beta + \gamma_1 = \delta \cdot \beta + \gamma_2,$$

whence by left cancellation (Corollary 8P), $\gamma_1 = \gamma_2$. ⊣

Digression Ordinal addition and multiplication are often introduced by transfinite recursion instead of by order types. In that case, the foregoing theorems can be used to establish the connection with order types, without need of Theorem 8L or Lemma 8M. The statements to be proved are those that we took in our development as definitions: The order type of an ordinal sum (or product) is the sum (or product) of the order types. That is, the equations

$$\overline{\alpha + \beta} = \bar{\alpha} + \bar{\beta} \qquad \text{and} \qquad \overline{\alpha \cdot \beta} = \bar{\alpha} \cdot \bar{\beta}$$

are, in this alternative development, to be proved. For addition, it suffices to give an isomorphism from $(\{0\} \times \alpha) \cup (\{1\} \times \beta)$ with lexicographic ordering onto $\alpha + \beta$ with epsilon ordering. The isomorphism f is defined by the equations

$$f(\langle 0, \gamma \rangle) = \gamma \qquad \text{for} \quad \gamma \text{ in } \alpha,$$
$$f(\langle 1, \delta \rangle) = \alpha + \delta \qquad \text{for} \quad \delta \text{ in } \beta.$$

Then it can be verified that $\operatorname{ran} f = \alpha + \beta$ and that f preserves order. For multiplication, the isomorphism g from $\alpha \times \beta$ with Hebrew lexicographic order onto $\alpha \cdot \beta$ with epsilon ordering is defined by the equation

$$g(\langle \gamma, \delta \rangle) = \alpha \cdot \delta + \gamma$$

for γ in α and δ in β. Again it can be verified that $\operatorname{ran} g = \alpha \cdot \beta$ and that g preserves order.

Logarithm Theorem Assume that α and β are ordinal numbers with $\alpha \neq 0$ and β (the base) greater than 1. Then there are unique ordinal numbers γ, δ, and ρ (the logarithm, the coefficient, and the remainder) such that

$$\alpha = \beta^\gamma \cdot \delta + \rho \quad \& \quad 0 \neq \delta \in \beta \quad \& \quad \rho \in \beta^\gamma.$$

Proof Since β-exponentiation is a normal operation, there is (by Theorem 8D) a largest γ such that $\beta^\gamma \subseteq \alpha$. Apply the division theorem to $\alpha \div \beta^\gamma$ to obtain δ and ρ such that

$$\alpha = \beta^\gamma \cdot \delta + \rho \qquad \text{and} \qquad \rho \in \beta^\gamma.$$

Note that $\delta \neq 0$ since $\rho \in \beta^\gamma \subseteq \alpha$. We must show that $\delta \in \beta$. If to the contrary $\beta \subseteq \delta$, then $\beta^{\gamma^+} = \beta^\gamma \cdot \beta \subseteq \beta^\gamma \cdot \delta \subseteq \beta^\gamma \cdot \delta + \rho = \alpha$, contradicting the maximality of γ. Thus we have the existence of γ, δ, and ρ meeting the prescribed conditions.

To show uniqueness, consider any representation $\alpha = \beta^\gamma \cdot \delta + \rho$ for which $0 \neq \delta \in \beta$ and $\rho \in \beta^\gamma$. We first claim that γ must be exactly the one we used in the preceding paragraph. We have

$$\begin{aligned} \beta^\gamma \subseteq \alpha &= \beta^\gamma \cdot \delta + \rho && \text{since } 1 \subseteq \delta \\ &\in \beta^\gamma \cdot \delta + \beta^\gamma && \text{since } \rho \in \beta^\gamma \\ &= \beta^\gamma \cdot \delta^+ && \\ &\subseteq \beta^\gamma \cdot \beta && \text{since } \delta \in \beta \\ &= \beta^{\gamma^+}. && \end{aligned}$$

Thus $\beta^\gamma \subseteq \alpha \in \beta^{\gamma^+}$. This double inequality uniquely determines γ; it must be the largest ordinal for which $\beta^\gamma \subseteq \alpha$. Once γ is fixed, the division theorem tells us that δ and ρ are unique. ⊣

Observe that in the logarithm theorem, we always have $\rho \in \beta^\gamma \subseteq \alpha$. An interesting special case is where the base β is ω. Then given any nonzero ordinal α, we can write

$$\alpha = \omega^{\gamma_1} \cdot n_1 + \rho_1,$$

where n_1 is a nonzero natural number and $\rho_1 \in \omega^{\gamma_1} \subseteq \alpha$. If ρ_1 is nonzero, then we can repeat:

$$\rho_1 = \omega^{\gamma_2} \cdot n_2 + \rho_2,$$

where n_2 is a nonzero natural number and $\rho_2 \in \omega^{\gamma_2} \subseteq \rho_1 \in \omega^{\gamma_1}$. Hence both $\rho_2 \in \rho_1$ and $\gamma_2 \in \gamma_1$. Continuing in this way, we construct progressively smaller ordinals $\rho_1, \rho_2, \ldots$. This descending chain cannot be infinite, so $\rho_k = 0$ for some (finite!) k. We then have α represented as an "ω-polynomial"

$$\alpha = \omega^{\gamma_1} \cdot n_1 + \cdots + \omega^{\gamma_k} \cdot n_k,$$

where $n_1, \ldots, n_k$ are nonzero natural numbers and $\gamma_k \in \gamma_{k-1} \in \cdots \in \gamma_1$. This polynomial representation is called the *Cantor normal form* of α. You are asked in Exercise 26 to show that it is unique.

Theorem 8R For any ordinal numbers,

$$\alpha^{\beta+\gamma} = \alpha^\beta \cdot \alpha^\gamma.$$

Proof We use transfinite induction on γ. In the limit ordinal case we will use the normality of α-exponentiation. But normality is true only for α greater than 1. So a separate proof is needed for the cases $\alpha = 0$ and $\alpha = 1$. We leave this separate proof, which does not require induction, to you (Exercise 27). Henceforth we assume that α is at least 2.

Suppose, as the inductive hypothesis, that $\alpha^{\beta+\delta} = \alpha^\beta \cdot \alpha^\delta$ for all δ less than γ; we must prove that $\alpha^{\beta+\gamma} = \alpha^\beta \cdot \alpha^\gamma$.

Case I $\gamma = 0$. Then on the left side we have $\alpha^{\beta+0} = \alpha^\beta$. On the right side we have $\alpha^\beta \cdot \alpha^0 = \alpha^\beta \cdot 1 = \alpha^\beta$, so this case is done.

Case II $\gamma = \delta^+$ for some δ. Then

$$\begin{aligned}
\alpha^{\beta+\gamma} &= \alpha^{\beta+\delta^+} \\
&= \alpha^{(\beta+\delta)^+} && \text{by (A2)} \\
&= \alpha^{\beta+\delta} \cdot \alpha && \text{by (E2)} \\
&= \alpha^\beta \cdot \alpha^\delta \cdot \alpha && \text{by the inductive hypothesis} \\
&= \alpha^\beta \cdot \alpha^{\delta^+} && \text{by (E2)} \\
&= \alpha^\beta \cdot \alpha^\gamma.
\end{aligned}$$

(The associative law was also used here.)

Case III γ is a limit ordinal. Then

$$\begin{aligned} \alpha^{\beta+\gamma} &= \alpha^{\sup\{\beta+\delta \mid \delta \in \gamma\}} && \text{by (A3)} \\ &= \sup\{\alpha^{\beta+\delta} \mid \delta \in \gamma\} && \text{by Theorem 8E} \\ &= \sup\{\alpha^{\beta} \cdot \alpha^{\delta} \mid \delta \in \gamma\} \end{aligned}$$

by the inductive hypothesis. On the other side,

$$\begin{aligned} \alpha^{\beta} \cdot \alpha^{\gamma} &= \alpha^{\beta} \cdot \sup\{\alpha^{\delta} \mid \delta \in \gamma\} \\ &= \sup\{\alpha^{\beta} \cdot \alpha^{\delta} \mid \delta \in \gamma\} \end{aligned}$$

by Theorem 8E again, this time applied to α^{β}-multiplication. Thus the two sides agree. ⊣

Theorem 8S For any ordinal numbers,

$$(\alpha^{\beta})^{\gamma} = \alpha^{\beta \cdot \gamma}.$$

Proof We again use transfinite induction on γ. And again we leave to you the case $\alpha = 0$ or $\alpha = 1$ (Exercise 27). Henceforth we assume that α is at least 2.

Suppose, as the inductive hypothesis, that $(\alpha^{\beta})^{\delta} = \alpha^{\beta \cdot \delta}$ for all δ less than γ; we must prove that $(\alpha^{\beta})^{\gamma} = \alpha^{\beta \cdot \gamma}$.

Case I $\gamma = 0$. Obviously both sides are equal to 1.

Case II $\gamma = \delta^+$. Then we calculate

$$\begin{aligned} (\alpha^{\beta})^{\gamma} &= (\alpha^{\beta})^{\delta^+} \\ &= (\alpha^{\beta})^{\delta} \cdot \alpha^{\beta} \\ &= \alpha^{\beta \cdot \delta} \cdot \alpha^{\beta} && \text{by the inductive hypothesis} \\ &= \alpha^{\beta \cdot \delta + \beta} && \text{by the previous theorem} \\ &= \alpha^{\beta \cdot \delta^+} \\ &= \alpha^{\beta \cdot \gamma} \end{aligned}$$

as desired.

Case III γ is a limit ordinal. Then

$$\begin{aligned} \alpha^{\beta \cdot \gamma} &= \alpha^{\sup\{\beta \cdot \delta \mid \delta \in \gamma\}} \\ &= \sup\{\alpha^{\beta \cdot \delta} \mid \delta \in \gamma\} && \text{by Theorem 8E} \\ &= \sup\{(\alpha^{\beta})^{\delta} \mid \delta \in \gamma\} && \text{by the inductive hypothesis} \\ &= (\alpha^{\beta})^{\gamma} \end{aligned}$$

as desired. ⊣

Example As with finite numbers, α^{β^γ} always means $\alpha^{(\beta^\gamma)}$. The other grouping, $(\alpha^\beta)^\gamma$, equals $\alpha^{\beta\cdot\gamma}$ by Theorem 8S. Now suppose that we start with ω and apply exponentiation over and over again. Let

$$\varepsilon_0 = \sup\{\omega, \omega^\omega, \omega^{\omega^\omega}, \ldots\}.$$

Then by Theorem 8E

$$\omega^{\varepsilon_0} = \sup\{\omega^\omega, \omega^{\omega^\omega}, \ldots\} = \varepsilon_0 .$$

The equation $\varepsilon_0 = \omega^{\varepsilon_0}$ gives the Cantor normal form representation for ε_0. More generally, the *epsilon numbers* are the ordinals ε for which $\varepsilon = \omega^\varepsilon$. The smallest epsilon number is ε_0. It is a countable ordinal, being the countable union of countable sets. By the Veblen fixed-point theorem, the class of epsilon numbers is unbounded.

Exercises

20. Show that every ordinal number is expressible in the form $\lambda + n$ where $n \in \omega$ and λ is either zero or a limit ordinal. Show further that the representation in this form is unique.

21. In Exercise 4 of Chapter 7, find the ordinal number of $\langle P, R\rangle$.

22. Prove parts (a) and (b) of Theorem 8Q by transfinite induction on α.

23. (a) Show that $\omega + \omega^2 = \omega^2$.
    (b) Show that whenever $\omega^2 \in\!\!\!\!\_\; \beta$, then $\omega + \beta = \beta$.

24. Assume that $\omega \in\!\!\!\!\_\; \alpha$ and $\alpha^2 \in\!\!\!\!\_\; \beta$. Show that $\alpha + \beta = \beta$. (This generalizes the preceding exercise.)

25. Consider any fixed ordinals α and θ. Show that

$$\alpha + \theta = \alpha \cup \{\alpha + \delta \mid \delta \in \theta\}.$$

26. Prove that the representation of an ordinal number in Cantor normal form is unique.

27. Supply proofs for Theorem 8R and Theorem 8S when α is 0 or 1.

28. Show that for any given ordinal number, there is a larger epsilon number.

29. Show that the class of epsilon numbers is closed, i.e., if S is a nonempty set of epsilon numbers, then $\sup S$ is an epsilon number.

CHAPTER 9

SPECIAL TOPICS

In this final chapter, we present three topics that stand somewhat apart from our previous topics, yet are too interesting to omit from the book. The three sections are essentially independent.

WELL-FOUNDED RELATIONS

Some of the important properties of well orderings (such as transfinite induction and recursion) depend more on the "well" than on the "ordering." In this section, we extend those properties to a larger class of relations.

Recall that for a partial ordering R and a set, D, an element m of D was said to be a minimal element if there was no x in D with xRm. This terminology can actually be applied to any set R:

Definition An element m of a set D is said to be an R-*minimal* element of D iff there is no x in D for which xRm.

Definition A relation R is said to be *well founded* iff every nonempty set D contains an R-minimal element.

If the set D in this definition is not a subset of fld R, then it is certain to contain an R-minimal element (namely any m in D − fld R). Thus the only sets D that matter here are the subsets of fld R.

Examples The finite relation

$$R = \{\langle 1,2\rangle, \langle 2,6\rangle, \langle 3,6\rangle\}$$

is well founded. Its field is $\{1, 2, 3, 6\}$, and all fifteen nonempty subsets of the field have R-minimal elements. The empty relation is also well founded, but for uninteresting reasons.

Example Consider any set S and the membership relation

$$\in_S = \{\langle x, y\rangle \in S \times S \mid x \in y\}$$

on S. Then we can show from the regularity axiom that $\in_S$ is well founded. For consider any nonempty set D. By regularity there some m in D with $m \cap D = \varnothing$, i.e., there is no x in D with $x \in m$. So certainly there is no x in D with $x \in_S m$. Hence m is $\in_S$-minimal in D. In fact the regularity axiom can be summarized by the statement: The membership relation is well founded.

The following theorem extends Theorem 7B, which asserted that a linear ordering was a well ordering iff it had no descending chains.

Theorem 9A A relation R is well founded iff there is no function f with domain ω such that $f(n^+)Rf(n)$ for each n.

Proof The proof is exactly the same as for Theorem 7B. If R is not well founded, then there exists a nonempty set A without an R-minimal element, i.e., $(\forall x \in A)(\exists y \in A)\, yRx$. So we can apply Exercise 20 of Chapter 6 to obtain a descending chain. ⊣

If R is a binary relation on A (i.e., $R \subseteq A \times A$) that is well founded, then we can say that R is a well-founded relation on A or that $\langle A, R\rangle$ is a *well-founded structure*.

Transfinite Induction Principle Assume that R is a well-founded relation on A. Assume that B is a subset of A with the special property that for every t in A,

$$\{x \in A \mid xRt\} \subseteq B \quad \Rightarrow \quad t \in B.$$

Then B coincides with A.

This theorem is the direct analogue of the transfinite induction principle for well-ordered relations (in Chapter 7). It asserts that any R-inductive subset of A must actually be A itself. In fact the earlier theorem is a special case of the above theorem, wherein R linearly orders A.

Proof The proof is the same as before. If B is a proper subset of A, then $A - B$ has an R-minimal element m. By the minimality, $\{x \in A \mid xRm\} \subseteq B$. But then by the special property of B (the property of being R-inductive), $m \in B$ after all. ⊣

Next we want to describe how any relation R (well founded or not) can be extended to R^t, the smallest transitive relation extending R.

Theorem 9B Let R be a relation. Then there exists a unique relation R^t such that:

(a) R^t is a transitive relation and $R \subseteq R^t$.
(b) If Q is any transitive relation that includes R, then $R^t \subseteq Q$.

Proof There are two equivalent ways to obtain R^t. The brute force method is to define R^* "from above" by

$$R^* = \bigcap\{Q \mid R \subseteq Q \subseteq \operatorname{fld} R \times \operatorname{fld} R \;\&\; Q \text{ is a transitive relation}\}.$$

The collection of all such Q's is nonempty, since fld $R \times$ fld R is a member. Hence it is permissible to take the intersection.

Then $R \subseteq R^*$ (because $R \subseteq Q$ for each of those Q's). R^* is a transitive relation (recall Exercise 34 of Chapter 3). And R^* is as small as possible, being the intersection of all candidates. Hence we may take $R^t = R^*$. Uniqueness is immediate; by (b) any two contenders must be subsets of each other.

Although the theorem is now proved, we will ignore that fact and give the construction of R^t "from below." Define R_n by recursion for n in ω by the equations

$$R_0 = R \qquad \text{and} \qquad R_{n+1} = R_n \circ R$$

and then define $R_* = \bigcup_{n \in \omega} R_n$. For example, we can describe R_n for small n as follows:

$$R_0 = R = \{\langle x, y\rangle \mid xRy\},$$
$$R_1 = R \circ R = \{\langle x, y\rangle \mid xRtRy \text{ for some } t\},$$
$$R_2 = R \circ R \circ R = \{\langle x, y\rangle \mid xRt_1Rt_2Ry \text{ for some } t_1 \text{ and } t_2\}.$$

In general,

$$R_n = \{\langle x, y\rangle \mid xRt_1R \cdots Rt_nRy \text{ for some } t_1, \ldots, t_n\}$$

and

$$R_* = \{\langle x, y\rangle \mid xR_ny \text{ for some } n\}.$$

Clearly $R \subseteq R_*$. To show that R_* is a transitive relation, suppose that xR_*yR_*z. Then xR_myR_nz for some m and n. We claim that $xR_{m+n+1}z$.

This is clear from the above characterization of R_n; the "three dots" technique can, as usual, be replaced by an induction on n. Hence $xR_* z$.

To see that R_* also satisfies the minimality clause (b), consider any transitive relation Q that includes R. To show that $R_* \subseteq Q$, it suffices to show that $R_n \subseteq Q$ for each n. We do this by induction on n. It holds for $n = 0$ by assumption. If $R_k \subseteq Q$ and $xR_{k+1}z$, then we have $xR_k yRz$ for some y, whence $xQyQz$ and (by transitivity) xQz. So $R_{k+1} \subseteq Q$ and the induction is complete.

We are now entitled to conclude that $R^t = R^* = R_*$. ⊣

Example Assume that S is a transitive set. Define the binary relation

$$\in_S = \{\langle x, y\rangle \in S \times S \mid x \in y\}$$

on S. Then for x and y in S,

$$\begin{aligned} x \in_S^t y &\Leftrightarrow x \in_S \cdots \in_S y \\ &\Leftrightarrow x \in \cdots \in y, \end{aligned}$$

where the intermediate points are automatically in S (by transitivity). Hence we have

$$x \in_S^t y \Leftrightarrow x \in \mathrm{TC}\, y.$$

Recall from Exercise 7 of Chapter 7 that TC y, the transitive closure of y, is the smallest transitive set that includes y. Its members are roughly the members of members of $\cdots$ of members of y. A little more precisely,

$$\mathrm{TC}\, y = y \cup \bigcup y \cup \bigcup\bigcup y \cup \cdots.$$

Theorem 9C If R is a well-founded relation, then its transitive extension R^t is also well founded.

Proof We will give a proof that uses the axiom of choice. Exercise 1 requests an alternative proof that does not require choice.

Suppose that, contrary to our expectations, R^t is not well founded. Then by Theorem 9A there is a descending chain f. That is, f is a function with domain ω and $f(n^+)R^t f(n)$ for each n.

The idea is to fill in this descending chain to get a descending chain for R, and thereby to contradict the assumption that R is well founded. Since $f(n^+)R^t f(n)$, we know that

$$f(n^+)Rx_1R \cdots Rx_k Rf(n)$$

for some $x_1, \ldots, x_k$. By interpolating these intermediate points between $f(n^+)$ and $f(n)$ (and doing this for each n), we get a descending chain g such that $g(m^+)Rg(m)$ for each m. ⊣

Corollary 9D If R is a well-founded relation on A, then R^t is a partial ordering on A.

Proof Certainly $R^t \subseteq A \times A$ and R^t is a transitive relation. So we need show only that R^t is irreflexive. But any well-founded relation is irreflexive; we cannot have xR^tx lest $\{x\}$ fail to have a R^t-minimal element. ⊣

Because of this corollary, transitive well-founded relations are sometimes called *partial well orderings.*

Transfinite Recursion Theorem Schema For any formula $\gamma(x, y, z)$ the following is a theorem:

Assume that R is a well-founded relation on a set A. Assume that for any f and t there is a unique z such that $\gamma(f, t, z)$. Then there exists a unique function F with domain A such that

$$\gamma(F \upharpoonright \{x \in A \mid xRt\}, t, F(t))$$

for all t in A.

Here γ defines a function-class G, and the equation

$$F(t) = \mathsf{G}(F \upharpoonright \{x \in A \mid xRt\}, t)$$

holds for all t in A. A comparison of the above theorem schema with its predecessor in Chapter 7 will show that we have added an additional variable to γ (and to G). This is because knowing what $F \upharpoonright \{x \in A \mid xRt\}$ is does not tell us what t is. The proof of the above theorem schema is much like the proofs of recursion theorems given in Chapters 4 and 7.

Proof As before, we form F by taking the union of many approximating functions. For any x in A, define seg x to be $\{t \mid tRx\}$. For the purposes of this proof, call a function v ***acceptable*** iff dom $v \subseteq A$ and whenever $x \in$ dom v, then seg $x \subseteq$ dom v and $\gamma(v \upharpoonright \text{seg } x, x, v(x))$.

1. First we claim that any two acceptable functions v_1 and v_2 agree at all points (if any) belonging to both domains. Should this fail, there is an R-minimal element x of dom $v_1 \cap$ dom v_2 for which $v_1(x) \neq v_2(x)$. By the minimality, $v_1 \upharpoonright \text{seg } x = v_2 \upharpoonright \text{seg } x$. But then acceptability together with our assumption on γ tells us that $v_1(x) = v_2(x)$ after all.

We can now use a replacement axiom to form the set $\mathscr{K}$ of all acceptable functions. Take $\varphi(u, v)$ to be the formula: $u \subseteq A \,\&\, v$ is an acceptable function with domain u. It follows from the preceding paragraph that

$$\varphi(u, v_1) \,\&\, \varphi(u, v_2) \quad \Rightarrow \quad v_1 = v_2.$$

Hence by replacement there is a set $\mathscr{K}$ such that

$$v \in \mathscr{K} \quad \Leftrightarrow \quad (\exists u \in \mathscr{P}A)\, \varphi(u, v).$$

Letting $\mathscr{K}$ be the set of all acceptable functions, we can explicitly define F to be the set $\bigcup\mathscr{K}$. Thus

$$(\star) \qquad \langle x, y\rangle \in F \quad \Leftrightarrow \quad v(x) = y \quad \text{for some acceptable } v.$$

Observe that F is a function, because any two acceptable functions agree wherever both are defined.

2. Next we claim that the function F is acceptable. Consider any x in dom F. Then there exists some acceptable v with $x \in \operatorname{dom} v$. Hence $x \in A$ and $\operatorname{seg} x \subseteq \operatorname{dom} v \subseteq \operatorname{dom} F$. We must check that $\gamma(F \upharpoonright \operatorname{seg} x, x, F(x))$ holds. We have

$$\begin{aligned} &\gamma(v \upharpoonright \operatorname{seg} x, x, v(x)) && \text{by acceptability,}\\ &v \upharpoonright \operatorname{seg} x = F \upharpoonright \operatorname{seg} x && \text{by } (\star) \text{ and } (1),\\ &v(x) = F(x) && \text{by } (\star), \end{aligned}$$

from which we can conclude that $\gamma(F \upharpoonright \operatorname{seg} x, x, F(x))$.

3. We now claim that dom F is all of A. If this fails, then there is an R-minimal element t of $A - \operatorname{dom} F$. By minimality, $\operatorname{seg} t \subseteq \operatorname{dom} F$. Take the unique y such that $\gamma(F \upharpoonright \operatorname{seg} t, t, y)$ and let $\hat{v} = F \cup \{\langle t, y\rangle\}$. We must verify that $\hat{v}$ is acceptable. Clearly $\hat{v}$ is a function and $\operatorname{dom} \hat{v} \subseteq A$. Consider any x in $\operatorname{dom} \hat{v}$. One possibility is that $x \in \operatorname{dom} F$. Then $\operatorname{seg} x \subseteq \operatorname{dom} F \subseteq \operatorname{dom} \hat{v}$ and from the equations

$$\hat{v} \upharpoonright \operatorname{seg} x = F \upharpoonright \operatorname{seg} x \qquad \text{and} \qquad \hat{v}(x) = F(x)$$

we conclude that $\gamma(\hat{v} \upharpoonright \operatorname{seg} x, x, \hat{v}(x))$. The other possibility is that $x = t$. Then $\operatorname{seg} t \subseteq \operatorname{dom} F \subseteq \operatorname{dom} \hat{v}$ and from the equations

$$\hat{v} \upharpoonright \operatorname{seg} t = F \upharpoonright \operatorname{seg} t \qquad \text{and} \qquad \hat{v}(t) = y$$

we conclude that $\gamma(\hat{v} \upharpoonright \operatorname{seg} t, t, \hat{v}(t))$. Hence $\hat{v}$ is acceptable, and so $t \in \operatorname{dom} F$ after all. Consequently, dom $F = A$.

4. F is unique, by an inductive argument like those used before. ⊣

We now proceed to show how transfinite recursion can be applied to the membership relation to produce the rank of a set (and thereby generate all of the ordinal numbers). Recall that in Chapter 7 we applied transfinite recursion to a well-ordered structure $\langle A, <\rangle$ to obtain a function E with domain A such that

$$E(a) = \{E(x) \mid x \in a\}$$

for each $a \in A$. It then turned out that ran E was (by definition) the ordinal number of $\langle A, <\rangle$.

Now we intend to perform the analogous construction using, in place of a well ordering, the membership relation on some set. After all, the regularity axiom assures us that the membership relation is always well founded. So we can apply transfinite recursion. What will we get? The answer is that if we do it right, we will get the rank of the set!

To be more specific, let S be the set $\langle 1, 0\rangle$, or

$$\{\{\{\varnothing\}\}, \{\varnothing, \{\varnothing\}\}\}$$

to use its full name. We can illustrate the formation of this set by Fig. 48. The sets appearing below S in this figure are the sets in TC S, the transitive closure of S.

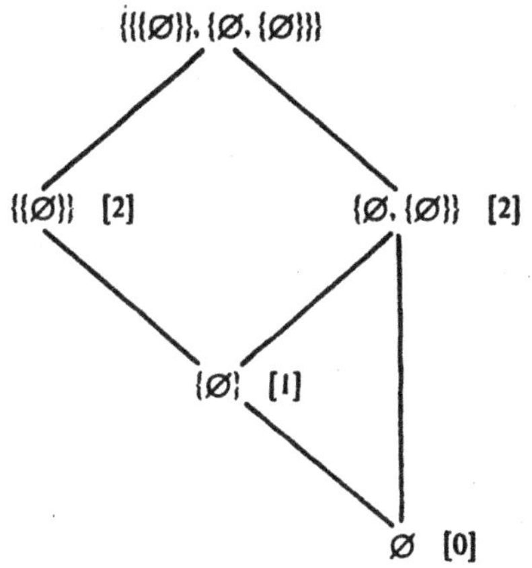

Fig. 48. The membership relation on $\mathrm{TC}\{\langle 1, 0\rangle\}$.

Let M be the membership relation on TC S:

$$M = \{\langle x, y\rangle \mid x \in y \in \mathrm{TC}\, S\}.$$

The pairs in this relation are illustrated by straight lines in Fig. 48. The relation M is sure to be well founded; in the present example it is also finite. We have seen that M^t is a well-founded partial ordering on TC S; in fact it is the partial ordering corresponding to Fig. 48. Since M^t is well founded, we can apply transfinite recursion to obtain the unique function E on TC S for which

$$E(a) = \{E(x) \mid xM^t a\}$$

for every a in TC S. (In the transfinite recursion theorem schema, take $\gamma(f, t, z)$ to be the formula $z = \operatorname{ran} f$. Then

$$\begin{aligned} E(a) &= \operatorname{ran}(E \restriction \{x \in \mathrm{TC}\, S \mid xM^t a\}) \\ &= E[\![\{x \mid xM^t a\}]\!] \\ &= \{E(x) \mid xM^t a\} \end{aligned}$$

as predicted.) The values of E are shown by the bracketed numerals in Figure 48. Observe that $\operatorname{ran} E = E[\![\mathrm{TC}\, S]\!] = 3$, which is indeed rank S. And furthermore $E(a) = \operatorname{rank} a$ for every a in TC S.

Was it just a lucky coincidence that for $S = \langle 1, 0\rangle$ we reproduced the rank function? (Would we have gone through the example if it were?) The following theorem says that the outcome is inevitable.

Theorem 9E Let S be any set and let

$$M = \{\langle x, y\rangle \mid x \in y \in \mathrm{TC}\, S\}.$$

Let E be the unique function with domain TC S such that

$$E(a) = \{E(x) \mid xM^{\text{t}}a\}$$

for each a in TC S. Then $\operatorname{ran} E = \operatorname{rank} S$. Furthermore $E(a) = \operatorname{rank} a$ for each a in TC S.

Proof This theorem is a consequence of Theorem 7V, which characterized rank a as the least ordinal strictly greater than rank x for every x in a. The heart of the proof consists of showing that

$$(\star) \qquad \operatorname{rank} a = \{\operatorname{rank} x \mid xM^{\text{t}}a\}$$

for all a in TC S.

To prove the "$\supseteq$" half of $(\star)$, we use Theorem 7V(a):

$$\begin{aligned} xM^{\text{t}}a &\Rightarrow x \in \cdots \in a \\ &\Rightarrow \operatorname{rank} x \in \cdots \in \operatorname{rank} a \\ &\Rightarrow \operatorname{rank} x \in \operatorname{rank} a. \end{aligned}$$

The "$\subseteq$" half of $(\star)$ is proved by transfinite induction on a over the well-founded structure $\langle \mathrm{TC}\, S, M^{\text{t}}\rangle$. Suppose then that $(\star)$ holds of b whenever $bM^{\text{t}}a$ and that $\alpha \in \operatorname{rank} a$. We must show that $\alpha = \operatorname{rank} x$ for some x with $xM^{\text{t}}a$. By Theorem 7V(b), $\alpha \in (\operatorname{rank} b)^+$ for some $b \in a$, so $\alpha \in\!\!\!\!\_\ \operatorname{rank} b$. If $\alpha = \operatorname{rank} b$, we are done, so suppose not. By the inductive hypothesis

$$\alpha \in \operatorname{rank} b \subseteq \{\operatorname{rank} x \mid xM^{\text{t}}b\}$$

so that we have

$$\alpha = \operatorname{rank} x \;\&\; xM^{\text{t}}b \;\&\; b \in a$$

for some x. Since $xM^{\text{t}}a$, the proof of $(\star)$ is complete.

From $(\star)$ and the equation $E(a) = \{E(x) \mid xM^{\text{t}}a\}$ we can conclude (from the uniqueness clause in recursion) that $E(a) = \operatorname{rank} a$ for all a in TC S.

By applying $(\star)$ to a set containing S (in place of S itself) we obtain

$$\operatorname{rank} S = \{\operatorname{rank} a \mid a \in \cdots \in S\} = \operatorname{ran} E$$

as claimed. ⊣

Theorem 9E gives one of the (many) possible ways of introducing ordinal numbers. It produces the rank operation directly (without reference to ordinals). Thus one could then define an ordinal number to be something that is the rank of some set. It then follows, for example, that rank $\alpha = \alpha$ for any ordinal α. And all this can be carried through without any mention of well-ordered structures! One can then proceed to develop the usual properties of ordinal numbers (e.g., those in Theorem 7M) by using the rank operation. The end result of such an approach would be equivalent to the more traditional approach followed in Chapter 7.

Exercises

1. Prove Theorem 9C without using the axiom of choice.
2. Give an intrinsic characterization of these relations R with the property that R^t is a partial ordering.
3. (König) Assume that R is a well-founded relation such that $\{x \mid xRy\}$ is finite for each y. Prove that $\{x \mid xR^ty\}$ is finite for each y.
4. Show that for any set S,

$$\mathrm{TC}\, S = S \cup \bigcup\{\mathrm{TC}\, x \mid x \in S\}.$$

NATURAL MODELS

In this section we want to consider the question of whether there can be a set M that is in a certain sense a miniature model of the class V of all sets. Consider a formula σ of the language of set theory; for example σ might be one of our axioms. (Recall from Chapter 2 the ways of constructing formulas.) We can convert σ to a new formula σ^M by replacing expressions $\forall x$ and $\exists x$ by $\forall x \in M$ and $\exists x \in M$. Then the proposition (true or false) that σ asserts of V is asserted by σ^M of just M.

Example Let σ be the pairing axiom,

$$\forall u\, \forall v\, \exists B\, \forall x[x \in B \quad \Leftrightarrow \quad x = u \text{ or } x = v].$$

Then σ^M is

$$(\forall u \in M)(\forall v \in M)(\exists B \in M)(\forall x \in M)[x \in B \quad \Leftrightarrow \quad x = u \text{ or } x = v].$$

In something closer to English, this says that you can take any u and v *inside the set* M, and be assured of the existence of a set B *inside the set* M whose members *belonging to the set* M are exactly u and v. If somebody has the delusion that M is the class of all sets, then σ^M asserts what he believes the pairing axiom to assert.

The formula σ^M is called the *relativization* of σ to M. We say that σ is *true in* M (or that M is a *model* of σ) iff σ^M is true. Now it is conceivable that there might be some set M that was a model for *all* of our axioms. It is also conceivable that no such M exists. In this section we will be particularly concerned with the question of whether V_α, for some lucky α, might be a model of all of our axioms.

Logical Notes (1) What we here call a model might also be called an "$\in$-model." There is a more general concept of a model $\langle M, E\rangle$ involving not only reinterpretation of $\forall$ and $\exists$ (to refer to M), but also a reinterpretation of $\in$ (by means of E). We will not delve into this more general concept; we only caution you that it exists. (2) There is a fascinating theorem of mathematical logic, called "Gödel's second incompleteness theorem." It implies, among other things, that if our axioms are consistent (as we sincerely believe them to be), then we can never use them to prove the existence of a model of the axioms. Such a model may exist (as a set), but a proof (from our axioms) that it is there does not exist. So do not expect such a proof in this section!

We will now go through the axioms one by one, checking to see whether its relativization is true in V_α for suitable α.

1. Extensionality. We claim that the relativization of extensionality to *any* transitive set (and V_α is always a transitive set) is true. That is, consider any transitive set M and any sets A and B in M. Our claim is that if

$$(\star) \qquad (\forall x \in M)(x \in A \iff x \in B),$$

then $A = B$. Well, if $x \in A$, then $x \in A \in M$, and by the transitivity of M we have $x \in M$. Hence we can apply ($\star$) to conclude that $x \in B$. Thus $A \subseteq B$, and similarly $B \subseteq A$. So $A = B$.

2. Empty set. We claim that the relativization of the empty set axiom to V_α is true provided that $\alpha \neq 0$. We must show that (for $\alpha \neq 0$),

$$(\exists B \in V_\alpha)(\forall x \in V_\alpha)\, x \notin B.$$

What should we take for B? The empty set, of course. Since $1 \in \alpha$, we have $\varnothing \in V_1 \subseteq V_\alpha$, so the empty set is in V_α. And certainly nothing in V_α (or elsewhere) belongs to the empty set.

3. Pairing. We claim that the relativization of the pairing axiom to V_λ is true for any limit ordinal λ. To prove this, consider any u and v in V_λ. Since $V_\lambda = \bigcup_{\alpha \in \lambda} V_\alpha$, we have $u \in V_\alpha$ and $v \in V_\beta$ for some α and β

less than λ. Either $\alpha \in \beta$ or $\beta \in \alpha$; by the symmetry we may suppose that $\alpha \in \beta$. Then $\{u, v\} \subseteq V_\beta$ and

$$\{u, v\} \in V_{\beta^+} \subseteq V_\lambda.$$

Thus in V_λ there is a set B (namely $\{u, v\}$) such that

$$(\forall x \in V_\lambda)(x \in B \quad \Leftrightarrow \quad x = u \text{ or } x = v).$$

And this is what we want.

4. Union. The union axiom is true in V_α for any α. For suppose that $A \in V_\alpha$. Then $A \subseteq V_\beta$ for some β less than α. Since V_β is a transitive set,

$$\begin{aligned} x \in \bigcup A \quad &\Rightarrow \quad x \in b \in A \qquad \text{for some } b \\ &\Rightarrow \quad x \in V_\beta. \end{aligned}$$

So $\bigcup A \subseteq V_\beta$, whence $\bigcup A \in V_\alpha$. Thus we have a set B in V_α (namely, $\bigcup A$) such that for any x in V_α (or elsewhere),

$$\begin{aligned} x \in B \quad &\Leftrightarrow \quad (\exists b)(x \in b \in A) \\ &\Leftrightarrow \quad (\exists b \in V_\alpha)(x \in b \in A) \end{aligned}$$

since V_α is transitive.

5. Power set. As with the pairing axiom, the power set axiom is true in V_λ whenever λ is a limit ordinal. To prove this, consider any set a in V_λ. Then rank $a < \lambda$, so that

$$\text{rank } \mathscr{P}a = (\text{rank } a)^+ < \lambda$$

and $\mathscr{P}a \in V_\lambda$. Thus we have a set B in V_λ (namely $\mathscr{P}a$) such that for any x in V_λ,

$$\begin{aligned} x \in B \quad &\Leftrightarrow \quad x \subseteq a \\ &\Leftrightarrow \quad \forall t(t \in x \quad \Rightarrow \quad t \in a) \\ &\Leftrightarrow \quad (\forall t \in V_\lambda)(t \in x \quad \Rightarrow \quad t \in a) \end{aligned}$$

since V_λ is transitive.

6. Subset. All subset axioms are true in V_α for any α. For consider any set c in V_α and any formula φ not containing B. We seek a set B in V_α such that for any x in V_α,

$$x \in B \quad \Leftrightarrow \quad x \in c \;\&\; \varphi^{V_\alpha}.$$

This tells us exactly what to take for B, namely $\{x \in c \mid \varphi^{V_\alpha}\}$. Since $B \subseteq c \in V_\alpha$, we have $B \in V_\alpha$.

7. Infinity. The infinity axiom is true in V_α for any α greater than ω. We have $\omega \subseteq V_\omega$, whence $\omega \in V_\alpha$ for any larger α. As with the other axioms, this implies that the infinity axiom is true in V_α.

8. Choice. The axiom of choice (version I) is true in V_α for any α. If R is a relation in V_α, then any subset of R is in V_α. In particular, the subfunction F of R provided by the axiom of choice (version I) is in V_α. (For some other versions of the axiom of choice, we need to assume that α is a limit ordinal. Although Theorem 6M proves that the several versions are equivalent, the proof to 6M uses some of the above-listed axioms. And those axioms may fail in V_α if α is not a limit ordinal.)

9. The regularity axiom is true in V_α for any α. Consider any nonempty set A in V_α. Let m be a member of A having least possible rank. Then $m \in V_\alpha$ and $m \cap A = \varnothing$. (Note that we do not need to use the regularity axiom in this proof, in contrast to the situation with all other axioms.)

We have now verified the following result.

Theorem 9F If λ is any limit ordinal greater than ω, then V_λ is a model of all of our axioms except the replacement axioms.

Our axioms (including replacement) are called the *Zermelo–Fraenkel* axioms, whereas the *Zermelo* axioms are obtained by omitting replacement. (Recall the historical notes of Chapter 1.) The above theorem asserts that V_λ is a model of the Zermelo axioms whenever λ is a limit ordinal greater than ω.

The smallest limit ordinal greater than ω is $\omega \cdot 2$. Informally, we can describe the ordinal $\omega \cdot 2$ by the equation

$$\omega \cdot 2 = \{0, 1, 2, \ldots, \omega, \omega^+, \omega^{++}, \ldots\}$$

listing the smaller ordinals. It is the ordinal of the set $\mathbf{Z}$ of integers under the peculiar ordering

$$0 < 1 < 2 < \cdots < -1 < -2 < -3 < \cdots.$$

V_ω contains the sets of finite rank. Any set in V_ω is finite, its members are all finite, and so forth. We can informally describe $V_{\omega \cdot 2}$ as

$$V_{\omega \cdot 2} = V_\omega \cup \mathscr{P}V_\omega \cup \mathscr{P}\mathscr{P}V_\omega \cdots.$$

In $V_{\omega \cdot 2}$ all the sets needed in elementary mathematics can be found. The real numbers are all there, as are all functions from reals to reals (compare Exercise 27 of Chapter 7). This reflects the fact that we needed only the Zermelo axioms to construct the reals. What sets are *not* in $V_{\omega \cdot 2}$? Well, the only ordinals in $V_{\omega \cdot 2}$ are those less than $\omega \cdot 2$, by Exercise 26 of Chapter 7. Since $\omega \cdot 2$ is obviously a countable ordinal, $V_{\omega \cdot 2}$ contains only ordinals that are countable, and not even all of those.

Lemma 9G There is a well-ordered structure in $V_{\omega \cdot 2}$ whose ordinal number is not in $V_{\omega \cdot 2}$.

Proof Start with an uncountable set S in $V_{\omega \cdot 2}$, such as $\mathbb{R}$ or $\mathscr{P}\omega$. By the well-ordering theorem, there exists some well ordering $<$ on S. Now $< \subseteq S \times S \subseteq \mathscr{P}\mathscr{P}S$ by Lemma 3B, so $<$ is also in $V_{\omega \cdot 2}$. (The rank of $<$ is only two steps above the rank of S.) Hence by going up two more steps, we have $\langle S, < \rangle$ in $V_{\omega \cdot 2}$. But the ordinal number of $\langle S, < \rangle$, being uncountable, cannot be in $V_{\omega \cdot 2}$. ⊣

Corollary 9H Not all of the replacement axioms are true in $V_{\omega \cdot 2}$.

Sketch of Proof Let σ be the formula of set theory: "For any well-ordered structure $\langle S, < \rangle$ there exists an ordinal α such that $\langle S, < \rangle$ is isomorphic to $\langle \alpha, \in_\alpha \rangle$." Then σ can be proved from the Zermelo–Fraenkel axioms—we proved it back in Chapter 7 (see Theorem 7D). But we claim that σ is not true in $V_{\omega \cdot 2}$. This claim follows from Lemma 9G, together with some argument to the effect that if it appears from inside $V_{\omega \cdot 2}$ that α is the ordinal number of $\langle S, < \rangle$, then it really is.

But if $V_{\omega \cdot 2}$ were a model for the Zermelo–Fraenkel axioms, it would have to be a model for any theorem that follows from those axioms. (This was the meaning of "theorem" back in Chapter 1.) So $V_{\omega \cdot 2}$ cannot be a model of the Zermelo–Fraenkel axioms. Since it is a model of the Zermelo axioms, it must be replacement axioms that fail in $V_{\omega \cdot 2}$. ⊣

Corollary 9I Not all of the replacement axioms are theorems of the Zermelo axioms.

Proof Any theorem of the Zermelo axioms must be true in any model of those axioms, such as $V_{\omega \cdot 2}$. But by the preceding result, not all replacement axioms are true in $V_{\omega \cdot 2}$. ⊣

The standard abbreviation for the Zermelo–Fraenkel axioms is "ZF" (or "ZFC" if we want to emphasize that the axiom of choice is included). The above corollary shows that there is a sense in which ZF is strictly stronger than the Zermelo axioms. It is our only excursion into the metamathematics of set theory.

Definition A cardinal number κ is said to be *inaccessible* iff it meets the following three conditions.

(a) κ is greater than $\aleph_0$.

(b) For any cardinal λ less than κ, we have $2^\lambda < \kappa$. (Here cardinal exponentiation is used.)

(c) It is not possible to represent κ as the supremum of fewer than κ smaller ordinals. That is, if S is a set of ordinals less than κ and if card $S < \kappa$, then the ordinal sup S is less than κ.

Cardinals meeting clause (c) are called *regular*; we will have more to say about such cardinals in the section on cofinality.

Examples Conditions (b) and (c) are true when $\kappa = \aleph_0$ (by Corollary 6K and Exercise 13 of Chapter 6). But of course condition (a) fails. Conditions (a) and (c) hold when $\kappa = \aleph_1$ (since a countable union of countable ordinals is countable). But condition (b) fails for $\aleph_1$. Conditions (a) and (b) hold when $\kappa = \beth_\omega$, but condition (c) fails since $\beth_\omega = \bigcup_{n\in\omega} \beth_n$. What is an example of a cardinal meeting all three conditions? Are there any inaccessible cardinals at all? We will return to these questions after the next theorem.

First we need the following lemma, which relates the beth numbers to the V_α sets.

Lemma 9J For any ordinal number α,

$$\text{card } V_{\omega+\alpha} = \beth_\alpha.$$

Proof We use transfinite induction on α.

Case I $\alpha = 0$. We must show that card $V_\omega = \aleph_0$. But this is clear, since V_ω is the union of $\aleph_0$ finite sets of increasing size.

Case II α is a successor ordinal β^+. Then

$$V_{\omega+\beta^+} = V_{(\omega+\beta)^+} = \mathscr{P}V_{\omega+\beta}$$

and its cardinality is

$$2^{\text{card } V_{\omega+\beta}} = 2^{\beth_\beta} = \beth_{\beta^+}$$

by the inductive hypothesis.

Case III α is a limit ordinal. Then $\omega + \alpha$ is also a limit ordinal and so $\omega + \alpha = \sup\{\omega + \delta \mid \delta \in \alpha\}$. Consequently (Exercise 9), $V_{\omega+\alpha} = \bigcup\{V_{\omega+\delta} \mid \delta \in \alpha\}$. Since card $V_{\omega+\delta} = \beth_\delta$, the cardinality of the union is at least $\beth_\alpha$. On the other hand, it is at most (card α) $\cdot \beth_\alpha$. Since card $\alpha \subseteq \alpha \subseteq \beth_\alpha$ (by the monotonicity of the beth operation), this product is just $\beth_\alpha$. ⊣

Lemma 9K Assume that κ is inaccessible.

(a) If α is an ordinal less than κ, then $\beth_\alpha$ is less than κ.
(b) If $A \in V_\kappa$, then card $A < \kappa$.

Proof We prove part (a) by transfinite induction on α. So suppose that the condition holds for all ordinals less than α, where α is less than κ.

Case I $\alpha = 0$. Then $\beth_0 = \aleph_0 < \kappa$ since κ is uncountable.

Case II $\alpha = \beta^+$ for some β. Then $\beth_\beta < \kappa$ by the inductive hypothesis, and $\beth_\alpha = 2^{\beth_\beta} < \kappa$ by inaccessibility.

Case III α is a limit ordinal. Then by the inductive hypothesis, $\beth_\gamma < \kappa$ for all $\gamma \in \alpha$. Then $\beth_\alpha = \sup\{\beth_\gamma \mid \gamma \in \alpha\} < \kappa$, since κ is not the supremum of fewer than κ (and card $\alpha < \kappa$) smaller ordinals.

This proves part (a). For part (b), suppose that $A \in V_\kappa$. Then $A \subseteq V_\alpha$ for some α less than κ. Hence we have

$$\text{card } A \leq \text{card } V_\alpha \leq \beth_\alpha < \kappa$$

by the preceding lemma and part (a). ⊣

Theorem 9L If κ is an inaccessible cardinal number, then all of the ZF axioms (including the replacement axioms) are true in V_κ.

Proof All of the Zermelo axioms are true in V_κ by Theorem 9F, since κ is an uncountable limit ordinal. The only axioms left to worry about are the replacement axioms.

Consider a set A in V_κ and any formula $\varphi(x, y)$ such that

$$(\forall x \in A)\, \forall y_1\, \forall y_2[\varphi(x, y_1)\ \&\ \varphi(x, y_2) \quad \Rightarrow \quad y_1 = y_2]$$

is true in V_κ. Define the function F by

$$F = \{\langle x, y\rangle \in A \times V_\kappa \mid \varphi(x, y) \text{ is true in } V_\kappa\}.$$

Notice that F *is* a function, by what we have said about φ. The domain of F is some subset of A. Let $B = \operatorname{ran} F$; thus for any y in V_κ

$$y \in B \quad \Leftrightarrow \quad (\exists x \in A)\, \varphi(x, y) \text{ is true in } V_\kappa.$$

What we must prove is that $B \in V_\kappa$, because then it will follow that

$$\exists B\, \forall y[y \in B \quad \Leftrightarrow \quad (\exists x \in A)\, \varphi(x, y)]$$

is true in V_κ.

To show that $B \in V_\kappa$, consider the set

$$S = \{\text{rank } F(x) \mid x \in \text{dom } F\}.$$

The ordinals in S are all less than κ (because $F(x) \in V_\kappa$), and card $S \le$ card dom $F \le$ card $A < \kappa$ by part (b) of the above lemma. So by the inaccessibility, the ordinal

$$\alpha = \sup S = \sup\{\text{rank } F(x) \mid x \in \text{dom } F\}$$

is less than κ. But any member $F(x)$ of B has rank no more than α, so $F(x) \in V_{\alpha^+}$. Hence $B \subseteq V_{\alpha^+}$ and so rank $B \underline{\in} \alpha^+ \in \kappa$. Thus $B \in V_\kappa$, as desired. ⊣

Now back to the question: Are there any inaccessible cardinals at all? If so, then by the above theorem, there are models of the ZF axioms. Because we cannot hope to prove the existence of such a model (due to Gödel's second incompleteness theorem, mentioned earlier in this section), it follows that we cannot hope to prove from our axioms that inaccessible cardinals exist.

On the other hand, we intuitively want the cardinal and ordinal numbers to go on forever. To deny the existence of inaccessible numbers would appear to impose an unnatural ceiling on "forever." With these thoughts in mind, Alfred Tarski in 1938 proposed for consideration as an additional set-theoretic axiom the statement:

For any cardinal number there is a larger inaccessible cardinal number.

This is an example of a "large cardinal axiom." Despite the fact that it literally concerns huge sets, one can prove from it new facts about natural numbers! The interest in various large cardinal axioms and their consequences motivates an important part of current research in set theory.

Exercises

5. Show that a set S belongs to V_ω iff TC S is finite. (Thus S belongs to V_ω iff S is finite and all the members of ... of members of S are finite. Because of this fact, V_ω is often called HF, the collection of *hereditarily finite* sets.)

6. Are the replacement axioms true in V_ω? Which axioms are *not* true in V_ω?

7. Let $\mathscr{F}$ be the collection of finite subsets of ω. Let g be a one-to-one correspondence between ω and $\mathscr{F}$ with the property that whenever $m \in g(n)$, then m is less than n. (For example, we can use $g(n) = \{m \in \omega \mid$ the coefficient of 2^m in the binary representation of n is $1\}$.) Define the binary relation E on ω by

$$mEn \quad \Leftrightarrow \quad m \in g(n).$$

Show that $\langle \omega, E \rangle$ is isomorphic to $\langle V_\omega, \in^\circ \rangle$.

8. The proof to Lemma 9G used the axiom of choice (in the form of the well-ordering theorem). Give a proof that does not use the axiom of choice.

9. Assume that λ is a limit ordinal. Show that

$$V_{\alpha+\lambda} = \bigcup\{V_{\alpha+\delta} \mid \delta \in \lambda\}.$$

10. Define the set

$$S = \omega \cup \mathscr{P}\omega \cup \mathscr{P}\mathscr{P}\omega \cup \cdots.$$

Prove that S is a model of the Zermelo axioms.

11. Assume that κ is inaccessible. Show that $\beth_\kappa = \kappa$ and card $V_\kappa = \kappa$.

COFINALITY

Any limit ordinal is the supremum of the set of all smaller ordinals; this merely says that $\lambda = \bigcup\lambda$ for any limit ordinal λ. But we do not need to take *all* smaller ordinals. We can find a proper subset S of λ such that λ is the supremum $\bigcup S$ of S. How small can we take S to be? That will depend on what λ is.

Definition The *cofinality* of a limit ordinal λ, denoted cf λ, is the smallest cardinal κ such that λ is the supremum of κ smaller ordinals. The cofinality of nonlimit ordinals is defined by setting cf $0 = 0$ and cf $\alpha^+ = 1$.

Thus to find the cofinality of λ, we seek a set S of smaller ordinals (i.e., $S \subseteq \lambda$) for which $\lambda = \sup S$. Such a set S is said to be *cofinal* in λ. There will be many sets that are cofinal in λ; for example, λ itself is such a set. There will not exist any smallest (with respect to inclusion) set S cofinal in λ. But there will be a smallest possible cardinality for S, and this smallest cardinality is by definition cf λ.

Example Obviously cf $\lambda \leq$ card λ, since λ is cofinal in itself. Sometimes equality holds; for example, cf $\omega = \aleph_0$. (We could not have cf $\omega < \aleph_0$, since a finite union of natural numbers would be finite.) On the other hand, sometimes cf $\lambda \neq$ card λ. For example, Theorem 9N will show that $\aleph_\omega$ (as a limit ordinal) has cofinality $\aleph_0$, being the supremum of $\{\aleph_0, \aleph_1, \aleph_2, \ldots\}$.

Definition A cardinal κ is said to be *regular* iff cf $\kappa = \kappa$, and is said to be *singular* if cf $\kappa < \kappa$.

The above example shows that $\aleph_0$ is regular whereas $\aleph_\omega$ is singular. Recall that an inaccessible cardinal is required to be regular.

Note on Style The definition of cofinality has an awkward three-part format. Can we not give a simple definition covering zero, the successor ordinals, and the limit ordinals in one sweep? Yes, we can do so by using the idea of the "strict supremum." The cofinality of any ordinal α is the least cardinal number κ such that there exists a subset S of α having cardinality κ and α is the least ordinal *strictly* greater than every member of S. (Exercise 12 asks you to verify that this is correct.)

Theorem 9M $\aleph_{\alpha+1}$ is a regular cardinal number for every α.

Proof Assume that $\aleph_{\alpha+1} = \sup S$, where S is a set of ordinals, each of which is less than $\aleph_{\alpha+1}$. Then each member of S has cardinality at most $\aleph_\alpha$. Therefore (compare Exercise 26 of Chapter 6) we have $\aleph_{\alpha+1} = \operatorname{card} \bigcup S \leq (\operatorname{card} S) \cdot \aleph_\alpha$. But this implies that $\operatorname{card} S \geq \aleph_{\alpha+1}$. ⊣

Since $\aleph_0$ is also a regular cardinal, there remain only the cardinals $\aleph_\lambda$ where λ is a limit ordinal.

Theorem 9N For any limit ordinal λ, $\operatorname{cf} \aleph_\lambda = \operatorname{cf} \lambda$.

Proof First of all, we claim that $\operatorname{cf} \aleph_\lambda \leq \operatorname{cf} \lambda$. We know that λ is the supremum of some set $S \subseteq \lambda$ with $\operatorname{card} S = \operatorname{cf} \lambda$. It suffices to show that $\aleph_\lambda = \sup\{\aleph_\alpha \mid \alpha \in S\}$. But this is Theorem 8E, applied to the aleph operation.

Second, we claim that $\operatorname{cf} \lambda \leq \operatorname{cf} \aleph_\lambda$. Suppose that $\aleph_\lambda$ is the supremum of some set A of smaller ordinals. Let

$$B = \{\gamma \in \lambda \mid \aleph_\gamma \text{ is the cardinality of some ordinal in } A\}.$$

Then $\operatorname{card} B \leq \operatorname{card} A$. To complete the proof it suffices to show that $\sup B = \lambda$. Any α in A has cardinality at most $\aleph_{\sup B}$, so $\alpha \in \aleph_{(\sup B)+1}$. Hence $\aleph_\lambda = \sup A \leq \aleph_{(\sup B)+1}$ and so $\lambda \mathrel{\underline{\in}} (\sup B) + 1$. Since λ is a limit ordinal, $\lambda \mathrel{\underline{\in}} \sup B$, whence equality holds. ⊣

For example, $\operatorname{cf} \aleph_\omega = \operatorname{cf} \omega = \aleph_0$. And $\operatorname{cf} \aleph_\Omega = \operatorname{cf} \Omega = \aleph_1$ (where Ω is the first uncountable ordinal). In particular, both $\aleph_\omega$ and $\aleph_\Omega$ are singular. For any limit ordinal λ,

$$\operatorname{cf} \aleph_\lambda = \operatorname{cf} \lambda \mathrel{\underline{\in}} \lambda \mathrel{\underline{\in}} \aleph_\lambda.$$

If $\aleph_\lambda$ is regular, then equality holds throughout. But does this ever happen? If it happens, λ is said to be *weakly inaccessible*. We cannot prove (from our axioms, if they are consistent) that any weakly inaccessible cardinals exist. If any inaccessible cardinals exist, then they are also weakly inaccessible (Exercise 15).

There is another way of characterizing cofinality that uses increasing sequences of ordinals instead of sets of ordinals. For an ordinal number α, define an *α-sequence* to be simply a function with domain α. For example, an ordinary infinite sequence is an ω-sequence, and a finite sequence is an n-sequence for some natural number n. For an α-sequence f, it is customary to write f_ξ in place of the Eulerian $f(\xi)$. An α-sequence f into the ordinals is called *increasing* iff it preserves order:

$$\xi \in \eta \quad \Rightarrow \quad f_\xi \in f_\eta.$$

An increasing α-sequence of ordinals is said to *converge* to β iff $\beta = \sup \operatorname{ran} f$, i.e.,

$$\beta = \sup\{f_\xi \mid \xi \in \alpha\}.$$

From a nonincreasing sequence we can extract an increasing subsequence (possibly with a smaller domain):

Lemma 9P Assume that f is an α-sequence (not necessarily increasing) into the ordinal numbers. There is an increasing β-sequence g into the ordinal numbers for some $\beta \underline{\in} \alpha$ with $\sup \operatorname{ran} f = \sup \operatorname{ran} g$.

Proof We will construct g so that it is a subsequence of f. That is, we will have $g_\xi = f_{h_\xi}$ for a certain increasing sequence h.

First define the sequence h by transfinite recursion over $\langle \alpha, \in_\alpha \rangle$. Suppose that h_δ is known for every $\delta \in \xi$. Then define

$$h_\xi = \text{the least } \gamma \text{ such that } f_{h_\delta} \in f_\gamma \text{ for all } \delta \in \xi,$$

if any such γ exists; if no such γ exists, then the sequence halts. Let $\beta = \operatorname{dom} h$, and define

$$g_\xi = f_{h_\xi} \qquad \text{for} \quad \xi \in \beta.$$

Then by construction,

$$\delta \in \xi \quad \Rightarrow \quad f_{h_\delta} \in f_{h_\xi} \quad \Rightarrow \quad g_\delta \in g_\xi,$$

so g is increasing. Also h is increasing, because h is one-to-one and h_ξ is the least ordinal meeting certain conditions that become more stringent as ξ increases.

We have $\operatorname{ran} g \subseteq \operatorname{ran} f$, so certainly $\sup \operatorname{ran} g \underline{\in} \sup \operatorname{ran} f$. On the other hand we have

$$f_\delta \underline{\in} f_{h_\xi} = g_\xi \qquad \text{for all} \quad \delta \underline{\in} h_\xi$$

by the leastness of h_ξ. In particular, $\xi \underline{\in} h_\xi$, so $f_\xi \underline{\in} g_\xi$. Hence if $\beta = \alpha$, then clearly $\sup \operatorname{ran} f \underline{\in} \sup \operatorname{ran} g$. But if β is less than α, it is only because there exists no γ for which f_γ exceeds every g_δ for $\delta \in \beta$. Again we conclude that $\sup \operatorname{ran} f \underline{\in} \sup \operatorname{ran} g$. ⊣

This lemma is used in the proof of the following theorem.

Theorem 9Q Assume that λ is a limit ordinal. Then there is an increasing $(\operatorname{cf} \lambda)$-sequence into λ that converges to λ.

Proof By definition of cofinality, λ is the supremum of $\operatorname{cf}\lambda$ smaller ordinals. That is, there is a function f from $\operatorname{cf}\lambda$ into λ such that $\lambda = \sup \operatorname{ran} f$. Apply the lemma to obtain an increasing β-sequence converging to λ for some $\beta \in\!\!\!\!\_\ \operatorname{cf}\lambda$. But it is impossible that $\beta \in \operatorname{cf}\lambda$, lest we have λ represented as the supremum of $\operatorname{card}\beta$ smaller ordinals, contradicting the leastness of $\operatorname{cf}\lambda$. ⊣

Corollary 9R For a limit ordinal λ, we can characterize $\operatorname{cf}\lambda$ as the least ordinal α such that some increasing α-sequence into λ converges to λ.

We know that $\operatorname{cf}\lambda \leq \lambda$. What about $\operatorname{cf}\operatorname{cf}\lambda$? Or $\operatorname{cf}\operatorname{cf}\operatorname{cf}\lambda$? These are all equal to $\operatorname{cf}\lambda$, by the next theorem.

Theorem 9S For any ordinal λ, $\operatorname{cf}\lambda$ is a regular cardinal.

Proof We must show that $\operatorname{cf}\operatorname{cf}\lambda = \operatorname{cf}\lambda$. We may assume that λ is a limit ordinal, since 0 and 1 are regular. By the preceding theorem, there is an increasing $(\operatorname{cf}\lambda)$-sequence f that converges to λ.

Suppose that $\operatorname{cf}\lambda$ is the supremum of some set S of smaller ordinals; we must show that $\operatorname{cf}\lambda \leq \operatorname{card} S$. (This will prove that $\operatorname{cf}\lambda \leq \operatorname{cf}\operatorname{cf}\lambda$ and will be done.) To do this, consider the set

$$f[\![S]\!] = \{f_\alpha \mid \alpha \in S\}.$$

This is a subset of λ having the same cardinality as S. It suffices to show that $f[\![S]\!]$ is cofinal in λ, since this implies that

$$\operatorname{cf}\lambda \leq \operatorname{card} f[\![S]\!] = \operatorname{card} S.$$

Consider then any $\alpha \in \lambda$. Since f converges to λ, we have $\alpha \in f(\beta) \in \lambda$ for some $\beta \in \operatorname{cf}\lambda$. Since $\operatorname{cf}\lambda = \sup S$, we have $\beta \in \gamma \in S$ for some γ. Hence

$$\alpha \in f(\beta) \in f(\gamma) \in f[\![S]\!].$$

So $f[\![S]\!]$ is indeed cofinal in λ. ⊣

Now we want to study the particular case where λ is an infinite *cardinal* number. In this case, we can give an "all cardinal" characterization of $\operatorname{cf}\lambda$ that makes no direct mention of ordinal numbers.

Theorem 9T Assume that λ is an infinite cardinal. Then $\operatorname{cf}\lambda$ is the least cardinal number κ such that λ can be decomposed into the union of κ sets, each having cardinality less than λ.

Proof We know that λ is the union of $\operatorname{cf}\lambda$ smaller ordinals, and any smaller ordinal α satisfies

$$\operatorname{card}\alpha \in\!\!\!\!\_\ \alpha \in \lambda = \operatorname{card}\lambda.$$

Hence λ can be represented as the union of $\operatorname{cf}\lambda$ sets of cardinality less than λ. It remains to show that we cannot make do with fewer than $\operatorname{cf}\lambda$ such sets.

So consider an arbitrary decomposition $\lambda = \bigcup\mathscr{A}$ where each member of $\mathscr{A}$ has cardinality less than λ. Let $\kappa = \operatorname{card}\mathscr{A}$; we seek to prove that $\operatorname{cf}\lambda \leq \kappa$. We can express $\mathscr{A}$ as the range of a κ-sequence of sets:

$$\mathscr{A} = \{A_\xi \mid \xi \in \kappa\}.$$

Thus $\lambda = \bigcup\{A_\xi \mid \xi \in \kappa\}$ and $\operatorname{card} A_\xi < \lambda$.

Define the cardinal $\mu = \sup\{\operatorname{card} A_\xi \mid \xi \in \kappa\}$. Then each $\operatorname{card} A_\xi \leq \mu$ and so $\lambda = \operatorname{card} \bigcup\{A_\xi \mid \xi \in \kappa\} \leq \mu \cdot \kappa$ (compare Exercise 26 of Chapter 6).

Case I $\lambda \leq \kappa$. Then $\operatorname{cf}\lambda \leq \lambda \leq \kappa$, so we are done.

Case II $\kappa < \lambda$. Then since $\lambda \leq \mu \cdot \kappa$, we can conclude that $\lambda = \mu$. Thus we have $\lambda = \sup\{\operatorname{card} A_\xi \mid \xi \in \kappa\}$, so λ is the supremum of κ smaller ordinals. Hence $\operatorname{cf}\lambda \leq \kappa$. ⊣

Cantor's theorem tells us that 2^κ is always greater than κ. The following theorem tells us that even $\operatorname{cf} 2^\kappa$ is greater than κ.

König's Theorem Assume that κ is an infinite cardinal number. Then $\kappa < \operatorname{cf} 2^\kappa$.

Proof Suppose to the contrary that $\operatorname{cf} 2^\kappa \leq \kappa$. Then 2^κ, and consequently any set of size 2^κ, is representable as the union of κ sets each of size less than 2^κ. The particular set to consider is ${}^\kappa S$, where S is some set of size 2^κ. (Thus $\operatorname{card} {}^\kappa S = (2^\kappa)^\kappa = 2^\kappa$.)

We have, then, a representation of the form

$${}^\kappa S = \bigcup\{A_\xi \mid \xi \in \kappa\},$$

where $\operatorname{card} A_\xi < 2^\kappa$. For any one ξ in κ,

$$\{g(\xi) \mid g \in A_\xi\} \subseteq S.$$

Furthermore we have proper inclusion, because $\operatorname{card}\{g(\xi) \mid g \in A_\xi\} \leq \operatorname{card} A_\xi < 2^\kappa$, whereas $\operatorname{card} S = 2^\kappa$. So we can choose some point

$$s_\xi \in S - \{g(\xi) \mid g \in A_\xi\}.$$

This construction yields a κ-sequence s, i.e., $s \in {}^\kappa S$. But $s \notin A_\xi$ for any $\xi \in \kappa$, by the construction. ⊣

Corollary 9U $2^{\aleph_0} \neq \aleph_\omega$.

Proof By König's theorem, $\operatorname{cf} 2^{\aleph_0}$ is uncountable. But $\operatorname{cf}\aleph_\omega = \aleph_0$. ⊣

Exercises

12. For any set S of ordinals, define the *strict supremum* of S, ssup S, to be the least ordinal strictly greater than every member of S. Show that the cofinality of any ordinal α (zero, successor, or limit) is the least cardinal κ such that α is the strict supremum of κ smaller ordinals.

13. Show that $\operatorname{cf} \beth_\lambda = \operatorname{cf} \lambda$ for any limit ordinal λ.

14. Assume that R is a well-founded relation, κ is a regular infinite cardinal, and $\operatorname{card}\{x \mid xRy\} < \kappa$ for each y. Prove that $\operatorname{card}\{x \mid xR^t y\} < \kappa$ for each y.

15. Prove that any inaccessible cardinal is also weakly inaccessible.

16. Assume that λ is weakly inaccessible, i.e., it is a limit ordinal for which $\aleph_\lambda$ is regular. Further assume that the generalized continuum hypothesis holds. Show that λ is an inaccessible cardinal.

17. (König) Assume that for each i in a set I we are given sets A_i and B_i with $\operatorname{card} A_i < \operatorname{card} B_i$. Show that $\operatorname{card} \bigcup_{i \in I} A_i < \operatorname{card} \times_{i \in I} B_i$. [*Suggestion*: If f maps $\bigcup_{i \in I} A_i$ onto $\times_{i \in I} B_i$, then select a point b_i in B_i not in the projection of $f[\![A_i]\!]$.]

18. Consider the operation assigning to each ordinal number α its cofinality $\operatorname{cf} \alpha$. Is this operation monotone? Continuous? Normal? (Give proofs or counterexamples.)

19. Assume that κ is a regular cardinal and that S is a subset of V_κ with $\operatorname{card} S < \kappa$. Show that $S \in V_\kappa$.

20. Consider a normal operation assigning t_α to each ordinal number α, and assume that λ is a limit ordinal. Show that $\operatorname{cf} t_\lambda = \operatorname{cf} \lambda$.

APPENDIX

NOTATION, LOGIC, AND PROOFS

In Chapters 1 and 2 we described how formulas of the language of set theory could be built up from component parts. We start with the indivisible "atomic" formulas, such as '$x \in S$' and '$x = y$'. Once we know what objects are named by the letters 'x', 'y', and 'S', we can consider the truth or falsity of such formulas.[1]

Given any formulas φ and ψ, we can combine them in various ways to obtain new ones. For example, $\ulcorner(\varphi \,\&\, \psi)\urcorner$ will be a new formula. The intended meaning of the ampersand can be captured by giving a truth table (Table 1). And similarly we can specify by this table that 'or' is to mean "one or the other or both." A simpler case is '$\neg$'; the truth value of $\ulcorner(\neg\varphi)\urcorner$ is determined by the last column of Table 1.

[1] In this appendix we utilize the convention of forming the name of an expression by the use of single quotation marks. For example, x might be a set, but 'x' is a letter used to name that set. We also use utilize corners, e.g., if φ is the formula '$x = x$', then $\ulcorner(\varphi \,\&\, \varphi)\urcorner$ is the formula '$(x = x \,\&\, x = x)$'.

TABLE 1

φ	ψ	$(\varphi \,\&\, \psi)$	$(\varphi \text{ or } \psi)$	$(\neg\varphi)$
T	T	T	T	F
T	F	F	T	F
F	T	F	T	T
F	F	F	F	T

Example We can show that the truth value, T or F, of $\ulcorner(\neg(\varphi \,\&\, \psi))\urcorner$ is always the same as the truth value of $\ulcorner((\neg\varphi) \text{ or } (\neg\psi))\urcorner$ by making a table (Table 2) illustrating every possibility.

TABLE 2

φ	ψ	$(\neg(\varphi \,\&\, \psi))$	$((\neg\varphi) \text{ or } (\neg\psi))$
T	T	F	F
T	F	T	T
F	T	T	T
F	F	T	T

Example The truth value of $\ulcorner(\varphi \,\&\, (\psi \text{ or } \theta))\urcorner$ is always the same as the truth value of $\ulcorner((\varphi \,\&\, \psi) \text{ or } (\varphi \,\&\, \theta))\urcorner$. The table for this must have eight lines to cover all possibilities. You are invited to contemplate the relationship of this example to Fig. 5 in Chapter 2.

The formula $\ulcorner(\varphi \Rightarrow \psi)\urcorner$ is to be read as "if φ, then ψ." The exact meaning is determined by Table 3. Note in particular that the formula is "vacuously" true whenever φ is false. The formula is sometimes read as "φ implies ψ." This usage of the word "implies" is a conventional part of mathematical jargon, but it is not exactly the way the rest of the world used the word.

TABLE 3

φ	ψ	$(\varphi \Rightarrow \psi)$	$(\varphi \Leftrightarrow \psi)$
T	T	T	T
T	F	F	F
F	T	T	F
F	F	T	T

Example The truth value of ⌜$(\varphi \Rightarrow \psi)$⌝ is always the same as the truth value of its contrapositive ⌜$((\neg\psi) \Rightarrow (\neg\varphi))$⌝. (You should check this.) Sometimes when it is desired to prove a formula of the form ⌜$(\varphi \Rightarrow \psi)$⌝, it is actually more convenient to prove ⌜$((\neg\psi) \Rightarrow (\neg\varphi))$⌝. And this suffices to establish the truth of ⌜$(\varphi \Rightarrow \psi)$⌝.

Example If θ is true, then the truth value of ⌜$((\neg\varphi) \Rightarrow (\neg\theta))$⌝ is the same as the truth value of φ. Sometimes to prove φ, we proceed in an indirect way: We show that ⌜$(\neg\varphi)$⌝ would lead to a result contradicting what is already known to be true, i.e., we use a "proof by contradiction."

Example The truth value of ⌜$(\neg(\varphi \Rightarrow \psi))$⌝ is the same as the truth value of ⌜$(\varphi \,\&\, (\neg\psi))$⌝. This is reflected in the fact that to give a *counterexample* to show that ⌜$(\varphi \Rightarrow \psi)$⌝ is false, we give an example in which φ is true and ψ is false.

To construct any interesting formulas, we must use (in addition to features already mentioned) the phrases "for all x" and "for some x." These phrases can be symbolized by '$\forall x$' and '$\exists x$', respectively. If φ is a formula (in which 'x' occurs but '$\forall x$' and '$\exists x$' do not), then the condition for ⌜$\forall x\, \varphi$⌝ to be true is that φ should be true no matter what 'x' names (in the universe of all sets). And the condition for ⌜$\exists x\, \varphi$⌝ to be true is that there is at least one thing (in the universe of all sets) such that φ is true when 'x' names that thing. These criteria do not reduce to simple tables. There is no mechanical procedure that can be substituted for clear thinking.

Example The truth value of ⌜$(\neg\forall x\, \varphi)$⌝ is the same as the truth value of ⌜$\exists x(\neg\varphi)$⌝. To deny that φ is true of everything is to assert that there is at least one thing of which φ is false, and of which ⌜$(\neg\varphi)$⌝ is consequently true. Similarly the truth value of ⌜$(\neg\exists x\, \varphi)$⌝ is the same as the truth value of ⌜$\forall x(\neg\varphi)$⌝.

Example Suppose that ⌜$\exists x\, \forall y\, \varphi$⌝ is true. Then, reading from left to right, we can say that there is some set x for which ⌜$\forall y\, \varphi$⌝ is true. That is, there is at least one set that, when we bestow the name 'x' upon it, we then find that no matter what set 'y' names, φ is true. This guarantees that the weaker statement ⌜$\forall y\, \exists x\, \varphi$⌝ is true. This weaker statement demands only that for any set (upon which we momentarily bestow the name 'y') there exists some set (which we call 'x') such that φ is true. The difference here is that the choice of x may depend on y, whereas ⌜$\exists x\, \forall y\, \varphi$⌝ demands that there be an x fixed in advance that is successful with every y. For example, it is true of the real numbers that for every number y there is some number x (e.g., $x = y + 1$) with $y < x$. But the stronger statement, that there is some number x such that for every number y we have $y < x$, is false. Or to give a different sort of example, it may well be true that every boy has some girl whom he admires, and yet there is no one girl so lucky as to be admired by every boy.

Next we want to give an example of a *proof*. Actually, the book is full of proofs. But what we want to do now is to illustrate how one constructs a proof, without knowing in advance how to do it. *After* one knows how the proof goes, it can be written out in the conventional form followed by contemporary mathematics books. But *before* one knows how the proof goes, it is a matter of looking at the assumptions and the conclusions and trying to gain insight into the connection between them. Suppose, for example, one wants to prove the assertion:

$$a \in B \quad \Rightarrow \quad \mathscr{P}a \in \mathscr{P}\mathscr{P}\bigcup B.$$

We will present the process of constructing the proof as a discussion among three mental states: Prover, Referee, and Commentator.

P: We assume that $a \in B$.

C: We might as well, because the assertion to be proved is vacuously true if $a \notin B$.

R: It is to be proved that $\mathscr{P}a \in \mathscr{P}\mathscr{P}\bigcup B$.

C: Always keep separate the facts you *have* available and the facts that you *want* to establish.

P: I have $a \in B$, but it is not obvious how to utilize that fact.

R: Look at the goal. It suffices to show that $\mathscr{P}a \subseteq \mathscr{P}\bigcup B$, by the definition of $\mathscr{P}$.

P: Well, in that case I will assume that $c \in \mathscr{P}a$.

R: Then the goal is to show that $c \in \mathscr{P}\bigcup B$.

C: The point is to consider an *arbitrary* member (call it c) of $\mathscr{P}a$, and show that it belongs to $\mathscr{P}\bigcup B$. Then we can conclude that $\mathscr{P}a \subseteq \mathscr{P}\bigcup B$.

P: $c \subseteq a$.

R: The goal is to get $c \subseteq \bigcup B$

C: This pair of sentences is a recasting of their previous pair, using definitions.

P: Since I am expected to prove that $c \subseteq \bigcup B$, I will consider any $x \in c$.

R: The goal now is to get $x \in \bigcup B$.

P: Looking back, I see that

$$x \in c \subseteq a \in B.$$

R: But you want $x \in \bigcup B$.

P: $x \in a \in B$, so I have $x \in \bigcup B$.

C: The job is done. The information Prover has is the same as the information Referee wants.

Now that the mental discussion is over, the proof can be written out in the conventional style. Here it is:

Proof Assume that $a \in B$. To show that $\mathscr{P}a \in \mathscr{P}\mathscr{P}\bigcup B$, it suffices to show that $\mathscr{P}a \subseteq \mathscr{P}\bigcup B$. So consider any $c \in \mathscr{P}a$. Then we have

$$\begin{aligned} x \in c \quad &\Rightarrow \quad x \in c \subseteq a \in B \\ &\Rightarrow \quad x \in a \in B \\ &\Rightarrow \quad x \in \bigcup B. \end{aligned}$$

Thus $c \subseteq \bigcup B$ and $c \in \mathscr{P}\bigcup B$. Since c was arbitrary in $\mathscr{P}a$, we have shown that $\mathscr{P}a \subseteq \mathscr{P}\bigcup B$, as desired. ⊣

We conclude this appendix with some comments on a task mentioned in Chapter 2. In that chapter there is a fairly restrictive definition of what a *formula* is. In particular, to obtain a legal formula, it is necessary to get rid of the defined symbols, such as '$\varnothing$', '$\bigcup$', '$\mathscr{P}$', etc. We will indicate a mechanical procedure for carrying out this elimination process.

Suppose then, we have a statement ⌜$\_\_\ \mathscr{P}t\ \_\_$⌝ in which '$\mathscr{P}$' occurs. We can rewrite it first as

$$\forall a(a = \mathscr{P}t \quad \Rightarrow \quad \_\_\ a\ \_\_)$$

and then rewrite '$a = \mathscr{P}t$' as

$$\forall x(x \in a \quad \Leftrightarrow \quad x \in \mathscr{P}t).$$

This reduces the problem to eliminating '$\mathscr{P}$' from statements of the form '$x \in \mathscr{P}t$'. And this is easy; '$x \in \mathscr{P}t$' can be replaced by '$x \subseteq t$' or by '$\forall y(y \in x \Rightarrow y \in t)$'.

This process is similar for other symbols. For the union symbol, we reduce the problem to eliminating '$\bigcup$' from '$x \in \bigcup t$'. And the definition of '$\bigcup$' tells us how to do this; we replace '$x \in \bigcup t$' by '$\exists y(x \in y \in t)$'.

SELECTED REFERENCES FOR FURTHER STUDY

Georg Cantor. *Contributions to the Founding of the Theory of Transfinite Numbers*, translated by P. Jourdain. Dover, New York, 1955.

This is a translation of two papers that Cantor originally published in 1895 and 1897. The papers treat cardinal and ordinal numbers and their arithmetic. Jourdain's introduction, first published in 1915, discusses Cantor's work in historical perspective.

Paul J. Cohen. *Set Theory and the Continuum Hypothesis*. Benjamin, New York, 1966.

This gives a highly condensed account of the main results in the metamathematics of set theory.

Frank R. Drake. *Set Theory, An Introduction to Large Cardinals*. North-Holland Publ., Amsterdam, 1974.

This book treats the implications that large cardinals have for the metamathematics of set theory, along with other related topics.

Abraham A. Fraenkel, Yehoshua Bar-Hillel, and Azriel Levy. *Foundations of Set Theory*, 2nd rev. ed. North-Holland Publ., Amsterdam, 1973.

This book considers the philosophical background for set theory and the foundations of mathematics. Different approaches to set theory are developed and compared. Among the topics discussed are the limitations of axiomatizations, the role of classes in set theory, and the history of the axiom of choice.

Thomas J. Jech. *The Axiom of Choice.* North-Holland Publ., Amsterdam, 1973.

Centered on the topic of its title, this book contains much material on the metamathematics of set theory.

K. Kuratowski and A. Mostowski. *Set Theory.* PWN—Polish Scientific Publ., Warsaw, and North-Holland Publ., Amsterdam, 1968.

In addition to the standard topics of set theory, this book has considerable material on the applications of set theory, particularly to topology. The footnotes provide a guide to the history of set-theoretic ideas.

Wacław Sierpiński. *Cardinal and Ordinal Numbers,* 2nd ed. PWN—Polish Scientific Publ., Warsaw, 1965.

This is a classic book on cardinal and ordinal arithmetic, the axiom of choice, and related topics. The approach is nonaxiomatic.

Jean van Heijenoort (editor). *From Frege to Gödel, A Source Book in Mathematical Logic, 1879–1931.* Harvard Univ. Press, Cambridge, Massachusetts, 1967.

This is a collection of forty-six fundamental papers in logic and set theory, translated into English and supplied with commentaries. There is an 1899 letter from Cantor to Dedekind and the 1902 correspondence between Russell and Frege. Zermelo's 1908 paper giving the first axiomatization of set theory is included, as are papers by Fraenkel, Skolem, and von Neumann.

LIST OF AXIOMS

Extensionality axiom

$$\forall A\ \forall B[\forall x(x \in A \ \Leftrightarrow\ x \in B) \ \Rightarrow\ A = B]$$

Empty set axiom

$$\exists B\ \forall x\ x \notin B$$

Pairing axiom

$$\forall u\ \forall v\ \exists B\ \forall x(x \in B \ \Leftrightarrow\ x = u \text{ or } x = v)$$

Union axiom

$$\forall A\ \exists B\ \forall x[x \in B \ \Leftrightarrow\ (\exists b \in A) x \in b]$$

Power set axiom

$$\forall a\ \exists B\ \forall x(x \in B \ \Leftrightarrow\ x \subseteq a)$$

Subset axioms For each formula φ not containing B, the following is an axiom:

$$\forall t_1 \cdots \forall t_k\ \forall c\ \exists B\ \forall x(x \in B \ \Leftrightarrow\ x \in c\ \&\ \varphi)$$

Infinity axiom

$$\exists A[\varnothing \in A \;\&\; (\forall a \in A)\, a^{+} \in A]$$

Choice axiom

$$(\forall \text{ relation } R)(\exists \text{ function } F)(F \subseteq R \;\&\; \operatorname{dom} F = \operatorname{dom} R)$$

Replacement axioms For any formula $\varphi(x, y)$ not containing the letter B, the following is an axiom:

$$\begin{aligned}\forall t_1 \cdots \forall t_k \forall A[(\forall x \in A)\forall y_1 \forall y_2(\varphi(x, y_1) \;\&\; \varphi(x, y_2) \;\Rightarrow\; y_1 = y_2)\\ \Rightarrow\; \exists B\, \forall y(y \in B \;\Leftrightarrow\; (\exists x \in A)\varphi(x, y))]\end{aligned}$$

Regularity axiom

$$(\forall A \neq \varnothing)(\exists m \in A)\, m \cap A = \varnothing$$

INDEX

Z

Printed and bound by CPI Group (UK) Ltd, Croydon, CR0 4YY
03/10/2024
01040414-0016